Storey's Guide to Raising Chickens

· STOREY'S GUIDE TO ·

RAISING
CHICKENS

BREED SELECTION · FACILITIES · FEEDING
HEALTH CARE · MANAGING LAYERS & MEAT BIRDS

Gail Damerow

Storey Publishing

The mission of Storey Publishing is to serve our customers by
publishing practical information that encourages
personal independence in harmony with the environment.

EDITED BY Deborah Burns and Sarah Guare
ART DIRECTION AND BOOK DESIGN BY Jeff Stiefel
TEXT PRODUCTION BY Liseann Karandisecky
INDEXED BY Samantha Miller

COVER PHOTOGRAPHY BY © Jason Houston, front; © blickwinkel/Alamy Stock Photo, back (top right);
© Juniors Bildarchiv GmbH/Alamy Stock Photo, spine & back (bottom); © Leslie Banks/Dreamstime,
back (middle left); © Leena Robinson/Alamy Stock Photo, back (top left)
INTERIOR PHOTOGRAPHY CREDITS appear on page 403.
ILLUSTRATIONS BY Bethany Caskey, except 126, 156, and 157 by Ilona Sherratt

Storey books are available for special premium and promotional uses and for customized
editions. For further information, please call 800-
793-9396.

Storey Publishing
210 MASS MoCA Way
North Adams, MA 01247
storey.com

Printed in China by R.R. Donnelley
10 9 8 7 6 5 4 3 2 1

Library of Congress Cataloging-in-Publication
Data on file

CONTENTS

PREFACE

Chickens were on earth long before humans and still have the same basic needs they always had — food, protection, and procreation. Unlike the jungle fowl from which they derived, domesticated chickens have become dependent on humans to help fulfill those basic needs.

Over the millennia, different breeds developed in different areas for different reasons. As a result, today's breeds range from those that are tiny enough to fit in the palm of your hand to those that are so tall they come nearly to your waist. Colors and patterns range from solid red, white, blue, or black to speckled, striped, and laced. The feathers might be long and thin, short and wide, or furlike and may appear not only on the chicken's body but also on its feet like a pair of boots, down its legs like trousers, beneath its beak like a beard, on the sides of its beak like a mustache, or on top of its head like a fancy Easter bonnet.

Depending on the traits for which these various breeds were bred, some chickens are nearly self-sufficient foragers; others squat by the trough waiting for the next meal. Some retain their innate sense of self-preservation; others don't have the sense to come in out of the rain. Some hens still have the instinct to collect eggs in a nest and hatch them into chicks; others have no interest in motherhood. Most cocks still mate the time-honored way; a few breeds require human intervention in order to fertilize eggs to produce more of their kind.

Meanwhile, we humans have diverse needs and desires. Some of us want lots of tasty eggs; some want meaty chickens that grow large or fast, or both large and fast; some enjoy nothing more than the beauty of brightly colored chickens frolicking in the sunshine; and some take pleasure in the simple companionship of these large land-based birds.

Our various lifestyles enter into the equation of our relationship with chickens. Some of us live on farms with plenty of land for our chickens to freely roam and few neighbors for them to bother. Others live in crowded communities where chickens must be more closely tended to avoid offending neighbors or becoming a meal for a neighborhood pet.

All of this diversity among chickens and humans gives rise to an incredible diversity in our purposes for keeping chickens, how many we keep, the breeds we choose, and the methods by which we shelter and maintain them. For this reason, no one can tell you the one right way to raise chickens, or offer you an established blueprint for keeping your own flock. The best anyone can do is explain the needs of fowl, offer possibilities for fulfilling those needs, and let you pick and choose the options that best fit your particular situation.

And that is the goal of this book.

May your chicken-keeping decisions result in happy, healthy birds that fulfill their purpose in your life and bring you abundant joy.

CHOOSING A BREED

The fun of raising chickens begins right from the start, when you get to choose which color, shape, and size to have. With so many options, you should have no trouble finding the perfect chicken — one that's both picture-pretty and ideally suited to your purpose. Your reasons for keeping chickens will influence your list of breeds to consider, and within each breed, you will encounter differences among varieties and strains that may or may not suit your purposes.

Breeds

No one knows exactly how many chicken breeds exist in the world. A *breed* is a genetically pure line having a common origin, similar conformation and other identifying characteristics, and the ability to reliably produce offspring with the same conformation and characteristics.

The latest edition of the American Poultry Association's *American Standard of Perfection* describes and depicts the one hundred-plus breeds currently recognized by the American Poultry Association (APA), a group that started out as the nation's premier organization for the poultry industry but has since narrowed its focus to exhibition. The American Bantam Association (ABA) publishes its own standard, which doesn't always agree with the APA

standard. Along with those listed in the two standards, other breeds are available in North America, and many more exist in the world. *Storey's Illustrated Guide to Poultry Breeds* offers color photographs and detailed descriptions of most breeds found in North America. (See the Recommended Reading list on page 402 for publishing information for all books referred to in this guide.)

All birds within a given breed share the same skin color, number of toes, carriage, and feathering. Skin color may be yellow like the skin of Cornish, New Hampshires, and Wyandottes, or white like that of Australorps, Orpingtons, and Sussex. Most breeds have four toes, but some, such as the Dorkings, Faverolles, and Houdans, have five. The carriage may be more horizontal, like a Plymouth Rock's, or more vertical, like the Shamo's.

Plumage offers yet more variety. Most roosters have pointed neck and saddle (lower back) feathers, but Sebright and Campine cocks are *hen feathered*, meaning they sport the rounded hackle and saddle feathers of a hen. Naked Necks have no feathers on their necks at all. Other breeds have feathers that form a beard (Faverolle, for example), boots down their legs and feet (Brahma), puffy topknots (Polish), or long, flowing tails (Yokohama).

One of the most important characteristics to consider in selecting chickens for a backyard flock is to choose birds you enjoy looking at.

Birds of most breeds have smooth, satinlike feathers, a result of tiny hooks, called barbicels, that hold a feather's webbing together. The feathers of Silkies, though, lack barbicels, making the birds look as if they're covered with fur. The feathers of Frizzles curl at the ends, giving the birds a permed look. Besides being a distinction of the Frizzle breed, frizzledness is a genetic condition that can be introduced into any breed.

For practical purposes, the various breeds may be grouped according to whether they are primarily *laying breeds*, *meat breeds*, *dual-purpose breeds*, or *ornamental*. The APA divides the breeds into two classifications: **bantam** and **large**. Bantam breeds are one-fifth to one-fourth the size of large breeds. Some bantam breeds are miniature versions of a corresponding large breed; others are distinctive breeds in their own right. Because bantams and their eggs are small, they are considered to be ornamental.

In contrast to the pointed neck and saddle feathers of most cocks (right), the Sebright cock (top) has the rounded neck and saddle feathers of a hen.

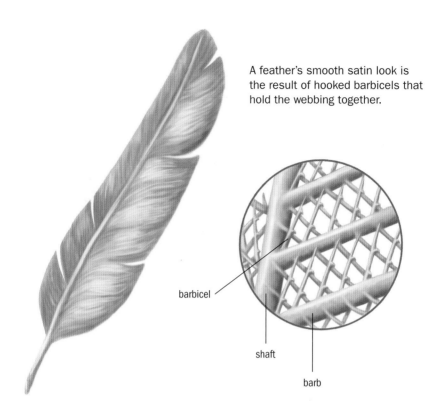

A feather's smooth satin look is the result of hooked barbicels that hold the webbing together.

barbicel

shaft

barb

LARGE BREED GROUPS

Asiatic (APA)	Brahma, Cochin, Langshan
Blue-Egg Layers	Ameraucana, Araucana
Continental (APA)	Barnevelder, Campine, Crevecoeur, Faverolle, Hamburg, Houdan, La Fleche, Lakenvelder, Polish, Welsumer
English (APA)	Australorp, Cornish, Dorking, Orpington, Redcap, Sussex
Game	Aseel, Malay, Modern Game, Old English Game, Shamo
Longtail	Cubalaya, Phoenix, Sumatra, Yokohama
Mediterranean (APA)	Ancona, Andalusian, Catalana, Leghorn, Minorca, Sicilian, Buttercup, Spanish
Other	Appenzeller, Fayoumi, Frizzle, Kraienkoppe, Maran, Naked Neck, Norwegian Jaerhon, Penedesenca, Sultan

EXAMPLES OF UNUSUAL FEATHERING

Frizzle

Naked Neck

Yokohama

Unusual feathering helps define such breeds and varieties as the the Frizzle, with its permed look; the Naked Neck, with its neck devoid of feathers; and the Yokohama, with its long, flowing tail.

Classes

Large and bantam breeds are subdivided into a number of classifications. While these classifications are established primarily to group breeds and varieties for show purposes, they are helpful in understanding the relationships among the various breeds.

Classifications for large breeds indicate their places of origin: American, Asiatic, Continental, English, Mediterranean, and Other (including Oriental). Each large breed is listed in only one class.

Bantams are classified according to specific characteristics: by whether or not they are game breeds; by comb style; and by the presence or absence of leg feathering. Among bantams, the same breed may be represented in different classes by distinctive varieties.

Varieties

Most breeds are subdivided into two or more varieties, usually based on plumage color. Some varieties are established based on feather placement or comb style. Plumage color ranges from a rainbow of solid colors to patterns such as speckled, barred, or laced. Wyandottes, for example, come in several varieties based on color, including solid hues such as buff, black, and white, as well as patterns such as gold or silver lacing.

Varieties defined by feather placement might have, for example, feathers on the legs or under the beak. Frizzle bantams may be clean legged or feather legged. Polish, Booted Bantams, and Silkies may be either bearded or nonbearded.

The most common comb style among chickens is the single comb, a series of upright sawtooth zigzags. Varieties defined by comb style might have buttercup, pea, rose, cushion,

Two varieties of Wyandotte: blue-laced red (top) and silver-laced (bottom)

walnut, strawberry, duplex (cup or V), or carnation combs. Among breeds with varieties defined by comb, Anconas and Rhode Island Reds each have two varieties — single comb and rose comb. Leghorns are an example of a breed that comes in different colors and different comb styles; among the possibilities are buff, black, and silver, with either single or rose combs.

Strains

A strain, or line, is a related family of birds bred with emphasis on specific traits. Strains bred by fanciers are derived from a single breed,

selected for what the owner perceives to be superior qualities. Whether or not these chickens may be called purebred is a matter of contention. Some people argue that they cannot be called purebred because chickens have no registry and therefore no papers as proof of lineage, so the term more accurately should be called *straightbred*.

Whatever you call them, your only guarantee is the owner's word. And it's not uncommon for these breeders to outcross to another breed to avoid close inbreeding or improve certain characteristics such as size or feather color.

FEATHER PATTERNS

barred

laced

Feather patterns include barred, laced, spangled, and penciled.

Commercial production lines are often hybrids — parented by a hen of one breed and a cock of another — developed for efficient egg or meat production. Commercial meat strains and brown-egg layers are usually true hybrids. Commercial white-egg layers, on the other hand, are not hybrids in the strictest sense, since they are bred from different strains of the same breed and variety — single-comb white Leghorn.

Whether hybrid or purebred, birds within a strain are so typical of the strain that an experienced person can recognize the strain at a glance. An established strain is usually identified by its developer's name — commonly a corporate name — for meat or egg production strains, and the breeder's name for noncommercial strains, such as those bred for exhibition, fly tying, and illegal cockfighting.

Learning about strains becomes important if you decide to specialize in a specific breed and discover your chickens do not entirely fit the published breed profiles relative to such things as temperament, rate of egg production, size or shape, and so on. You then have the choice of seeking a more suitable strain or developing your own strain through selective breeding.

spangled

penciled

LARGE BREEDS
(See key below for meanings of symbols)

Breed	Varieties	Eggs	Meat	Ornamental	Forager	Climate	Temperament
Ameraucana	8	*	†			Hardy	D
Ancona	2	**			FF	Hardy	Fl
Andalusian	1	*		§	FF	Heat tolerant #	Fl
Appenzeller	3	*		§	FF	Cold hardy	Fl
Araucana	8	*		§		Hardy	D
Aseel	5		†	§		All	D (A)
Australorp	1	*	†			Cold hardy	D
Barnevelder	1	*	†		F	Hardy	D
Brahma	5		†	§	F	Cold hardy	D
Buckeye	1	*	†		F	Cold hardy	D (A)
Campine	1	*		§	F	#	D
Catalana	1	*	†		F	Heat tolerant #	Fl
Chantecler	2	*	†		F	Cold hardy	Fl
Cochin	18		†	§	F	Cold hardy	D
Cornish	12		††	§		Temperate	D (A)
Crevecoeur	1			§		Temperate	Fl
Cubalaya	4			§	F	Heat tolerant	A
Delaware	1	*	††	§	F	Hardy	D
Dominique	1	*	†		F	Hardy	D
Dorking	5	*	†		F	#	D
Faverolle	5	*	†	§		Hardy	D (A)
Fayoumi	1	**			FF	Heat tolerant #	Fl
Frizzle	2			§		Temperate	D
Hamburg	6	*		§	F	Temperate	Fl
Holland	1	*	†		F	#	D
Houdan	2	*	†	§	FF	Temperate	D
Java	3	*	†	§	F	Cold hardy	D
Jersey Giant	2		†	§		Cold hardy	D
Kraienkoppe	2			§	FF	Cold hardy	Fl
La Fleche	1	*	†	§	F	Hardy	Fl
Lakenvelder	1	*		§	F	#	Fl
Langshan	3	*	†	§	F	Cold hardy	D
Leghorn	16	**			F	Heat tolerant #	Fl
Malay	7		†	§	F	Heat tolerant	Fl
Maran	9	*	†		F	#	Fl
Minorca	5	**	†		FF	Heat tolerant #	Fl
Modern Game	18			§	F	Heat tolerant	A

LARGE BREEDS
(See key below for meanings of symbols)

Breed	Varieties	Eggs	Meat	Ornamental	Forager	Climate	Temperament
Naked Neck	6	*	†	§	F	Temperate	D
New Hampshire	1	*	†		F	#	D
Norwegian Jaerhon	2	**			F	Cold hardy #	A
Old English Game	34			§	FF	Hardy #	A
Orloff	6		†	§		Cold hardy	D
Orpington	4	*	††		F	Cold hardy #	D
Phoenix	7			§		Temperate	D
Polish	10	*		§		Temperate	D
Redcap	1	*	†	§	FF	Hardy	Fl
Rhode Island Red	2	*	†		F	#	D (A)
Rhode Island White	1	*	†		F	Hardy	D
Shamo	8	*	††	§		Temperate	D (A)
Sicilian Buttercup	1			§		Heat tolerant	Fl
Spanish	1			§	FF	Heat tolerant	Fl
Sultan	3			§		Temperate	D
Sumatra	1			§		Heat tolerant	Fl (A)
Sussex	7	*	†	§	F	#	D
Welsumer	1	*	†		FF	Cold hardy #	D
Wyandotte	18	*	†	§	F	Cold hardy	D (A)
Yokohama	5			§		Temperate	D

Key to Symbols

*	lay well
**	lay exceptionally well
†	especially suitable for meat production
††	are exceptional
§	stand out for aesthetic qualities
F	breed forages well
FF	will forage for a high percentage of sustenance
#	breed having cocks with large combs prone to frostbite
D	generally docile and suitable for backyard situations and as pets
(A)	breed known to have aggressive individuals
Fl	breed that tends to be flighty to a greater or lesser extent
A	aggressive breed that is best left to experienced keepers

BREED BY COMB STYLE

Rose Comb

Rhode Island White,* Wyandotte

Rose Comb (spiked)

Ancona,* Dominique, Dorking,* Hamburg, Leghorn,* Minorca,* Redcap, Rhode Island Red,* Rosecomb bantam

Cushion Comb

Chantecler

Walnut Comb

Kraienkoppe, Orloff, Yokohama

Single Comb

Ancona,* Australorp, Barnevelder, Blue Andalusian, Campine, Catalan, Cochin, Delaware, Dorking,* Fayoumi, Faverolle, Frizzle,** Holland, Java, Jersey Giant, Lakenvelder, Langshan, Leghorn,* Maran, Minorca,* Modern Game, Naked Neck, New Hampshire, Norwegian Jaerhon, Old English Game, Orpington, Phoenix, Plymouth Rock, Rhode Island Red,* Rhode Island White,* Spanish, Sussex, Welsumer

Carnation Comb

Penedesenca

V Comb

Appenzeller, Crevecoeur, Houdan, La Fleche,
Polish, Sultan

Buttercup Comb

Sicilian Buttercup

Strawberry Comb

Malay

Ameraucana, Araucana, Aseel, Brahma, Buckeye,
Cornish, Cubalaya, Shamo, Sumatra

Pea Comb

  * Breeds having varieties based on differing comb styles.
 ** The condition of frizzledness may appear in combination
    with any comb style.

Breed Selection

Way back in time, all of today's many breeds, varieties, and strains had a common origin: the wild red jungle fowl of Southeast Asia. Over tens of thousands of years, chicken keepers selectively bred their flocks to favor different combinations of characteristics related to economics, aesthetics, and other factors, such as aggressiveness or duration of the cocks' crow.

The availability of all these various breeds, varieties, and strains ranges widely from common (and therefore inexpensive) to extremely rare (and therefore quite dear). While crossbred production strains are as common as dirt, pure strains of some breeds can be hard to come by. Not every hen that lays a blue egg is an Ameraucana or Araucana, for example, and most hens sold as New Hampshires are crossed with Rhode Island Reds or some other production breed. Indeed, among the brown-egg-laying "pure" New Hampshire hens I once bought, some laid blue eggs.

Layer Breeds

If you have never seen or tasted homegrown eggs, you will be amazed at their superior color and flavor. All hens, unless they are old or ill, lay eggs. Breeds known as "layers" lay nearly an egg a day for long periods at a time. Other breeds might enjoy longer rest periods between bouts of laying, or else go broody — when they have such a strong nesting instinct, they'd rather hatch eggs than lay them. Still others may be just as prolific as the typical layer breeds but eat more feed per dozen eggs. Efficient laying hens share four desirable characteristics:

- They lay a large number of eggs per year.
- They have small bodies.
- They begin laying at 4 to 5 months of age.
- They are not inclined to nest.

The best layers average between 250 and 280 eggs per year, although individual birds may exceed 300. Compared to larger hens, small-bodied birds need less feed to maintain muscle mass. Purebred hens rarely match the laying abilities and efficiency of commercially bred strains but can still be efficient enough for a backyard flock. Since a hen stops laying once she begins to *set* (nest), the best layers don't readily go broody.

Some of the Mediterranean breeds, especially Leghorn, are particularly efficient layers. These breeds, and a few others specializing in egg production, tend to be high strung, however, and therefore not much fun to work with, especially if you take up chicken keeping for relaxation. Another characteristic of Mediterraneans is that they lay white-shelled eggs.

Commercial brown-egg hybrids lay nearly as well as Leghorn-based strains and are not as flighty, because they're derived from breeds in the American classification, which tend to be more laid-back than the Mediterraneans. A popular brown-egg strain is the Hubbard Golden Comet, a buff-colored bird called "the brown-egg layer that thinks like a Leghorn." You can expect 180 to 240 eggs per year from a commercial-strain brown-egg layer.

Purebred brown-egg layers generally lay fewer eggs than commercial hybrids but still lay respectably enough for a backyard flock. Shell color ranges from pale tan to dark brown. The darkest eggs come from Barnvelders, Marans, Penedesencas, and Welsumers, although none of these breeds lays spectacular numbers of eggs.

Many breeds originally developed for laying did quite well in their time, but their laying abilities don't stack up to the production rate of scientifically bred modern strains. Then, too, many breeds that once had superior laying

LAYER BREEDS

Ancona

Leghorn

Marans

The most efficient layers are small bodied and flighty — like the Ancona or Leghorn — and lay eggs with white shells.

The heavier-bodied Marans lays fewer eggs, but with dark brown shells, making it popular for backyard flocks.

abilities are now bred for exhibition, where production is less important than appearance and therefore not high on the list of traits considered necessary for breeding-stock selection. If you have your heart set on a fancy breed that doesn't lay well, your options are to look for a strain bred for production as well as appearance (which is rare), expand your flock to obtain the requisite number of eggs (requiring the expense of more space and feed), or develop your own laying strain (an option that may take years but could be fun as well as rewarding).

The most efficient laying breeds all tend to be nervous or flighty. Kept in small numbers in uncrowded conditions, with care to avoid stress (such as being chased by dogs or children) and extra time spent ensuring their comfort around people, these breeds can work fine in a backyard setting.

After a few years, layers become *spent*, meaning they slow down in production. At that point they don't have much meat on their bones, since their energy has been concentrated on laying. A good layer fleshes out slowly and never would have made a good meat bird in the first place. If your poultry interest leans toward grilled chicken, consider a meat breed instead.

Meat Breeds

People who raise chickens for meat enjoy better-tasting, more healthful, and safer poultry than is generally available at the supermarket. Any healthy chicken may be prepared for dinner, although some breeds are more suited to meat production than others. Efficient meat strains share four characteristics:

- They grow and feather rapidly.
- They reach target weight in minimum time.
- They are broad breasted.
- They have white feathers for clean picking.

The more quickly a bird grows to butchering weight, the more tender it is and the cheaper it is to raise. The most efficient meat strains were developed from a cross between Cornish and an American breed such as New Hampshire or Plymouth Rock. The 1- to 2-pound (0.5 to 0.9 kg) Cornish hen (as a commercial marketing ploy, sometimes called a Cornish game hen even though it's not a game breed) is nothing more than a 4-week-old Rock-Cornish hybrid. A commercial meat bird eats just 2 pounds of feed for each pound of weight gained. A hybrid layer, by comparison, eats three to five times as much for the same weight gain.

Raising a hybrid meat flock is a short-term project. You buy a batch of Cornish-cross chicks, feed them to butchering age, dispatch them into the freezer, and enjoy the fruits of your labor for the rest of the year. Since these birds aren't around long, performance as a meat bird (the ability to grow quickly on the least possible amount of feed) takes precedence over appearance.

Purebreds are not as efficient as hybrids at converting feed to meat, but some are heavy bodied enough to make respectable meat birds. Because of their slower growth, their meat is more flavorful than that of a fast-growing hybrid. Breeds originally developed for meat include Brahma, Cochin, and Cornish. Although the Jersey Giant grows to be the largest of all breeds, it does not make an economical meat bird because it first puts growth into bones, then fleshes out, reaching 6 months of age before yielding a significant

MEAT BREEDS

Cornish

Shamo

Orpington

Meat breeds are broad breasted and tend to be more laid-back than layers; Shamos, although friendly with people, are aggressive among themselves and toward other chicken breeds.

DUAL-PURPOSE BREEDS

Plymouth Rock

Rhode Island Red

Faverolle

Dual-purpose breeds are ideal for family self-sufficiency because they lay better than meat breeds and grow bigger than layer breeds.

amount of meat. Many backyard chicken keepers, regardless of their chosen breed or purpose in having them, hatch an annual batch of chicks and put the extra cockerels into the freezer.

Dual-Purpose Breeds

If you want the best of both worlds — eggs and meat — you have two choices: keep a year-round laying flock and raise a batch of meat birds on the side, or compromise by keeping one dual-purpose breed. Dual-purpose chickens don't lay as well as laying hens and don't grow as fast or as big as meat birds, but they lay better than meat birds and grow faster and larger than laying hens.

Dual-purpose breeds are the classic backyard chickens. Their chief advantage over a laying breed is that young excess males and spent layers are full breasted and otherwise have an appreciable amount of meat on their bones. Their advantage over a meat breed is that the hens lay a reasonable number of eggs for the amount of feed they eat.

Most breeds in the American and English classifications are dual-purpose, although many others, including some of the breeds typically considered ornamental, are equally versatile. Among these dual-purpose breeds, some are slightly more efficient at producing eggs, while others grow bigger and tend to go broody, tilting them more toward use as meat birds over layers. These characteristics vary not only from breed to breed, but differ from strain to strain within the same breed. As among egg-laying breeds, dual-purpose strains developed for show are generally prettier than they are useful.

A few hybrids have been developed as efficient dual-purpose birds. The most popular of these hybrids are the Black Sex Link and the Red Sex Link (so-called because the chicks' sex may be determined by the color of their down). The Red Sex Link lays about 250 eggs a year. The Black Sex Link lays slightly fewer but larger eggs and weighs 1 pound (0.5 kg) more at maturity. If your purpose in keeping a dual-purpose flock is for self-sufficiency and to that end you plan to hatch your own future replacement chicks, hybrids are not the way to go, since they do not breed true.

Low-Maintenance Breeds

Some breeds are inherently more self-reliant than others. Chickens that have been bred in confinement for generations are generally less aggressive foragers than breeds that have been allowed to exercise their foraging instinct. In the South, for example, you commonly see Old English Games wandering along country lanes. As the closest domesticated kin to the ancient wild jungle fowl, they are not as plump or prolific as the dual-purpose breeds but compensate by being almost entirely maintenance-free.

Breeds that are not aggressive foragers, or should not be required to forage, include those with heavy leg feathering or large crests. Leg feathers inhibit scratching the ground to turn up food. Crests offer head protection in cold weather but inhibit vision, making crested breeds easier prey. Crests are also known to freeze in wet winter weather. In freezing weather, breeds with tight combs such as cushion, pea, or rose cope with the cold better than breeds with large single combs. If you plan to pasture your birds, choose a breed suitable for your prevailing climate.

Two other characteristics that enhance a breed's self-sufficiency are strong reproductive instincts and plumage color. Breeds that have

LOW-MAINTENANCE BREEDS

Houdan

Old English Game

Penedesenca

Low-maintenance breeds are good foragers that have other-than-white feathers for camouflage, tend toward broodiness, and are suitable for almost any local climate; crested breeds like the Houdan should not be left out in wet, freezing weather.

retained their instinct to brood require no human intervention to reproduce, in contrast to their specialized industrial cousins, whose brooding instinct has been taken away. Some breeds, notably the Cornish, have been so distorted in the quest for a broad-breasted meat bird that the cocks have difficulty mounting a hen. Feather colors other than white blend more easily into the surroundings, offering birds protection from predators.

Ornamental Breeds

In contrast to production birds — chickens that are kept for meat or eggs — *ornamental birds* are kept primarily for aesthetic reasons, and many keepers like to show them off at poultry exhibitions. While the same breed might be raised for both show and production, rarely will you find exhibition and production qualities in the same strain.

The ideal shape, or type, for each breed is described by the relevant standard. A chicken coming close to the ideal for its breed is said to be "true to type" or "typey." Production strains are generally less typey than exhibition strains, since their owners emphasize economics rather than aesthetics. By the same token, the typier exhibition strains tend to be less efficient at producing meat and eggs, since their owners emphasize appearance over production. When you think about it, that doesn't make much sense, because a breed's type is directly related to its original production purpose, but exhibitors tend to select for showy, exaggerated type to the detriment of production.

Even among exhibition strains, not all birds are created equal, since different exhibitors choose different characteristics to emphasize or exaggerate. The *American Standard* states, for example, that a Sebright should have a short,

well-rounded back, yet I have seen more than enough Sebrights with long, straight backs and have shown under a judge who preferred them that way — proving the old adage that beauty is in the eye of the beholder.

Endangered Breeds

In 1868, Charles Darwin published an inventory of chicken breeds that existed at the time — all 13 of them. Most of the breeds we know today have been developed since then. Breeds and varieties proliferated in the United States between 1875 and 1925, fueled by interest in both unusual exhibition birds and dual-purpose backyard flocks.

This incredible genetic diversity has since fallen victim to the whims of fowl fads. The decline began in the 1930s, when new zoning ordinances prohibited the raising of chicken flocks in many backyards, and emphasis began shifting to large-scale commercial production strains requiring higher-energy feed and climate-controlled housing. Among the hardest-hit breeds were North America's two oldest: Dominique, the oldest American breed, and Chantecler, the oldest Canadian breed.

In an ever-speedier downward spiral, birds that had been valued for their appearance declined in popularity, and poultry shows became less frequent; as the number of poultry shows dwindled, interest in chickens declined further. The Depression added its impact — people could no longer afford to keep chickens just "for pretty."

Interest in exhibiting regained some of its popularity in the affluent 1950s, but by then some of the more exotic breeds had all but disappeared. In the spring of 1967, poultry fancier Neil Jones warned *Poultry Press* readers that a serious effort was needed to preserve

ORNAMENTAL BREEDS

White-faced Black Spanish

Cochin

Polish

Ornamental breeds offer a wide range of interesting features.

ENDANGERED BREEDS

Fayoumi

Orloff

Redcap

Some breeds are quite rare and face extinction without a serious conservation effort.

these endangered ornamental breeds. Jones's warning led to the birth of the Society for the Preservation of Poultry Antiquities (SPPA), the mission of which is to perpetuate "and improve" rare breeds. To this end, the SPPA publishes an annual critical list of historical show breeds, designating as rare those breeds and varieties not widely available and seldom seen at poultry shows. (See the Resources pages in the appendix for the SPPA's contact information and that of all other agencies referred to in this book.)

Dual-purpose breeds experienced a comeback during the self-sufficiency movement of the 1960s and 1970s, then declined at an alarming rate as people abandoned country life in their scramble for paying jobs. Those of us who stuck with our classic breeds took our domestic chickens for granted (after all, they'd been around since Grandma's time). Unaware of the dynamics at work, most of us never dreamed they were becoming irreplaceable.

I count myself among the unwary but guilty who did not fully appreciate the value of my birds. During the 1970s, I had a wonderful flock of New Hampshires, which in the 1980s I dispersed before moving cross-country, thinking I would replace them when I got resettled. I never again found a strain that equaled those New Hamps in uniform appearance, rapid growth, steady laying ability, and laid-back disposition. During my fruitless quest, I heard countless similar stories.

The Livestock Conservancy periodically conducts a survey to identify the most endangered old-time production breeds significant to the United States, locate existing flocks, and tally their numbers. Their survey is based on respondents answering ads in a few publications and queries sent to hatcheries and to members of the National Poultry Improvement Plan handling nonindustrial breeds. Their sample base is quite small, and the number of volunteer responses even smaller, so quite likely the survey misses countless family flocks quietly scratching in backyards throughout North America. But even though the numbers may not be entirely accurate, the fact remains that old-time agricultural breeds are losing ground.

Rare Breeds Canada (RBC) — dedicated to conserving, evaluating, and studying heritage, rare, and minor breeds of Canadian farm animals — also periodically publishes a conservation list, which is considerably shorter than those of the SPPA and The Livestock Conservancy. Like The Livestock Conservancy, RBC's mission is to preserve endangered agricultural breeds by increasing numbers, to assist people in finding particular breeds, and to educate the public about the need for breed preservation.

Some people argue that poultry breeds become extinct for lack of interest and therefore no longer have a purpose in the modern world. Others argue that losing these irreplaceable breeds depletes the overall genetic pool, resulting in the loss of such valuable traits as disease resistance, the brooding instinct, superior taste, and an ability to live naturally. They feel these heritage breeds must be preserved to safeguard the survival of poultry into the future. Since most noncommercial breeds and varieties are endangered to a greater or lesser extent, raising any breed — or supporting people who raise a pure breed by buying chicks, eggs, or meat birds—helps ensure their survival.

BANTAMS

Bearded d'Anvers

Nankin

Japanese

A breed having no large counterpart is considered to be a true bantam.

Bantams

Bantams are miniature chickens generally weighing 2 pounds (0.9 kg) or less. Their history closely follows that of the Industrial Revolution and the movement of families away from farms. Folks who didn't want to give up their chickens turned to miniatures that require little backyard space, don't eat much, don't mind being confined, and respond well to human relationships. Interest in keeping bantams boomed in the affluent 1950s, when raising chickens in the backyard was considered good family fun; interest waned in later decades, and then came back stronger than ever at the turn of the 21st century.

Nearly every breed and variety of large chicken has a bantam version that's one-fifth to one-fourth its size. A few breeds and varieties come only in bantam size. Some people make a distinction between bantams that have a large counterpart and those that do not, calling the former *miniatures* and the latter *true bantams*. The *Bantam Standard*, published by the American Bantam Association (ABA), lists many more varieties than are listed in the APA *Standard*.

Banties are popular as pets, as exhibition birds, and as ornamentals that add character to the yard or garden. Although their eggs are smaller than those of larger breeds, some banty strains are prolific layers. The chunkier breeds may not rival commercial Rock-Cornish in growth rate, but they look and taste just as good, if not better.

Broody Breeds

If you have your heart set on seeing your hens hatch their own chicks, you'll want a breed that is known for hens that brood successfully. Except among commercially bred strains, the majority of hens will brood, some more successfully than others. Two outstanding broodies are Silkie and Aseel; hens of both breeds are often used to hatch eggs laid by other breeds.

As a general rule, heavy hens tend to make good broodies that can handle large numbers of eggs, but a really heavy hen with a loaded nest may break some eggs. Feather-legged breeds may have a problem with leg feathers accidentally flicking eggs out of the nest.

Because laying stops when setting starts, the instinct to gather eggs in a nest and keep them warm for 21 days until chicks hatch has, over the years, been selectively bred out of layer strains. The light, flighty laying breeds tend not to go broody, and when they do, they are not reliable.

Even among breeds that are not typically broody, the occasional hen will take a notion to hatch out some chicks. The accompanying table lists breeds that are least likely to produce broody hens.

May Brood	
Appenzeller	Rhode Island Red
Faverolle	Welsumer
Malay	Wyandotte
Maran	

Rarely Brood	
Ancona	Minorca
Andalusian	Norwegian Jaerhon
Barnvelder	Orloff
Campine	Polish
Catalana	Redcap
Crevecoeur	Rhode Island White
Hamburg	Rosecomb (bantam)
Hybrid layers	Sebright (bantam)
La Fleche	Serama (bantam)
Lakenvelder	Sicilian Buttercup
Leghorn	Sultan

Chickens as Pets

Chickens kept as pets may be any breed that appeals to you and has a calm disposition. If you like the looks of a big, heavy meat breed like the Jersey Giant or Cochin, and you have the space to accommodate them, by all means go ahead and enjoy having a few roam your backyard, since the economy of meat production won't be an issue. If space is limited, at the other end of the scale is the Serama, the tiniest of all chickens, weighing as little as three-quarters of a pound (0.3 kg) and found in every possible color. Even the breeds typically known to be flighty can work in small numbers if you start them as chicks and spend a lot of time with them, although some breeds tend to be too nervous to trust around small children.

Breeds to avoid are those typically described as aggressive — it's no picnic to be bending over filling a feed hopper and have a belligerent rooster mount a rear attack or, worse, fly in your face. Mean individuals occasionally appear in nearly any breed and are more typically cocks than hens. Despite anything you might hear about taming an ornery rooster, your best bet is to get rid of it before you, a family member, a neighbor, or a young child gets seriously injured. The calmer breeds tend to be more easygoing in urban or suburban confinement.

Where the nearest neighbor is just over the fence, consider the noise factor, especially if you intend to keep a rooster. Cocks of the larger breeds have a deep crow that doesn't travel far, while the smaller breeds have a high-pitched crow that travels a good distance and can be annoying close up, especially to someone other than the owner. The difference is similar to the woof-woof of a big dog compared to the yip-yip-yipping of a small one.

Jersey Giant

Serama

From the largest to the smallest, many breeds make ideal pets.

House chickens are becoming increasingly more common as domestic pets. If that is your aim, you certainly want a docile breed. You'll also want to make sure no one in your family is allergic to chicken dander. The easiest way to do that is for your family to spend time visiting someone else's chickens before making your own commitment. And, since chickens can't easily be housebroken, you'll need a plan for dealing with chicken "accidents."

In narrowing down your breed choice, if you can't decide which breed you like best, and you don't plan to raise future generations for show or sale, get one of each. My own first chickens came with the first house I owned, its backyard prepopulated with a "Heinz 57" flock. My already keen interest in chickens deepened as I had the fun of learning to identify the breed of each bird I had acquired.

Breeds for Feathers

Colorful chicken feathers are used for making jewelry and home decorations, trimming hats and other clothing, and tying fishing flies. Different crafts require different kinds of feathers. For some crafts, wing and tail feathers are preferred. For fly tying — the most lucrative feather market — only the hackle and saddle feathers of cockerels have value.

The hackle is ideal for fly tying because it's lightweight and floats on water, like the insect the tie is designed to imitate. The ideal hackle is long — as long as 12 inches — narrow, and free of webbing, and it has a strong, flexible shaft. The best feathers come from fast-growing hard-feathered breeds in colorful varieties, such as barred Plymouth Rock, blue Andalusian, buff Minorca, silver-penciled Wyandotte, and crele Penedesenca. The Coq de Leon from Spain is an ancient chicken specifically bred for fly tying; its feathers are so suited to the purpose that even hen feathers may be used.

Flocks are selectively bred for feather color and shape, and specially fed for optimum feather production. When the feathers are prime for harvest, the cockerel is killed and his hackles removed in a cape with the skin attached. Saddle feathers are sometimes individually harvested, but more often are removed as a saddle patch or as a unit together with the cape. Capes and patches are then dried for use or sale.

BUYING REPLACEMENTS VERSUS HATCHING YOUR OWN

An important factor in deciding on which breed to raise is whether or not you want to be self-sustaining and hatch your own future replacement chicks (in which case your only options are the pure breeds), or whether you will be content to buy chicks when time comes to replenish your flock (in which case you might consider hybrids). Some Extension agents admonish backyard poultry keepers to purchase new chicks, rather than hatch eggs from their own flocks. The rationale is that repurchasing chicks breaks the disease cycle that otherwise accumulates in a chicken yard or incubator. In my experience just the opposite is true: If you take great care to keep your flock healthy, you're better off hatching future chicks from your own birds than bringing in new ones later and running the risk of introducing problems.

You can turn a good profit marketing feathers or selling products made from them. First you have to find out what kind of feathers are in demand for the market that interests you and what those feathers are worth. Then learn how to harvest, process, and package feathers for that market. The most successful feather sellers are associated in some way with the craft for which their feathers are used.

Of course, you need to find out which chickens grow the priciest feathers, and in some cases you might breed and sell the chickens themselves. I've been visited by a fly tyer looking for suitable birds to include in his own breeding program. I've made scores of feather earrings, have woven feathers into tapestries, and have furnished feathers to other weavers. In my living room I keep a ceramic vase filled with an ever-changing bouquet of colorful poultry feathers.

Purebred versus Hybrid

Whether you keep purebreds or hybrids depends in good part on whether you wish to incubate your flock's eggs. Purebreds, also called straightbreds, will breed true, meaning their chicks will grow up to be pretty much like the parents. Hybrids won't breed true. Because they result from matings between different breeds (or highly specialized strains within a single breed), the only way to get more chickens exactly like them is to reproduce the same cross. So a major issue in deciding between hybrids and straightbreds is whether or not you wish to perpetuate your flock by hatching eggs from your own chickens.

It's true that hybrids are more efficient than purebreds at egg or meat production, but they also require expensive high-quality feed. By contrast, many purebreds do well on forage and table scraps as supplements to commercial rations, and they also enjoy a longer productive life. On the other hand, if you plan to raise meat birds to stock your freezer, hybrids will get you there quicker than purebreds — but won't taste nearly as good.

The decision of whether to keep purebreds or hybrids may depend on whether you intend to show. Except for 4-H shows involving production birds, most shows require entries to conform to breed descriptions in the *Standard*. Show birds should be purebred, although some

PLACATING NEIGHBORS

Neighbors are less likely to take exception to your chickens if you avoid breeds that are especially noisy or tend not to stay in their own yards.

Noisy Breeds		Flying Breeds	
Ancona	Leghorn	Ancona	Lakenvelder
Andalusian	Modern Game	Andalusian	Leghorn
Cornish	Old English Game	Campine	Old English Game
Cubalaya	Spanish	Fayoumi	Most bantams
		Hamburg	

breeders secretly cheat by crossing their birds with other breeds to improve such things as plumage color or comb type.

If you're raising chickens as pets or just for fun, you might want a mix of interesting breeds. For this purpose, some hatcheries sell ornamental assortments.

Which First — Chicks or Eggs?

Once you know what kind of chickens you want, the next step is to decide whether to purchase eggs to hatch, newly hatched chicks, started birds, or full-grown chickens. Each option has distinct advantages and disadvantages.

REGULATIONS

Before bringing home your first birds, check your local zoning laws and other ordinances. Regulations may limit or prohibit chicken-keeping activities in your area and may pertain to birds bought, sold, traded, shown, shipped, bred, or hatched. Two common regulations restrict how many chickens you may keep and how far they must be from your property line. Obtain information from your town or county zoning board or Extension agent and from your state poultry specialist or veterinarian.

Even if specific laws don't pertain, consider the possibility that grumpy neighbors may file a nuisance suit if your chickens make too much noise or get into their garden and scratch up the petunias. If that seems likely, you'll fare better by avoiding breeds that are noisier than most or tend to fly more than most.

Hatching Eggs

Hatching eggs are fertilized eggs that will hatch if incubated for 21 days. Starting out with a small incubator and a dozen or so eggs can be a fun and educational project, but finding fertile eggs of your chosen breed may be a challenge. And if you do find them, the project may end in disappointment if the eggs are suboptimal for hatching, the incubator's temperature or humidity isn't properly set, the power goes out, or any number of other things go wrong.

Once you overcome the challenges of hatching, you have a new set of challenges in caring for freshly hatched chicks. Unless you are particularly adventuresome, or have prior experience running an incubator, consider starting out with live birds.

Baby Chicks

Chicks are a surer bet than hatching eggs and are usually cheaper than older birds. If you want your chickens to be pets, chicks will bond with you more easily than started or mature birds. Baby birds shipped any distance travel much better than eggs for hatching and are less likely than older birds to bring disease into your yard. Chicks come in two options: sexed and unsexed.

Unsexed chicks — also called *as-hatched* or *straight run* — are mixed in gender exactly as they hatch, or approximately 50 percent cockerels (males) and 50 percent pullets (females). Some people swear hatcheries stack the deck by throwing in extra cockerels, since the mix often comes out more like 60/40 or even 75/25, but some hatches just naturally turn out to be nearly all cockerels, while others are nearly all pullets.

Sexed chicks are sorted so you get exactly as many pullets or cockerels as you want. Within a given breed, sexed pullets cost the most, straight run next, and sexed cockerels the least. Cockerels have the least value because a flock needs fewer roosters than hens, or none at all if you or your neighbors don't want to hear crowing.

If you're establishing a laying flock, you can be sure to get the number of hens you need by buying sexed pullets, but starting with chicks means you'll have to wait several months before gathering your first eggs. You do not need a rooster to get eggs, and roosters can sometimes be rough on the hens, although many people feel that including at least one cock in the flock doubles the enjoyment of having chickens.

For a dual-purpose flock, you might start out with a batch of straight-run chicks and raise the surplus cockerels for the freezer. If your poultry project is strictly for meat, you can save money on chicks and grow out the birds faster by getting all cockerels.

If you want your birds to compete in shows, the downside to starting with chicks is you won't know if you have a potential winner or quality breeder until the chicks mature. And regardless of your purpose, chicks require more care at the outset than started or mature birds.

Started Birds

Started birds, when you can find them, are partially grown and make a good deal if you don't want the bother of brooding chicks. For a laying flock, *started pullets* have two advantages: you won't spend much time feeding unproductive birds, since they'll soon begin laying; and because the birds are just coming into lay, you'll have them for the longest possible productive life compared to full-grown hens.

Started birds are also a good option if you plan to show. They're less expensive than proven show birds, but also less likely to have serious faults than chicks, since birds with serious faults are culled early. Just make sure you are not acquiring the culls.

CHICKS SHIPPED BY MAIL

When you order chicks by mail, open the box in front of the mail carrier to verify any claim you may have for losses. Introduce the chicks to household pets to let them know the chicks are yours and shouldn't be touched. Provide the chicks with heat and water as soon as possible after their arrival.

Mature Birds

Full-grown birds are the most expensive but offer the fewest surprises, since you see exactly what you're getting. Two unpleasant surprises you can get unwittingly are disease and excessive age. The older a bird gets, the longer it is exposed to potential diseases and the more likely it is to carry one. That goes double if the bird has been traveling the show circuit.

Excessive age can be a serious problem if you're buying mature birds for laying or breeding. Eggeries commonly sell spent hens to unsuspecting people wishing to start their own backyard laying flocks. These unwary folks believe they're getting the best hens from a commercial operation specializing in egg production.

If you think about it, though, a commercial farm isn't likely to sell its best layers. The most you can hope for from a place like that is cheap stewing hens, but even then they'd have little meat on their bones.

Examining Birds

To avoid getting stuck with chickens that are unhealthy or past their prime, learn to recognize a healthy chicken and to tell the difference between a young one and an old one. Although most sellers won't try to palm off unhealthy birds, and most will freely tell you how old their birds are, the occasional unscrupulous seller sees the wonderstruck novice buyer as an opportunity to turn unwanted birds into cash.

Determining Health

When you buy grown chickens, look for bright eyes; smooth, shiny feathers; smooth, clean legs; and full, bright combs. When you buy chicks, make sure they are bright-eyed and perky. If they come by mail, open the box in front of the mail carrier to verify your refund or replacement claim in case any have died.

A well-kept bird of any age is parasite free. You can check for parasites by peeking under the wings and around the vent — external parasites may be visible; internal parasites often cause diarrhea that sticks to vent feathers.

If you visit the seller in person, listen for coughing or sneezing in the flock — when a few chickens catch cold, chances are good the whole flock is coming down with it. Old-time poultry keepers whistle whenever they near a flock; the birds quiet down and listen to see what's making the whistling sound, and any coughs and sneezes are easier to hear.

One way to be sure you are getting healthy birds is to purchase from a flock that's enrolled in the National Poultry Improvement Plan (NPIP), which certifies flocks to be free of several serious diseases. As part of their biosecurity agreement, NPIP members will not allow you to visit their flocks. And you may not find an NPIP member who has the kind of birds you want, because a lot of poultry breeders don't want to get tied up in government bureaucracy, although that does not automatically mean their chickens are unhealthy.

Determining Age

To make sure you aren't getting an old, worn-out bird, carefully look it over. You can never be certain of a chicken's exact age (you can hardly, for example, check its teeth as you would a horse), but you can always tell a young bird from an old one.

Cockerels and pullets tend to look like gangly teenagers compared to the more rounded, finished look of a cock or hen of the same breed. Cockerels and pullets have smooth legs. Older birds have rough scales on their legs.

── FEATURES OF A HEALTHY CHICKEN ──

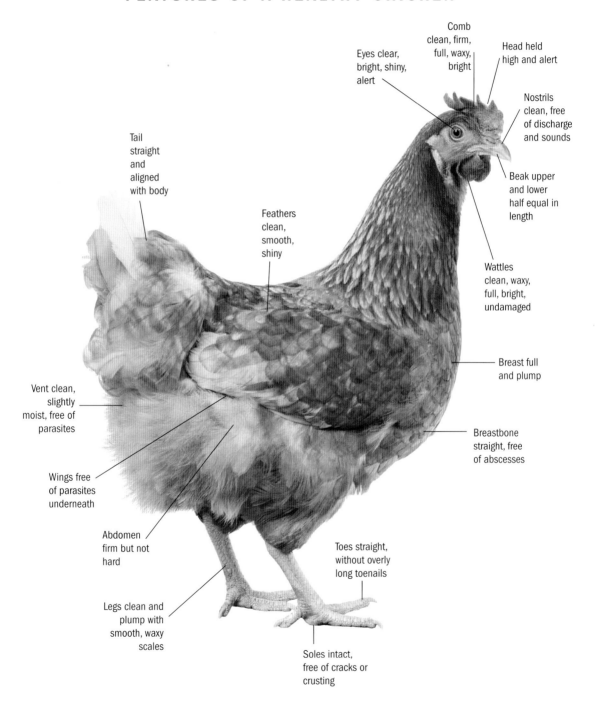

Comb
clean, firm,
full, waxy,
bright

Head held
high and alert

Eyes clear,
bright, shiny,
alert

Nostrils
clean, free
of discharge
and sounds

Beak upper
and lower
half equal in
length

Tail
straight
and
aligned
with body

Feathers
clean,
smooth,
shiny

Wattles
clean, waxy,
full, bright,
undamaged

Breast full
and plump

Vent clean,
slightly
moist, free of
parasites

Breastbone
straight, free
of abscesses

Wings free
of parasites
underneath

Abdomen
firm but not
hard

Toes straight,
without overly
long toenails

Legs clean and
plump with
smooth, waxy
scales

Soles intact,
free of cracks or
crusting

Some pullets and all cockerels have little nubs where their spurs will grow. Cocks, and some hens, have long spurs; the longer the spur, the older the bird.

To confirm your findings, pick up the bird and examine it by feel. The breastbone is fairly flexible in a young bird, quite rigid in an older bird. The muscle is soft in a young bird, firm in an older bird. The skin is papery thin and somewhat translucent in a young bird, thick and tough in an older bird. A young bird will, in general, feel light compared to the solid, heavy feel of an older bird.

Getting Started

The best place to buy birds depends on what you want. If it's a commercial hybrid strain, your only choice is a hatchery. Unfortunately, some hatcheries churn out large numbers of low-quality chicks. The same can be true of chick brokers — feed stores and mail-order houses that market chickens from outside sources, often with little knowledge of or concern for the birds' condition or bloodlines.

Some hatcheries specialize in exhibition breeds, but rarely do they sell prize-winning strains. If you want quality purebreds, look for a serious breeder who keeps records on breeding, production, and growth. If possible, make a personal visit so you can ask questions, examine records, and see the conditions under which the birds live.

Your county Extension agent should know who keeps chickens in your area, and may know a 4-H member with chicks or chickens for sale. The county fair poultry show is a good place to meet people who own chickens; if you don't see what you want there, perhaps one of the exhibitors knows someone who has that breed. Many parts of the country have a regional poultry club whose members can be helpful in getting you started.

If you can't find someone local who has the birds you want, seek a reputable seller willing to ship. Many breeds have a national association whose members are dedicated to their chosen breed and a directory of members with birds for sale. The Society for the Preservation of Poultry Antiquities publishes a membership directory. The American Bantam Association website maintains a list of links to members' websites. Several websites have a place where sellers list chickens for sale.

The bimonthly magazine *Backyard Poultry* has a breeders' directory. The newspaper *Poultry Press* offers monthly commentary on who's winning at shows and who has birds for sale. Canada has a similar paper called *Feather Fancier*. Ask exhibitors and judges at poultry shows for tips on who to deal with and — just as important — who to stay away from.

If you're looking for a classic production strain, The Livestock Conservancy or Rare Breeds Canada can help you find a producer. Seek one who specializes in the specific chickens you want, has worked with the same flock for a long time, and has taken the trouble to trace the flock's history to verify that it is an original strain.

The Right Time of Year

A good time to visit poultry breeders and examine their flocks is late November or early December, when young birds are nearly grown and old birds have finished molting. A good time to raise chicks is March or April, when the weather is turning warm but is still cool enough to discourage diseases.

Large breeds started in December and bantams started in March will feather out in time for fall exhibitions. Spring pullets will start laying in the fall and will continue laying throughout the winter.

Flock Size

One of the most common mistakes made by novice chicken owners is getting too many birds too fast. An extreme example is a young couple I knew who had the noble idea of setting up a chicken zoo where they would display every known breed. Before their facilities were ready, they went around buying chickens and crowding them together in a holding pen. The exciting venture soured when, within a few months, most of the chickens had sickened and died.

Decide how many birds you want or need, build your facilities accordingly (or a little larger, in case you catch Chicken Fever and have to expand), acquire the number of birds you planned to have, and keep your flock that size. When you buy chicks, get at least 25 percent more than you want to end up with to allow for natural deaths and the elimination of any that turn out to be undesirable as they grow.

If you're starting a laying flock, decide how many eggs you want and size your flock accordingly. As a rough average, you can expect two eggs a day for each three hens in your flock. Since hens don't lay at a steady rate year-round, you may sometimes have more eggs than you can use, and at other times too few.

If you plan to breed show birds, a mature trio or quartet of birds will give you a nice start. A trio consists of one cock and two hens; a quartet is one cock and three hens.

Unless you're raising cockerels for meat or feathers, most of the chickens in your flock should be hens; a good average is one cock per 12 to 20 hens. If you have an excess of roosters, they'll fight. If you don't need fertilized eggs for hatching, or if the local zoning ordinance doesn't allow roosters, you don't need cocks at all — but you'll miss out on their amusing antics.

THREE GOLDEN RULES OF FLOCK SIZE

1. Keep no more chickens than you have space for.
2. Keep no more chickens than you have time to care for.
3. Keep no more chickens than you can afford to maintain.

FOWL DISPOSITION

All chicken breeds and varieties originated with ancient jungle fowl. In many ways modern chickens are still much like their ancestors, having retained some of their natural instincts, such as scratching the ground for food — something you'll see chicks doing when they're only a few hours old. In other ways they differ; some of today's modern breeds no longer have the instinct to make a nest and hide their eggs and incubate them for 21 days until chicks appear. But chickens are like people — no two are exactly alike, and as soon as you make a statement about what all chickens do or don't do, one comes along to prove you wrong.

Still, as soon as you get your chickens home, you will begin to notice certain distinctive characteristics that may surprise you. **Each chicken has a unique personality.** Even if all your chickens are of a single breed and look nearly identical, you will easily be able to tell them apart by their individual personalities.

Your chickens communicate with each other, and with you, using sounds that convey specific meanings. Before long you'll become adept at "talking chicken" yourself.

Each individual bird has a unique tone of voice — even with your eyes closed, you can tell precisely which one is making the sounds you hear. And, just like people, you'll recognize that some chickens are more chatty than others.

Fowl Language

Chickens make a lot of different sounds, and every one of them means something. Anyone who spends much time around chickens can tell by the sounds they make when they are frightened, contented, or cautious or expressing a whole range of other emotions.

Some scientists insist that the idea of chickens communicating through the sounds they make is mere *anthropomorphizing* — attributing human behavior to an animal. They cling to this notion because communicating through language is supposedly a major distinction between humans and animals. A few progressive scientists — most likely those who grew up with chickens — spend their lives studying the sounds chickens make and seeking to understand what they mean.

Chicks make a pleasant sound that says they feel safe and warm.

Chicken Talk

In the 1960s, a German physician named Erich Baeumer identified 30 distinct sounds made by chickens. At about the same time, Nicholas E. Collias of the University of California at Los Angeles identified 24 calls made by red jungle fowl, from which most of our chicken breeds originated. The discrepancy may be attributed to the specific sounds each man identified as being distinct from other sounds.

To use a human example of the difficulty of identifying separate words, if you put your finger in front of your mouth and softly make the sound "shh," you communicate a request for silence. If, on the other hand, you more forcefully hiss a short "shh!" you insist on instant silence. In both cases the sound "shh" means hush, but inflection conveys important differences in meaning. Where one person might consider them to be two distinct words with different meanings, another might consider them to be the same word uttered with different intensity.

Animal behaviorist Chris Evans of Macquarie University in Sydney, Australia, is another researcher devoted to studying chicken communications. He points out that the conveying of information by the sounds chickens make reveals a complex and sophisticated system paralleling human language. Evans recognizes three similarities between chicken talk and human language:

- The ability to distinguish specific sounds
- The use of sounds to denote environmental events, such as the discovery of food or the approach of a predator
- The production of sounds for the benefit of others of the same species

Evans does not, however, imply that chickens have a language comparable to that of humans. For one thing, their vocabulary is extremely limited. For another, chickens don't — as far as we know — discuss abstract concepts or past or future events, but limit their communications to the present.

A significant feature of human language is that it must be learned. A chick isolated from other chickens will grow up to make typical chicken sounds. On the other hand, chickens associating with other chickens have a richer repertoire, indicating that some degree of learning takes place.

To date, no one has developed a definitive list of all the sounds chickens make or determined with certainty what each sound means to the chickens. Still, plenty of words in the vocabulary of chicks, hens, and cocks are clearly recognizable by anyone taking time to listen.

Baby Talk

A chick peeps even before it hatches from the egg, and shortly after hatching, it makes a number of different sounds by which you can tell if it is content or unhappy. The happy sounds tend to swing upward in pitch; the unhappy sounds descend in pitch.

Pleasure peep is a soft irregular sound chicks use to maintain contact with each other and their mother. *Its meaning:* "I'm right here."

Pleasure trill is the soft, rapidly repeated sound chicks make when they've found food or are nestling under the hen, happy to have a warm, safe place to sleep. *Its meaning:* "Life is good."

Distress peep is a loud, sharp, group of sounds chicks make when they're cold or hungry. *Its meaning:* "I'm miserable."

Panic peep is a loud, penetrating peep of a chick that's scared or lost. The sound is similar to the distress peep but louder and more insistent. *Its meaning:* "Help!"

Fear trill is the sharp, rapidly repeated sound made by a chick that sees something strange or potentially threatening, such as a small unfamiliar object or a hand reaching toward it. *Its meaning:* "Don't hurt me."

Startled peep is the sharp, surprised cry of a chick that's been grabbed abruptly. *Its meaning:* "Whoa!"

In communicating with chicks, make sounds that are low pitched, brief, soft, and repetitive to attract, calm, and comfort them. Sudden, high-pitched, long, and loud sounds (such as the noise made by active children and some machinery) frighten them.

Listen to Mama

When a chick starts peeping before it hatches from the egg, a setting hen will respond to the unhatched chick. Through this early communication, chicks learn to recognize the sound of their mother's voice. After the chicks hatch, the hen uses three specific calls to keep them together, help them find food, and warn them of danger.

Cluck is a short, low-pitched repetitive sound made by a hen with chicks. Some setting hens start clucking well before their eggs are due to hatch, especially when off the nest briefly to eat or eliminate. Most setting hens start clucking when their chicks peep prior to or during the hatch. The frequent cluck of a mother hen, sometimes accompanied by the ruffling of her feathers, encourages her chicks to follow. *Its meaning:* "Stay close."

Food call is a high-pitched sound repeated more rapidly than the cluck. Sometimes a clucking hen, upon encountering some tasty tidbit, will segue from clucking to the *tuck-tuck-tuck* food call that inspires chicks to come running and look for food, which the hen indicates by pecking the ground, picking up and

COLOR RECOGNITION

Chickens sometimes segregate into groups of like feather color while foraging, leading one to wonder if their organization by color is deliberate. Nicholas E. Collias of UCLA showed that chickens do recognize each other by color. He took chicks brooded by a red hen, a black hen, and a white hen and put them into a room with three hens of the same color as the mother hens. Most of the red hen's chicks went to the red hen, the black hen's chicks to the black hen, and the white hen's chicks to the white hen.

dropping bits, or breaking an item into smaller pieces the chicks can handle more easily. Once in a great while, a hen without chicks, or a chick itself, will make this sound. *Its meaning:* "I found something tasty to eat."

Hush sound is a soft, vibrating sound, something like *errrr*, that warns chicks of potential danger and causes them to flatten to the ground and be quiet. When the chicks are young and staying close to the hen, they dive under her, and she spreads her wings to cover them. As they get older and begin to stray, they may flatten into the grass on hearing their mother sound the hush note, which she may repeat periodically if she perceives continuing danger. *Its meaning:* "Be still."

Hen Sounds

Some hens are considerably more talkative than others. Hens that are free to roam around their premises are noisier, in my experience, than hens that are more closely confined, such as for breeding or exhibition. And some breeds are just naturally more talkative than others.

My hens have a huge and fascinating repertoire of sounds, not all of which I have succeeded in deciphering, mainly because they stop to look at me when I peek in to see what they're doing. One hen occasionally repeats an unusual single-syllable sound I can best describe as a *howl*. It's so loud it carries farther than a cock's crow. She doesn't seem to be doing anything particular while making the sound, and in decades of keeping chickens of many breeds, I've never heard any other chicken make that sound.

Laying cackle is a series of short, sharp sounds made by a hen after she lays an egg and is leaving the nest; therefore, it might more properly be called the nest-leaving cackle. Some hens don't cackle at all, some cackle only briefly, and others carry on far too long. It's tempting to

PRECOCIAL PEEPERS

Chickens, guinea fowl, turkeys, and other barnyard species each have their own distinct vocabulary. So how is it that guinea keets or turkey poults hatched by a chicken hen so readily recognize the hen as mom?

Any animal that can feed itself almost from the moment of birth is considered *precocial*, which certainly describes barnyard hatchlings. The word "precocial" comes from the Latin word *praecox*, meaning mature before its time. (A synonym for precocial is *nidifugous* — from the Latin word *nidus*, meaning "nest," and *fugere*, meaning "to flee.")

The chief characteristic of precocial birds is spryness soon after hatching, and as a result they may easily get separated from their mom. They don't have much time to learn to recognize the sound of her call, which is essential to their survival.

Another characteristic of precocial birds is that they communicate with their setting-hen mama just before they hatch; the chicks in the shell *peep*, and the broody hen *clucks*. These chicks learn to recognize the sounds their mother makes while they are still in the shell — even if she doesn't speak their native language. Entering the world with the ability to quickly find their way back to Mom, the precocial peepers are ready to hit the ground running.

think they're bragging about having just laid an egg, but chances are the cackle is designed to scare away any predator that may have sneaked up while the hen was occupied in the nest, and to put other chickens on notice that she may need help should a predator in fact be there. *Its meaning:* "Danger may be near."

Broody hiss is a hissing sound, something like the hiss of a snake, made by a setting hen that's annoyed at having been disturbed on the nest to indicate she's wary and has her guard up. *Its meaning:* "Stay away."

Broody growl is a harsh sound, more serious and intense than a hiss, made by a disrupted hen on the nest. It may also be sounded by a hen with chicks in protest to a cock intent on mating; a low-ranking hen approached by a higher-ranking hen; or any hen on seeing a small, familiar animal such as a cat or rat. The sound is not particularly loud, but it indicates defensiveness and mistrustfulness. It is accompanied by feather ruffling to increase the intimidation and may be accompanied by a peck — for instance to a human hand reaching under the hen to retrieve a fresh egg. *Its meaning:* "Don't mess with me."

Singing is the sound of happy hens. The notes are usually rapidly repeated but are sometimes drawn out. The purpose of singing is likely self-amusement, akin to a human's humming while doing dishes or singing in the shower. If I linger in the barn after feeding, I am sure to be serenaded by a chorus of cheerfully singing hens. *Its meaning:* "All is well."

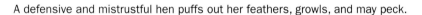

A defensive and mistrustful hen puffs out her feathers, growls, and may peck.

Social Sounds

Chickens use a variety of sounds to maintain social contact. The general purpose is likely to ensure flock cohesion and to keep individuals from straying into the jaws or clutches of a predator.

Contentment call is a low-pitched sound repeated by both cocks and hens when they are safe and comfortable. They make this sound while actively moving around but not intensely foraging. It probably works to keep them in touch with one another so none are left behind as they travel. *Its meaning:* "Let's stick together."

Nesting call is a series of rapid, excited gabbles used by a hen looking for a site she feels is suitable for laying eggs. A cock will make a similar but more intensely excited sound to show a hen a potential nest site, which might be a gap between bales of straw or a nook behind an opened door. While he gabbles, he nestles into the spot as if he is going to lay an egg himself. A lot of times the cock is ignored, but occasionally a hen will check out the spot he's found and create a duet by responding with her own song. The sound is more common to pullets and cockerels but also comes from mature hens that resume laying after a rest period. *Its meaning:* "Here's a good place to lay an egg."

Roosting call is a low-pitched, rapidly repeated sound made at nightfall when chickens are ready to roost. A large flock can make quite a racket, but it doesn't last long. The function of this call is to ensure that all the chickens roost together for safety's sake. *Its meaning:* "Let's sleep here."

Cock Talk

Roosters have a colorful vocabulary covering a wide range of activities. They seem to enjoy being the center of attention, as many of their sounds attract attention to themselves.

Food call, similar to that of a hen calling her chicks, is used by a cock to call hens to him. He'll use this excited, rapid *tuck-tuck-tuck* sound to tell the hens he has found something tasty, such as a patch of grain thrown on the ground. He might repeatedly pick up and drop a bug or a piece of fruit, a practice known as *tidbitting*, or hold a piece for a hen to take from his beak. Sometimes a cock will give a less excited food call on encountering feathers and other debris raked together during yard cleanup. *Its meaning:* "Come see what I found."

Courtship croon is the low sound a cock makes when he circles a hen while flicking one wing against the ground. Sometimes a cock will attract a hen to himself with the food call, even with no food evident, and when a hen gets near enough, he'll start the courtship song and dance. Of course, he can't often be so deceitful or the hens will soon catch on and stop being fooled. *Its meaning:* "Let's mate."

Flying object alert is a sound a cock makes when he sees a high-flying bird or airplane overhead. He makes this sound while turning his head to look upward with one eye. Some of the other chickens may look up to see what he's looking at. *Its meaning:* "Something's up there, but it doesn't look dangerous."

Startled note is a short sound a cock makes when startled or surprised. The intensity of the sound varies, or the sound may be repeated, depending on how startled the bird is. The sound might be triggered, for example, by asphalt shingles heaved down by a roofing crew or by any sudden nearby noise that disturbs a rooster at rest. *Its meaning:* "What was that?"

Crowing is an assertion of maleness. A rooster flapping his wings and stretching his neck in a mighty crow is akin to Tarzan's beating his chest and shouting out his familiar jungle call. Crowing has so many interesting facets that it has a section all its own later in this chapter. *Its meaning:* "I'm in charge here."

Predator Alarms

Cocks and hens use a variety of sounds to warn each other of potential danger. Different sounds are used to distinguish between possible danger and immediate danger and between a predator in the air and one on the ground.

Caution call consists of a few quick notes briefly repeated, made by a chicken that sees, or thinks it sees, a predator in the distance. A house cat wandering by might trigger this sound. It is not a particularly loud or insistent call and doesn't last long unless the predator becomes a threat. *Its meaning:* "Keep an eye on that intruder."

Alarm cackle is a more insistent caution call announcing the approach of an apparent predator on the ground or perhaps perched in a nearby tree or on a fence post. It consists of a brief series of short, sharp sounds followed by one loud, high-pitched sound. *Kukukukuh-KACK! Kukukuh-KACK!* Other chickens take notice, and some may join the cackling while stretching their necks to get a better look and moving around in an agitated way, as though not quite sure if or where to run. These sounds increase in intensity the longer the assumed predator is in the flock's sight and may continue after the creature has gone. The same sound may come from a hen that's been disturbed while on the nest. *Its meaning:* "Something's out there, might be dangerous."

EYEING THE SKY

The eyes on the sides of a chicken's head give it a larger range of peripheral vision but a smaller range of binocular vision, compared to birds and other creatures (including humans) whose eyes at the front focus on objects with both eyes. By contrast, a chicken has a right-eye system and a left-eye system, each with different and complementary capabilities.

The right-eye system works best for activities requiring recognition, such as identifying items of food. The left-eye system works best for activities involving depth perception, which is why a chicken watching an approaching hawk is likely to peer warily at the raptor out of its left eye.

Air raid is a loud warning cry made by a cock, or occasionally a hen, that spots the approach of a raptor. The chicken makes this sound while looking up with one eye and flattening its head and tail in a crouch to make itself less conspicuous. Without looking up, the other chickens run for cover. False alarms occasionally occur, but too many false alarms produce the same result as the boy who cried "Wolf!" Although the alarm may be triggered by anything suddenly appearing above — a leaf fluttering down from a tree, a butterfly flitting by, or a windblown feather — chickens don't sound this alarm every time a tree loses a leaf. Crows, buzzards, and light planes frequently fly over our farm, and our chickens learn to differentiate them from predatory hawks and eagles, although a suddenly appearing falling leaf or passing crow still sometimes triggers an alert. *Its meaning:* "Take cover!"

Distress Calls

A chicken that's been pecked or caught raises a fuss. The noise communicates the bird's surprise but may also be intended to unnerve the aggressor.

Startled squawk is a moderately loud cry of pain by a chicken that's suddenly pecked by another chicken. Depending on the pecked chicken's temperament, its position in the peck order, and how hard it's been pecked, the squawk may be shrill or barely audible. I've heard a similar single brief, loud squawk made by a chicken that had been foraging at the edge of a forest and was pounced on by a fox. *Its meaning:* "Ow!"

Distress squawks are loud, long, repeated sounds made by a chicken that's been captured and is being carried away, especially if it's carried upside down by its legs. The squawking

may be intended to frighten the aggressor into letting go but also warns other chickens of immediate danger. The other chickens may run and hide, although a courageous cock, or occasionally a hen, may try to rescue the distressed bird by attacking the person or animal carrying it away. *Its meaning:* "Let go!"

Communicating with Chickens

Cocks and hens have a larger vocabulary than that outlined in the previous pages. As a general rule, brief, soft, repetitive notes of low frequency are comfort calls. Loud harsh sounds with high frequencies are alarm cries. Harsh sounds emphasizing low frequencies are threats.

An extremely dedicated person with lots of time on his or her hands and skill at imitating sounds (especially speaking other languages) could learn to communicate effectively with chickens on their own terms — I mean apart from the usual chicken-keeper talk, such as calling *chick, chick, chick* at treat time and having your chickens run over for an anticipated snack.

Once, during a snowstorm, my husband and I discovered a strange cock in our backyard. We had no idea where he came from, but in the snow (which is uncommon here in Tennessee) he had no shelter and could find nothing to eat. We tried to catch him, but he would have nothing to do with us.

At night he huddled in a tree, exposed to the cold wind. Since a cock doesn't sleep with his head tucked under his wing, like a hen, he must have been pretty miserable. My husband fetched a ladder and tried to get him down from the tree, but the bird flew out of reach, then fluttered to the ground and ran into the dark. Next morning he was still there, although obviously colder and hungrier.

While my husband went to get a handful of grain to try to coax him in, I stepped out the back door and in my best chicken voice gave the *tuck-tuck-tuck* food call. The rooster stretched his neck to listen, then raced toward the door just as my husband arrived in time to help me capture the bird and bring him in to warmth, shelter, and food. As I carried the rooster inside, my astonished husband turned to me and asked, "What did you say to him?!"

When the Cock Crows

The chief characteristic of roosters is their crowing, and everyone seems to have an opinion about it. Some people love it, some hate it. Which side you fall on has a lot to do with whether or not you own the rooster doing the crowing.

Why a rooster crows is a question without a definitive answer, because no one can get into the bird's mind to find out what he's thinking. Cocks certainly crow at first light, perhaps to announce the dawning of a new day or perhaps to proclaim, "I'm still here." They may also crow during the dark of night, sometimes triggered by the sound of movement or a passing light, such as from a car or a switched-on porch light. They might believe an intruder is approaching, but instead of hushing up and laying low, they put on a loud show to warn off the intruder.

Cocks crow in the daytime as well, presumably to put potential challengers on notice. During Europe's Thirty Years' War (1618–1648), marauding soldiers carried cocks to help them find livestock that villagers had hidden in the forest. When the soldiers' cock called out, the villagers' roosters responded with "This territory is occupied," thus giving away themselves and their comrades.

Anatomy of a Crow

Nicholas E. Collias studied the crowing of the four species of jungle fowl: the green jungle fowl (*Gallus varius*) of Java; the red jungle fowl (*G. gallus*) of India; the gray jungle fowl (*G. sonnerati*), also of India; and the Ceylon jungle fowl (*G. lafayettei*) of Sri Lanka. He found that all crowing has considerable harmonic structure but each species has unique characteristics, differing in number of notes, length of crowing, accent on different notes, structure and pitch of notes, and interval between notes.

The green jungle fowl has a two-note crow that is higher in pitch than the crowing of the other species. The Ceylon jungle fowl has a three-note crow that differs from the others in having a long interval between the first and second notes. The red and gray jungle fowl both issue four notes, but the gray puts more energy into the second note, and the red puts the most energy into the third note.

Collias describes the crow of the red jungle fowl — from which most of our domestic chicken breeds originate — as being loud and complex. He confirms that a rooster crows to advertise his presence on his territory to other males and also to attract females. Two red jungle fowl cocks may engage in a crowing duel at their territorial boundaries or when competing for a hen. A dominant cock will respond to the crowing of his chief rivals, even when they are out of sight on the edge of his territory, but usually will ignore crowing by the young subordinate cockerels he tolerates in his flock. A cock beaten in a fight stops crowing near the victor altogether.

While crowing, the cock moves his head in a specific sequence. During the first note, he holds his head horizontal and stretches his neck up and forward. On the second and third

A CROW BY ANY OTHER NAME

The crowing sound is designated differently in different languages and in all languages is onomatopoeic, meaning it mimics the crowing sound. In English, of course, it's *cock-a-doodle-doo*. In German, it's *kikeriki*; in French, *cocorico*; in Spanish, *kikiriki*; in Dutch, *kukeleku*; in Finnish, *kukkokiekuu*; in Norwegian, *kykkeliky*; in Swedish, *kuckeliku*; in Greek, *kikiriku*; in Russian, *kukareku*; in Portuguese, *cocorico*; in Hebrew, *kookooreekoo*; in Japanese, *kokekokkoo*; in Wolof (spoken in Senegal), *kookoriikook*; in Hausa (spoken in Nigeria), *k'ik'irik'i*; in Korean, *kokiyo*. Chinese cocks say *gu-gu-gu*, and in Mandarin Chinese, *'o'o'o*. I find it fascinating that Chinese and Korean roosters crow in three syllables, but I don't know enough about languages or chickens in that part of the world to understand why.

The interesting thing about most of the words used to designate crowing is that they start with a *k* sound, while some emphasize *i* and *e* sounds and others emphasize *o* and *u*. A linguist will tell you the differences relate to cultural perception, but that's not the whole story. The pitch of the crow corresponds to the size of the breed. Some countries traditionally favor heavy breeds that have a deeper-sounding crow (the *o* and *u* sounds), while other countries favor lighter breeds that produce a more piercing crow (the *i* and *e* sounds).

notes (the latter, in red jungle fowl, being the loudest), he sways his head and neck back. On the fourth and final note, he again swings his head and neck forward.

Where one cock may utter the second and third as two separate notes, another may combine them into one note, producing three notes or energy peaks, rather than four. The crows of individual cocks differ not only in the number and length of notes, but also in their pitch and the clarity of their tones. Through these variations chickens recognize different individuals.

Learning to Crow

A rooster supposedly says *cock-a-doodle-doo*, from the Irish Gaelic *cuc-a-dudal-du*. One day, I sat and listened until I heard every one of my roosters crow. Half of them said *cock-a-doodle-doo*; the other half said *a-cock-doodle-doo*. I've had roosters that got right to the point with *cock-doodle-doo*. Cockerels learning to crow often cut it even shorter, sounding less like they're making a pronouncement and more like they're asking a question: *cock-a-erk?* These early attempts can be pretty funny, but with practice the young fellows eventually get it right. The age at which cockerels start crowing depends on how rapidly they mature, which is partially breed dependent, since some breeds mature significantly faster than others. Baeumer reported that chicks may be induced to crow at just a few days old by injecting them with a hefty dose of male hormone.

Since a cock inherits his style of crowing, all cocks within a given family sound somewhat similar. But each individual adds his own distinctive touch. If you have two or more roosters, you'll be able to recognize each one by the sound of his crow.

Just as some breeds are selectively bred for good laying ability or rapid growth, others are selected for the sound or duration of their crow. At the end of a crow is a faint sound you don't normally hear unless you're up close and listening carefully. In a heavy cock with a deep call, it may be loud enough to hear as part of the crow. It's also clearly audible in breeds known as longcrowers. You can make a similar sound by trying to push out the last little bit of air from your lungs.

Some cocks crow louder than others, but to my knowledge no one has ever attempted to correlate amplitude with breed, because of the difficulty of obtaining recordings that are exactly the same distance from the origin of the sound.

However, a Scottish man was cited for antisocial behavior in 2006 because his rooster crowed too early and too often and the sound supposedly exceeded the 30-decibel limit set by the World Health Organization.

THE CHICKEN'S VOICE BOX

Unlike a human, whose voice box consists of vocal cords within the larynx at the top of the windpipe (the trachea), the chicken's voice box has no vocal cords and consists of the syrinx at the bottom of the windpipe, where the trachea splits at an upside-down Y-shaped junction to create the bronchi that go into the two lungs. In addition to the lungs, the chicken's respiratory system includes nine air sacs attached to the lungs.

Crowing and all the other sounds a chicken makes require a cooperative effort among the tracheal muscles, syrinx, air sacs, and respiratory muscles. Contraction of muscles in the abdomen and thorax (the part of the body between the head and the abdomen) forces air from the air sacs into the bronchi and syrinx, while tracheal muscles work to alter the syrinx's shape to create various sounds. Each individual chicken produces a unique set of sounds, and the astute flock owner can identify each bird in the flock by the sounds it makes.

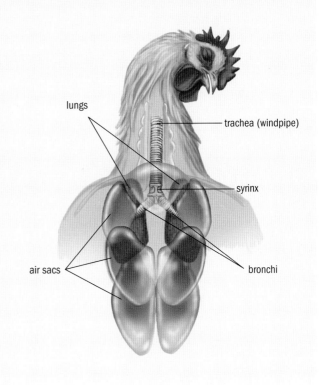

lungs

trachea (windpipe)

syrinx

air sacs

bronchi

Crowing and Dominance

A feisty little bantam cock I had attacked people's legs without provocation. As soon as he learned we anticipated his attacks and were ready to ward them off, he developed a unique tactic. Whenever someone went into the chicken yard, he hustled to the far side of the coop and crowed repeatedly. Recognizing the sound of his crow, we knew he was too far away to attack, and we let down our guard. After a time he would stop crowing, scurry around the coop, and launch a surprise attack.

In an attempt to correlate crowing to fighting, a group of researchers at the University of New Mexico in Albuquerque compared the comb lengths of 20 adult male red jungle fowl to the acoustic qualities of their crows, including the mean fundamental frequency and dominant frequency at peak amplitude. They found that roosters producing low-dominant-frequency crows have longer combs than those with high-dominant-frequency crows. They postulated that since comb length is a reflection of both testosterone levels and good health, it correlates with a cock's fighting ability.

And that fighting ability is reflected in his position in the pecking order. A previous study had found a correlation between the fundamental frequency of a cock's crow and his peck-order status — cocks higher in the pecking order have higher-pitched crows than subordinate cocks.

These studies imply that if an intruding male's crow is roughly equivalent in quality to the territorial male's crow, the intruding cock's next step in assessing the territorial male would be a closer inspection of his comb size. By crowing, a cock puts other males on notice, asserting his dominance without engaging in unnecessary fights.

Frequency of Crowing

One morning I heard a strange crow coming from a pen of replacement pullets. It didn't sound like our cock Brewster, and sure enough, on investigation I spotted a golden cockerel that had mistakenly been put in with the pullets. Little Goldie didn't crow often, so I was surprised some days later to hear him crowing nonstop. He had wandered into the goat stall and was making quite a spectacle of himself for some of Brewster's hens that were scratching in the bedding. Suddenly, Brewster appeared in the doorway. Goldie immediately stopped bragging and pretended he was just one of the girls.

How often a rooster crows depends, in part, on how secure he feels. Contests to see whose rooster can crow the most within a certain period of time rely a good deal on making a cock feel secure enough to boast. Since the 1950s, rooster-crowing contests have been popular at rural county fairs and other festivals. The allotted time ranges from 10 or 15 minutes to 30 minutes, and the cocks are encouraged to crow by participants clapping their hands, flapping their arms, and otherwise making themselves look foolish.

Longcrowers

Certain breeds are prized for the duration of the cocks' crow. They are relatively unknown in the United States but are common in other countries. These breeds generally have an upright stance, long legs, and long necks. They likely evolved from Japanese longcrowers, which in turn have their origins in the Shamo breed.

The call of a Japanese longcrower lasts 15 seconds or more. It starts out sounding like the kind of crow we're all familiar with, but

SOMETHING TO CROW ABOUT

After attending a contest at the Indiana State Fair, a journalist with Indianapolis's *NUVO Newsweekly* reported that the winning rooster crowed 60 times in 15 minutes, while the nearest challenger crowed 59 times. "After the contest," he said, "in contrast to the sparse crowing during the competition, they crowed up a storm." Apparently each rooster, on realizing he wouldn't be attacked by the others, became intent on proclaiming his victory.

The longest-running contest takes place annually in Rogue River, Oregon, as an event started in 1953 by a group of merchants called the Rogue River Booster Club. One of the members had heard that coal miners in Wales held crowing contests during holidays, so the group decided to offer a cash prize to the owner of the rooster that crowed the most within 30 minutes.

Despite rainy weather, 75 cocks were entered into the first contest, which was won by Hollerin' Harry, a rooster that crowed 71 times and won his owner $50. "I was a sophomore in high school then, so fifty dollars was a lot of money," the cock's owner, Don Martin, told the *Rogue River Press* in 2003. Hollerin' Harry "was a big black rooster, and I still have a scar where his talon impaled my hand."

To inspire roosters to crow, "You'd put 'em in a cage the night before and cover them till noon the next day, and they'd think the sun just came up," Don says. "The idea was that when one set out the challenge, then they'd all get going. You were also supposed to get in front of your cage and strut your stuff and crow."

Following that first contest, an auction was held for contestants who wished to rid themselves of their noisy fowl. First on the block was Silent Sam, having lived up to his name. The cock's disgusted owner made an opening bid of 10 cents and then watched, astounded, as the bidding went to $50. Other rooster owners rushed to cash in, only to see the bottom drop out of the bidding.

Rogue River hosted a second contest later that same year, with the prize money of $100 going to Beetle Baum. Having crowed 109 times in 30 minutes, Beetle Baum reigned supreme for 25 years. In 1978, White Lightning set a new record by crowing 112 times. Of White Lightning's success, his owner, Willie Beck, said, "I just kept him away from any other roosters or chickens for six weeks before the contest and brought him to the contest in a large sack."

the final note is sustained (like a drawn-out train whistle) before petering out as the cock appears to run out of breath. The three distinct parts of the crow are called *dashi* (the beginning), *hari* (the stretch), and *hiki* (the finish).

Japan recognizes three major longcrower breeds. The Tomaru (black crower) is noted for its rich two-tone call that deepens toward the end. In pitch, it is intermediate between the calls of the other two breeds. The Koeyoshi (good crower), supposedly developed by crossing the Tomaru with the Plymouth Rock, has a deeper voice. Koeyoshi cocks take 12 to 18 months to mature and usually don't start crowing until they are about 8 months old. The Totenko (red crower) is noted for its long tail, as well as for the duration of its high-pitched crow. You can find audio examples of all three by doing a keyword search on the Internet.

Germany has its Bergische Kraeher, supposedly the oldest German breed, imported during the Middle Ages from the Balkans,

The typical longcrower has an upright stance, long legs, and a long neck.

WHEN A HEN CROWS

Among longcrowers, crowing hens are considered valuable as breeders. In other breeds, the crowing of a hen implies that she's either diseased or getting on in years. An exception is in a flock with no rooster, in which case a hen may take on the masculine role, including crowing.

A hen has two ovaries, but only the left one produces eggs, while the right one remains undeveloped. If the left ovary becomes inactive due to atrophy or disease, the testicular tissue of the right ovary is stimulated into functional activity, resulting in the hen's getting a dose of the male hormone responsible for crowing. Sometimes an aging hen will crow during nonlaying periods, when male hormones exert greater influence than female hormones.

Researchers in Lithuania investigated the role of female hormones in crowing by identifying the gender of week-old incubated eggs and injecting the male eggs with a form of estrogen, while injecting the female eggs with an estrogen inhibitor. All the chicks matured normally, except that in the cocks the rate of crowing, duration of crowing, and strength of crowing were significantly reduced, and four of the seven treated hens regularly produced brief, weak crowlike sounds.

where its nearest relative is the Bosnian crower. In Kosovo the Drenica breed was fairly common in the Drenica province before the late-20th-century war and subsequent displacement of rural people into cities; folks who remained in the country opted for breeds that lay better. The best cocks of this breed reportedly crow for up to 60 seconds.

In Russia, the Yurlov crower was developed during the second half of the 19th century. Although its crow stretches out like that of other longcrowers, the call typically lasts a mere 7 to 9 seconds. Similarly, the crow of Turkey's Denizli breed averages 10 to 15 seconds, although some individuals may crow as long as 35 seconds. To ensure the survival and purity of this breed, the Turkish government maintains a breeding station where cocks are selected based on voice.

The duration of crowing is preserved through generations by constant selection for the longest-crowing males. Unfortunately, the better the crowing ability of the breeding stock, the lower the fertility of the eggs and the more readily the delicate chicks succumb to various diseases. So, although longcrower breeds appear in many countries, they remain relatively rare.

To Minimize Crowing

If you keep chickens in a populated area, you might be tempted to believe that the sole function of crowing is to annoy the neighbors. The question often comes up: How do you keep a rooster from crowing?

Sorry, but no 100 percent foolproof way has been found to prevent roosters from crowing. Decrowing surgery is not a ready answer. Assuming you could find a vet to do it, the operation is expensive, risky, and not always successful. Caponizing (severing or removing the cock's gonads) minimizes crowing, but a capon is useless for breeding. At any rate, most people, including veterinarians, find the surgical procedures of decrowing and caponizing to be distasteful if not downright inhumane.

To minimize crowing, let the chickens out only during reasonable hours. Overnight, either close shutters on coop windows to keep out passing lights, or leave a light on in the coop to reduce the disturbance caused by passing lights. Softly playing a radio helps keep roosters from crowing in response to sounds coming from outside the coop, and insulating the coop walls and surrounding the coop with shrubbery will muffle any crowing that does take place.

Since a cock stretches his neck to crow, putting him overnight in a ventilated box or cage small enough to prevent a good stretch will discourage crowing. Covering the container to keep out light also helps. I once visited a Cornish bantam breeder who brought his roosters into the house at night, putting them into cages stacked in an unused shower stall (a basement works well, too). The cocks still occasionally crowed, but his neighbors couldn't hear it.

Of course, you have to let the roosters out during the day, and they're going to crow no matter what. If neighbors still complain, you have only two remaining, but drastic, options. You could get rid of the rooster — thereby losing his functions of maintaining social cohesion and fertilizing eggs. Or you could move to a neighborhood where chickens are welcome.

Peck Order

By about 6 weeks of age, chicks spar to establish their place in the pecking order, which governs a flock's social organization and thus reduces tension and stress. In a flock containing both sexes, the peck order involves a complex hierarchy on three levels: among all the males, among all the females, and between the males and the females.

In general the cocks are at the top of the peck order, then hens, then cockerels, and finally pullets, although cockerels will work their way through the hens as they mature, and similarly, maturing pullets will work their way up the ladder. A new bird added to the flock must also work its way up but won't necessarily start at the bottom.

Challenges

Once the flock establishes a peck order, a bird of lower rank infringing on the space of one of higher rank will earn a glare from the higher-ranking bird — as if to say, "I can't believe your impudence" — causing the lower-ranking bird to move on. Fighting is thus kept to a minimum and mainly involves challenges to the top cock. The older he is, the more often he'll be challenged by younger upstarts.

Some of the interesting things you'll learn by observing peck-order activities are that dominant cocks mate more often than lower-ranking cocks, but submissive hens mate more often than dominant hens because they are more easily intimidated and therefore crouch more readily. And among birds with various comb styles, single-comb birds rank higher than birds with other comb styles.

Keeping the Peace

You can reduce stress among your chickens by helping them maintain a stable peck order in the following ways:

- Give your chickens plenty of room to roam so the lowest-ranking birds have space to get away from those of higher rank.
- Design your facilities with enough variety so your timid birds can find places to hide.

- Provide enough feeders and drinkers for the number of chickens you keep; otherwise, higher-ranking birds will chase away lower-ranking birds.
- If you have more than one cock, furnish one feeding station per cock and position feeders and waterers so no bird has to travel more than 10 feet (3 m) to eat or drink. Well-placed troughs allow each cock to set up his own territory and gather a group of hens around him; fighting is further minimized if no bird has to pass through another's territory to reach feed and water.
- When you move chickens, do not combine birds from different groups. Doing so adds to the stress of moving and increases peck-order fighting.

- Avoid introducing new chickens into your flock, which causes a reshuffling of the peck order. Constantly introducing new birds that disrupt the peck order causes stress that can lead to feather pulling, vent picking, and other forms of cannibalism.
- If you do introduce new birds, reduce bright lighting to make the unfamiliar birds less conspicuous.
- If your chickens constantly fight, look for management reasons such as poor nutrition, insufficient floor space, or inadequate ventilation.
- Never cull a bird just because it is lowest in the pecking order — as long as you have at least two birds, one will always be lowest in rank.

Most fights to determine peck-order status end as soon as one cock backs down.

Cockfighting

Quite the opposite of attempting to reduce stress by minimizing fights, *cockers* (fighting-cock owners) deliberately maximize aggression. Some poultry historians believe cockfighting played at least as large a role in the domestication of chickens as the production of eggs and delicious meat — maybe even a larger role. By the time of the Roman Empire, chickens had been bred along specialized lines, with the Romans emphasizing birds that provided a good return for farmers, and the Greeks focusing on fighting ability. To this day, certain lines are bred to fight.

These so-called sporting fowl are housed year-round in barrels or small individual A-frames, where they have little protection from the elements. Only the toughest survive, making the game breeds incredibly hardy.

Some individual cocks — and some entire breeds — have a natural inclination to fight, although they generally fight only until one backs down. But that's not always the case. I had two roosters — of a breed not known for a strong inclination to fight — that persisted in beating up each other. We constructed a separate pen to house one of the cocks with a few hens, but the two continued to batter each other through the wire partition, obviously intent on killing one another.

This behavior is more typical of the game breeds that have been around for centuries, although even among these breeds the strains bred for show have had much of the fight bred out of them. The characteristics of a fighting breed include a big-boned body with heavy muscling (for strength), long neck and legs (for reach), hard feathering (for armor), and a hardy constitution (for resilience). The cocks are fed a specialized diet and given regular exercise. The result is a lean, tough, sinewy bird.

Cocks in the fighting pit are paired by weight and have been trained not to back down but to keep fighting, sometimes to the death. To make matters worse, the cocks are fitted with razorlike spurs to augment their already formidable natural spurs.

Once considered one of America's national sports, cocking is now considered barbaric and inhumane; it is illegal in the United States and its territories. However, a Nevada corporation is promoting a bloodless variation designed to "make this ancient sport legally acceptable … and allow gamecock breeders to continue to legally test and perfect their breed." Called *game cock boxing*, it involves covering the spurs with foam rubber gloves and fitting each cock with a vest that electronically records hits, so cocks can spar without causing injury.

Fowl Intelligence

The term "dumb cluck" in referring to a stupid person is an insult to chickens. For far too long chickens have been considered not too bright, a perception that has gradually changed over the past few decades. In the 1960s, German physician Erich Baeumer wrote a little book — the title translates as *The Stupid Chicken? Behavior of Domestic Chickens* — in which he demonstrated that chickens are a lot brighter than most people believed at the time.

Since then, the status of chickens in general has improved to the point where some have moved from the coop to the house — and I don't mean the hen house. Chickens have joined parrots and parakeets as house birds. I met my first house chicken in the 1970s. This hen slept at night in a basket in her owner's bedroom, traveled in the car happily tucked in her little basket, and enjoyed watching television. I have since heard from several other house-hen owners that chickens love TV.

A mother hen "home-schools" her chicks.

Although a chicken needs to spend daily time outdoors doing what chickens do — sunbathing and dust bathing, scratching in dirt, and snacking on such tasty treats as creepy crawlies and tender green things — more and more people find that a single hen of a calm breed makes an entertaining but challenging house pet. The limiting factor is the difficulty of house-training a chicken.

I have brooded lots of newly hatched chicks in my house — at one time I was known as the lady who keeps chickens in her living room — but I never had a chicken as a house pet. I did once have a rooster that was smart enough to come into the basement in the wintertime to warm himself by the wood stove whenever I was dumb enough to leave the basement door open.

Self-Control

That a chicken can recall the past and anticipate the future has been proven by British researchers. In 2003, Siobhan Abeyesinghe and her colleagues at the Silsoe Research Institute determined that chickens are capable of exercising self-control, which requires resisting immediate gratification in anticipation of a future benefit.

To determine if chickens are capable of self-control, they offered hens a choice between an immediate but small payoff and a larger payoff available after a delay. The impulsive hen choosing the less-delayed reward obtained less value, while the hen waiting for a more valuable reward was able to maximize her gain by showing self-control.

Hens were trained to peck colored keys giving them a choice between access to feed almost immediately (impulsive) but only for a short time and waiting several seconds (self-control) to gain access to feed for a longer period that allowed them to eat more. A significant number of the hens held out for more feed, proving chickens are capable of understanding that a current choice has future consequences.

Training a Chicken

Training a chicken is simple but not easy. It requires a consistent, methodical approach and lots of patience. It involves carefully watching the bird for the behavior you desire, letting it know at the precise moment it has done what you want it to do, rewarding it in a timely manner, and repeating the exercise until the bird gets it right every time.

This type of training is known as operant conditioning and is the way chickens and other animals normally learn how to behave, whether they are being deliberately trained or are learning to survive in their natural environment. The technique was perfected by the late Keller and Marian Breland, who founded the field of applied animal psychology, and Bob Bailey, who married Marian after Keller passed away.

The Brelands and the Baileys developed a system of training dog trainers by teaching them to train chickens. They chose chickens for this purpose because chickens are readily available, learn fast, and lack complex social interactions. A chicken is behaviorally pretty simple — focusing most of its attention on eating, not being eaten, and making more chickens — so altering its behavior is relatively simple. On the other hand, a chicken moves fast, offering a challenge to the experienced animal trainer and the novice chicken owner alike.

Until 1990, the Brelands demonstrated the results of their training method at the IQ Zoo in Hot Springs, Arkansas, where chickens and other trained animals performed tricks with little or no human intervention. In one exhibit, a chicken named Casey pecked a small baseball bat to hit a home run, then rounded the bases of a scaled-down baseball field. In another exhibit, a chicken enclosed in a fiberglass box played tic-tac-toe against human visitors.

The method perfected by the Brelands involves obtaining a desired behavior by using positive reinforcement, or a reward. A positive reinforcer may be anything a chicken wants, seeks, or needs — most commonly food. The idea is that if you reinforce a behavior, it's more likely to occur again. If you don't reinforce it, it's less likely to recur.

Reducing or eliminating an undesirable behavior is done through nonreinforcement, or the withholding of a reward. It is not the same as punishment, which is difficult to apply to get the response you want. Even when punishment is successful as a training tool, it represses (rather than eliminates) the undesired behavior exhibited.

To modify a chicken's behavior, first determine exactly what behavior you want, then shape the chicken's behavior by breaking down your training sessions into baby steps that eventually lead to your goal. Start with a simple step the bird can easily handle and, in subsequent training sessions, gradually escalate toward your goal behavior.

To offer a timely reward, you have to know your chicken so well you can tell what it's going to do before the bird does it; otherwise your reward will be late, and the bird won't associate it with the desired behavior (or may associate the reward with an undesired behavior). If your chicken seems unable to grasp the concept, most likely your timing is off. Follow the established technique, and your chicken's behavior should steadily improve. Keep your training periods short (10 to 15 minutes) and consistent, and remain calm. If you feel yourself getting upset or frustrated, end the session early.

The technique developed by the Brelands and the Baileys involves the use of a clicker to let the chicken know the precise moment it has done what you want. Using a clicker lets you avoid the inevitable delay between the chicken's accomplishing the desired behavior and your letting it know it has earned a reward. Not all clicker training uses positive reinforcement.

Books and videotapes by the Brelands and the Baileys describe and depict their operant conditioning technique in detail. Although nothing is available solely and specifically on training chickens, the same general principles apply as are used for other animals. Do a keyword search for "training chickens" on the Internet, and you will find lots of information and some amusing video clips.

Clicker training can produce astonishing results, such as the ability to skateboard.

3

SHELTER

If you're looking for the definitive, perfect all-purpose chicken shelter, *dream on*. The design that best fits your needs must take into consideration your geographic location and weather patterns, your available land, how many chickens you plan to keep, the breed or breeds you choose, and the purpose for which you intend to keep them — not to mention how much you want to spend. If you live in a populated area, you'll avoid neighborhood hassles if your structure blends in with the surroundings. Zoning or other building restrictions may further narrow down your facility options.

When deciding which of the many management methods might work best for you, the first step is to analyze your available ground to determine how much space you can devote to your chickens and how that space would best be used. A good way to gather ideas is to find successful chicken keepers in your area — or correspond with people who live in a similar climate — and pick their brains as to what works for them and what doesn't work at all. Don't be readily swayed by someone who just finished designing the "perfect chicken shelter." Unless the thing has been used for at least a year, and preferably longer, the drawbacks haven't had time to surface.

Free Range

The term "free range" has been applied to both yarding and pasturing to imply the birds have complete freedom to roam an outdoor area that provides some degree of forage, although in both cases the birds are confined at least by a fence and therefore are not entirely free. True free-range chickens are the ones you see running along the lane when you drive through the countryside.

Nonconfinement was a common management practice until the 1950s, when increasing urbanization began limiting available space for chickens to roam and new neighbors didn't appreciate having their freshly planted petunias and tomatoes scratched up. At the same time, industrialization dictated that chickens be maintained under controlled conditions to maximize egg and meat production and minimize costs.

Those birds you see today along country lanes, if they belong to anyone, may have access to a shelter, yet may choose to roost in trees as their ancestors did and therefore have little or no protection from bad weather and predators. Most likely, these chickens descended from Old English bantams or a similar hardy, self-reliant breed that reproduces easily enough to make up for the inevitable

---

losses. As picturesque as it might be to see chickens running free, scratching wherever they please, and roosting in trees, no responsible owner would deliberately turn chickens loose to annoy the neighbors and face certain early death from predators of one form or another.

Fenced Range

A fenced yard gives chickens a safe place to get the sunshine, fresh air, and exercise they need to remain healthy. As many advantages as fenced confinement offers, it has one big disadvantage — chickens can quickly destroy the ground cover by pecking at it, scratching it up, pulling it up, and covering it with droppings. The smaller the yard, the quicker it will turn to either hardpan or mud, depending on your climate. Therefore, your first consideration when designing your perfect chicken shelter is its impact on your land. By planning your land use carefully, you can easily avoid creating a situation that soon becomes unsightly and unsanitary. Where space for a yard is truly limited, and you have only a few pet chickens, one way to avoid the problem is to level the small yard area and cover it with several inches of clean sand. Go over the sand every day with a grass rake to smooth out dusting holes and remove droppings and other debris. Some folks choose to use gravel instead of sand, but droppings get packed in the spaces between bits of gravel, and eventually the mess has to be removed and replaced with a load of fresh gravel.

SHELTERING METHODS

Sheltering methods are as varied as people who keep chickens and range in style from complete confinement to total freedom. What each method is called depends on who is trying to sell you on the idea as being new and different. If you enroll in an organic certification program, you'll need to abide by the program's definitions, which may or may not be the same as those in general use. The basic options are:

- No confinement (*free range*) — seen most often in rural areas

- Confinement to a portable shelter with a fenced foraging area (*pastured, range fed, day range*) — used on farms with available pasture where the fence or shelter can be relocated periodically

- Confinement within a floorless portable shelter — used in family gardens (*ark, chicken tractor*), and on farms (*pastured poultry*) with enough land for the shelter to be moved frequently

- Confinement to a permanent building with an outdoor fenced yard (*yarding*) — the traditional method for housing homestead poultry and other small backyard flocks

- Confinement within a permanent building (*loose housing*) — generally used for raising broilers or breeders or maintaining a flock during cold or wet weather

- Cage confinement (*hutch, ark*) — most often used in urban and suburban areas and for show chickens

The bare patch in this yard shows where chickens have pecked and scratched the ground near their shelter.

The larger your yard, the better chance you'll have of maintaining some vegetation in it. Since chickens are most active near their shelter, denuding will start around the entrance and work progressively outward. In a nice roomy yard, the ground cover may continue to grow in areas farthest from the doorway. Between the barren area and the grassy area may grow a band of weeds so tough or unpalatable the chickens won't eat them and can't trample them. To keep the vegetated areas tidy and safe, you'll need to mow occasionally. How often depends on the climate, time of year, and number of chickens.

Chickens will range farther if their yard is not entirely open but includes trees and shrubs offering shade from the hot sun, relief from blowing winds, and protection from flying predators. Trees and shrubs are also an attraction because they may drop leaves and fruits, as well as harbor insects that chickens like to snack on.

One of my chicken yards includes a single oak tree about 75 feet (23 m) from their shelter, and the chickens spend hot summer days lolling in its shade. Across the driveway is an apple orchard, and the chickens persist

in ducking under or flying over the fence to scratch and dust under trees as far as 350 feet (107 m) from their shelter. (It's no coincidence that, even though we never spray, our apples have few worms.)

If your proposed large chicken yard has no trees, provide one or more small range shelters scattered throughout the yard to offer shade, a windbreak, and a refuge from flying predators. Including a waterer in the shelter will ensure its use. Protective trees or constructed shelters encourage chickens to forage away from their home shelter, reducing their impact around the doorway.

A covered dooryard or sizable overhang helps prevent muddy conditions around the doorway. A concrete pad at the doorway, frequently scraped off with a hoe or flat shovel, reduces the tracking of mud and manure into nests. Alternatives to concrete are mulch and gravel — both less expensive but more difficult to maintain.

If your soil is neither sandy nor gravelly, site your yard at the top of a hill or on a slope, where puddles won't collect when it rains. A south-facing slope, open to full sunlight, dries fastest after a rain.

SIMPLE RANGE SHELTER

A range shelter to protect the birds from sun and wind can be as simple as a roof on posts. This design can be made from a sheet of plywood covered with rolled roofing or, for lighter weight, from metal or fiberglass corrugated roofing.

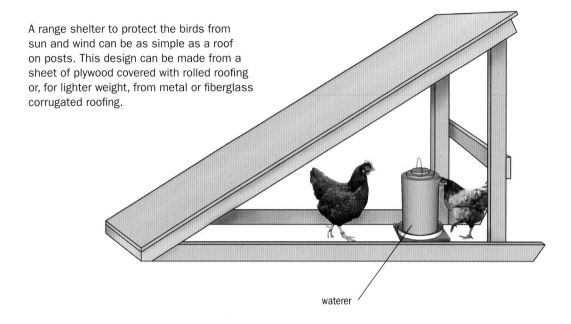

waterer

TREE CAVEATS

Trees in the yard help keep chickens healthier and happier but offer a few distinct disadvantages:

- Trees provide a place for a hawk to land while selecting his next meal, as do fence posts. (Measures for dealing with hawks are discussed in chapter 5.)

- A tree close to the inside of the fence provides a handy way for chickens to fly up and out, and an overhanging tree outside the fence lets predators go up and in; make sure tree branches do not overhang the perimeter fence.

- Chickens like to roost in trees at night, leaving them open to predation by owls and other nighttime stalkers; however, training your chickens to go into their shelter at night is not difficult.

As a guideline to establishing your chicken yard, a nice spacious area provides 8 to 10 square feet (0.7 to 1 sq m) per chicken. If the perimeter fence is 100 feet (30 m) or more from the shelter and the yard has no trees, station a basic range shelter about every 60 feet (18 m).

Yard Rotation

One way to cope with the problem of a messy chicken yard is to divide the yard into paddocks (small fields) and let the chickens into only one paddock at a time. Rotation reduces the concentration of pathogens and parasites in the soil by periodically giving each paddock a rest. This system works only if you can successfully keep the chickens out of any resting paddocks, which means no flying over or ducking under fences. How often to rotate from one paddock to another depends on how fast the vegetation is destroyed, which is a function of such things as your climate and the number of chickens you keep.

One rotation scheme calls for dividing the yard into two separate paddocks, each with its own entrance from the coop. While the chickens are in one paddock, recondition and reseed the other paddock, which you might do every six months or once a year, depending on what you plant and the time of year it grows best in your area. The chickens may still destroy the ground cover in the paddock they use, but the rested paddock will be sanitized, thanks to rest, sunshine, and plant growth.

If you have plenty of space, set up four paddocks so you can rotate the flock more often. A six-way rotation is even better. Size each paddock as if it were the only yard, allowing the requisite 8 to 10 square feet (0.7 to 1 sq m) per chicken.

Range Rotation

Instead of a yard rotation, which involves sequentially turning chickens into a series of paddocks surrounding a fixed shelter, range rotation involves moving a portable shelter on pasture. It's a good option for a short-term project of raising broilers, but for long-term chicken keeping, it requires plenty of land and labor. Leave the portable house in one place too long and the pasture will be destroyed; without expensive and time-consuming renovation, the bare spots will grow up to weeds.

Various sources publish information on how often the shelter should be moved, but the brutal truth is you have to learn to judge for yourself based on how rapidly the chickens destroy the patch of pasture they're on. When we pasture broilers, we start out moving them every three days, and spring rains nicely revive the patches they've been on. As the birds grow, they trample plants faster and poop more, while at the same time dry weather slows plant regrowth, so the shelter must be moved more often. Toward the end, moving the broilers twice daily is just barely enough to preserve the pasture.

Some growers figure pasture renovation goes with the territory. Others are willing to devote the time necessary to observe changing conditions, determine when moving is necessary, and get the job done in a timely manner. Keen observation is needed because the speed with which a given group of birds will devastate vegetation depends on their size, how crowded they are, the climate, the season, the weather, and the type of pasture they are on. It also depends on whether or not they are entirely confined within the portable shelter. Surrounding the shelter with a movable fence gives the chickens more room to move around

YARD ROTATION

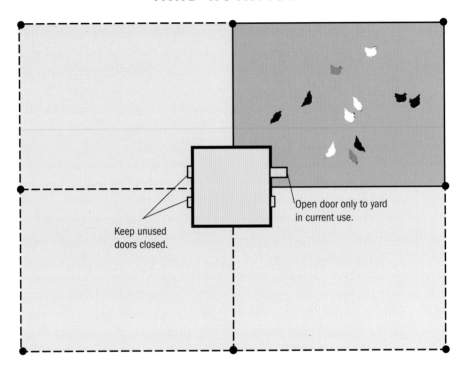

Keep unused
doors closed.

Open door only to yard
in current use.

To rotate yards without moving the housing, put chicken-size
doors on different sides of the coop.

and therefore lengthens the amount of time they may remain in one place.

The major drawback to pasture rotation is that most portable shelters, being light enough to be moved easily, offer little protection in cold weather. Another potential problem is that a pasture isn't as nicely graded as a lawn, and all the bumps and dips create a significant challenge in making a portable shelter tight enough to exclude predators. Both these issues may be resolved by having a sturdier, and more expensive, portable shelter on a trailer chassis,

but in winter's snow and ice, you'll still have to rotate it to prevent a manure buildup that will be toxic to spring pasture.

As a general guideline, plan on providing at least 110 square feet (10 sq m) per bird (which, on an acreage basis, means you may pasture up to one hundred chickens on 0.25 acre [0.1 hectare]). Since broilers don't hang around long, you can grow the same number in a little less than half the space; plan on 45 square feet (4 sq m) per meat bird (or about 250 broilers on a quarter acre). Another good

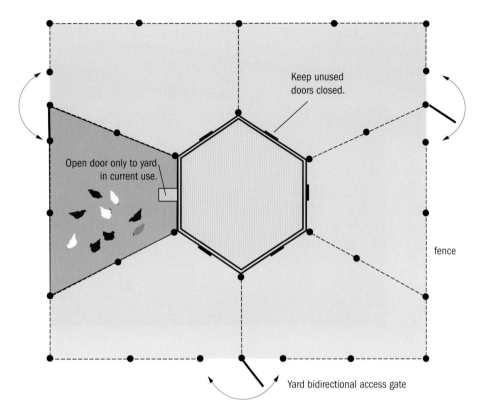

Keep unused
doors closed.

Open door only to yard
in current use.

fence

Yard bidirectional access gate

A six-way rotation offers more options during seasons when
vegetation grows especially fast or not at all.

rule of thumb is to keep only as many chickens as you can rotate without revisiting the same ground within a given year.

The so-called chicken tractor concept uses the same principle as range rotation in a confined shelter, but the shelter is moved around a garden expressly so the chickens will destroy weeds, eat cutworms and other pests, and fertilize the soil. Although active chickens will scratch in the dirt, hence the name chicken tractor, standard broilers and other inactive types tend instead to compact the soil. Like pasture shelters, portable garden shelters must be moved often enough to prevent the chickens from foraging in their own droppings. And in northern areas the birds will need alternative housing that offers protection from rough winter weather.

Portable Shelters

Portable shelters have become popular because they are less expensive than permanent housing, are not taxed as property improvements in some areas, and may be moved periodically to give chickens healthful ground and fresh

The handles at both ends of this shelter make it easy for two people to move it (the handles in the middle are for lifting off the siding to gain access).

FLOORED OR FLOORLESS?

Portable shelters are used either with or without access to additional range. Those with a floor are often combined with a movable fence, while those without a floor are more often used for total confinement. A shelter without access to range is little better than a cage: it does not allow birds that are low in the peck order to get away from aggressive ones, and the active birds have insufficient space to do their avian thing. Chickens that can spend days outside their shelter have more room to forage, scratch, gobble up insects, and enjoy dust baths. Because they are less crowded, they experience less stress.

forage. To aid in moving them, these shelters come in four basic styles:

- With handles, to be moved by hand
- On skids, to be moved by hand if light enough; otherwise by a truck, tractor, or draft animal
- On wheels without an axle, to be moved by hand; the wheels may be incorporated into the shelter design or on a separate dolly
- On wheels with an axle (*chicken mobile home*, *henmobile*, or *eggmobile*), to be moved by a truck, trailer, or draft animal

Whatever design you choose, the shelter must be so easy to move that you'll move it as often as necessary. A shelter that is movable by one person is more likely to get moved in a timely fashion than one requiring one or more helpers who may not always be around when you need them.

Lightweight Shelters

A lightweight shelter that's movable by hand works fine in a mild climate or for a warm-season poultry project. Such a shelter is inexpensive enough to allow experimentation and revision as needed. The most common construction issue is making it so flimsy it twists apart after a few moves. A basic understanding of structural strength comes in mighty handy in designing one of these shelters. Overbuilding, on the other hand, can make the shelter too heavy or cumbersome to move with ease.

Basic lightweight designs are peaked, boxy, or rounded. Variations include hoops and hexagons. Compared to a flat roof, a rounded or pointed roof sheds more rainwater, and a pointed roof is more comfortable in the hot summer sun. Where winds are strong, the shelter should be low — no more than 4 feet (1.2 m) high — and/or well staked to the ground.

A floorless shelter is cheaper to construct than one with a wire or solid floor. It has the advantage that you can enclose the chickens entirely within, and they'll have access to pasture or garden soil without being exposed to flying predators. Take care to block ground-level dips that provide access to four-legged predators and rainwater with this sort of shelter, and be sure to move it often to keep the chickens off built-up droppings.

A 6-by-10-foot (1.8 by 3 m) shelter, moved daily, may be used to raise up to three dozen broilers; a 4-by-8-foot (1.2 by 2.5 m) camper-shell size will handle about 15. By attaching nest boxes to the outside (where they don't take up interior space), you could enclose about the same number of laying hens.

A shelter with a solid floor is more rigid and offers greater protection from ground predators, provided you close the door each night. Because the birds have no access to pasture inside the shelter, they'll need to run loose during the day. They may be confined to a pasture area in one of two ways — use a portable fence that moves with the shelter, or let the chickens range within a permanently fenced pasture. Either way, you'll need to move the shelter at least once a week to avoid killing the pasture beneath it; the plants underneath will start to yellow but should recover in about a month. In contrast to a floorless shelter, which acquires a clean "floor" every time it's moved to new ground, a floored shelter requires bedding that must be periodically refreshed or cleaned out and replaced, increasing the labor involved.

SHELTER-MOVING ISSUES

When moving a shelter without a floor, take care that no chickens get trampled. Chickens can learn to move with the shelter, especially when they see fresh pasture coming up. But initially they will try to stay where they were, and as the shelter moves they bunch up and may get crushed. Until they catch on to the idea of moving, have a helper shoo them ahead as the shelter moves.

An issue for chickens ranging outside their shelter during the day is that they can get confused about finding their home after it's been moved. Help them along by not moving the shelter far outside the previous range and by rounding up stragglers that insist on bedding down at the old place.

A wire floor lets droppings fall through, solving the issue of cleaning a solid floor, while providing more predator protection than no floor. But wire should not be used to house heavy breeds, as it's hard on their feet.

Heavy-Duty Shelters

For a harsh climate a shelter on skids, or on an axle or two, may be built sturdier than a lightweight hand-moved shelter and may be

MUST-HAVE FEATURES

Whether portable or stationary, any successful chicken shelter has these features:

- Provides adequate space for the number of birds
- Is well ventilated
- Is free of drafts
- Maintains a comfortable temperature
- Protects the chickens from wind and sun
- Keeps out rodents, wild birds, and predatory animals
- Offers plenty of light during the day
- Has adequate roosting space
- Includes clean nests for the hens to lay eggs
- Has sanitary feed and water stations
- Is easy to clean
- Is situated where drainage is good

insulated more easily. A single-axle shelter needs a dolly wheel or jack at the front to level the thing while it's in place, and any shelter with wheels must be braked, blocked, or tethered to prevent rolling. A shelter built on an axle has a floor and may be moved easily early in the morning or late in the evening when the chickens are inside.

A shelter on skids may be constructed with or without a floor. A floorless shelter would have a gap at the front and back, between the skids, that leaves the structure wide open to predators. If the front and back walls are brought down to ground level to close off these gaps, they will scrape the ground during a move. Instead, close off the gaps with hinged boards at the front and back that can be lifted during a move.

With the exception of mobility, a heavy-duty shelter on wheels or skids has pretty much all the same features as a permanent shelter. Most such shelters lack lights and automatic waterers, although with a little ingenuity you could include both.

Permanent Shelters

A permanent shelter, or fixed housing, offers a number of advantages over portable housing:

- It is easier to service and maintain.
- It can be built larger and of heavier construction.
- It more easily protects chickens from predators.
- It can be made for comfort in bad weather.

Because the yard outside a fixed shelter readily turns to packed dirt or mud and the accumulated manure eventually leads to disease, permanent housing works best for a small flock,

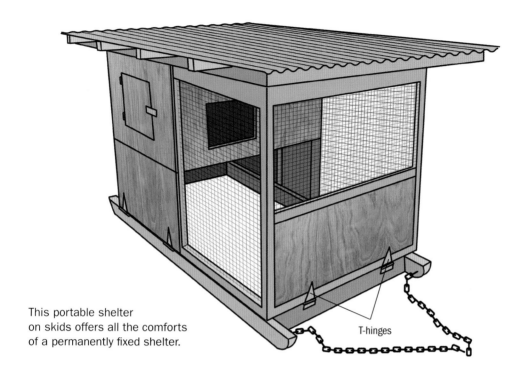

This portable shelter
on skids offers all the comforts
of a permanently fixed shelter.

T-hinges

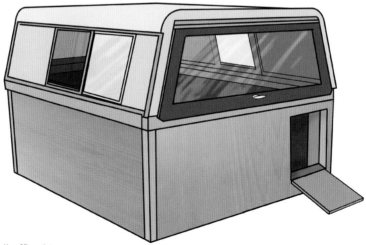

A lightweight camper shell affixed to a
plywood corral may be insulated against
cold weather and built on skids for
easy moving.

in a dry climate, and with a good system of yard rotation. The structure may be a repurposed building, such as a child's playhouse or toolshed, with the addition of perches and nesting boxes. Or it may be built from scratch, complete with automatic ventilation, security lights, waterers, and door openers. Whatever the design, your structure should blend in with the surroundings to keep it from looking out of place and keep complaining neighbors at bay, especially if you live in a populated area. The shelter should fit in with existing structures in size, type of construction, and style and color of the siding and roofing.

Simple, open housing is easier to clean than a coop with numerous nooks and crannies, although the latter offers more hiding places for birds that are lowest in the peck order. If the shelter is high enough for you to stand in, you'll be more inclined to clean it as often as it needs cleaning. If you prefer a low coop — for reasons of economy, to retain your flock's body heat in a cold climate, or to keep the structure from

BASIC COOP DESIGN

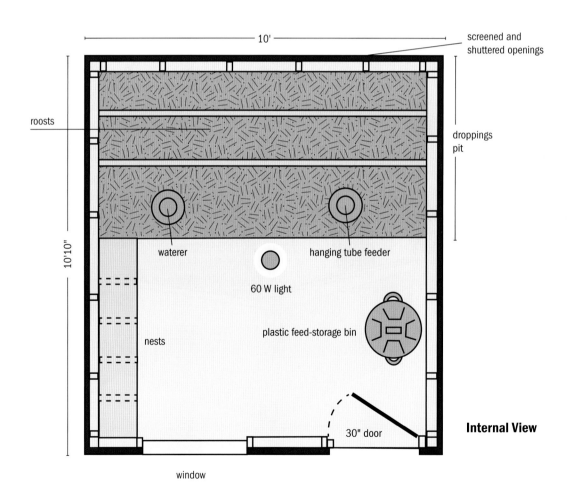

roosts

screened and shuttered openings

droppings pit

waterer

hanging tube feeder

60 W light

plastic feed-storage bin

nests

30" door

Internal View

window

10'

10'10"

blowing over in a high wind — design it like a chest freezer, with a hinged roof that opens so you can stand erect during cleaning.

Space

How much space chickens need — and, therefore, how big your shelter should be — depends on their age and breed. Chickens need more space as they grow, and some breeds are naturally more active than others. Adequate space allows chickens to engage in normal avian behavior, minimizes stress, and helps prevent health and behavioral problems. The more room your chickens have, the healthier and more content they'll be.

All sorts of rules of thumb are offered for sizing coops, many of them calling for variations on pounds of chicken per square foot (0.1 sq m) of space. Whenever I see one of these formulas, I work it out to square feet per chicken. I understand these formulas are established to optimize the available funds, yet

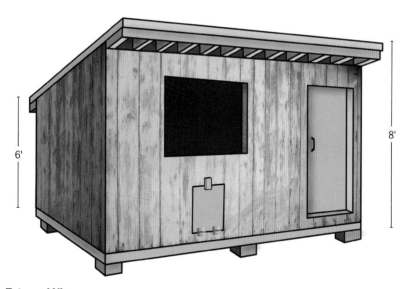

External View

This basic coop plan features roosts over a droppings pit for good sanitation, a window for light, and screened and shuttered openings on the north side to control ventilation. To expand the interior floor space, build the nests on the outside of the coop (see page 80).

LARGE COOP DESIGN

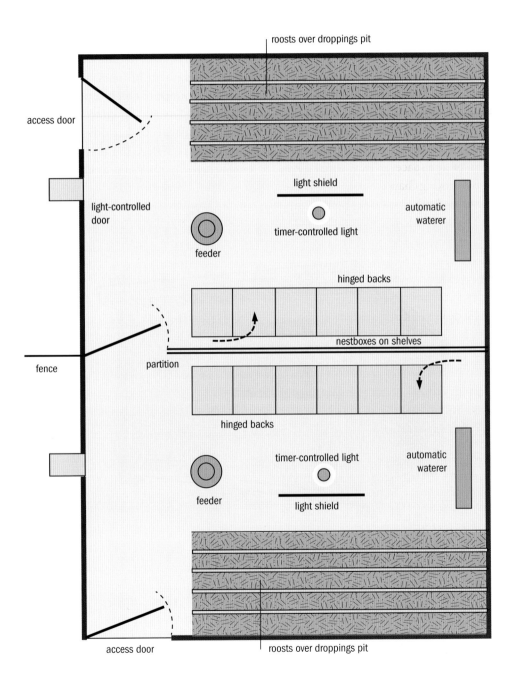

roosts over droppings pit

access door

light shield

light-controlled door

automatic waterer

timer-controlled light

feeder

hinged backs

nestboxes on shelves

fence

partition

hinged backs

timer-controlled light

automatic waterer

feeder

light shield

roosts over droppings pit

access door

This layout can offer two separate areas for raising different breeds or birds of different ages, or the two areas may be combined for one large flock.

I invariably feel sorry for chickens living under such crowded conditions.

By comparison, the minimums I follow are often criticized as being too spacious, yet my chickens rarely experience the problems I hear about from other chicken keepers. Given adequate space, the lowest birds in the peck order have an easier time of it; if you have more than one rooster, each will gather its own group of hens, and they'll all be less apt to fight.

The minimum space requirements shown in the accompanying chart are the least amount of indoor space birds should have when they can't or won't go outside for an extended period of time. Even in the best climate, chickens may sometimes prefer to remain indoors because of rain, extreme cold, or extreme heat.

A flock that never has access to an outside run will do better if given more space than these minimums. Total confinement without proper ventilation carries the risk of respiratory illness due to ammonia from droppings and dust from litter, skin, and feathers. On the other hand, a confinement house with good ventilation and properly managed litter may be more healthful than a yard covered in mud and manure.

Given a well-maintained yard, birds that spend most of their time outdoors, coming in mainly at night to roost, do quite nicely with less space. To encourage chickens to spend most of their daytime hours outdoors, even in poor weather, give them a covered area adjoining the coop where they can loll out of the rain, wind, and sun. Encouraging your chickens to stay out in the fresh air has two advantages: they will be healthier, and their shelter will stay clean longer.

MINIMUM SPACE REQUIREMENTS

Birds	Age	Open House		Confined Housing		Caged	
		sq ft/Bird	Birds/sq m	sq ft/Bird	Birds/sq m	sq ft/Bird	Birds/sq m
Heavy	1 day–1 week	—	—	0.5	20		
	1–8 weeks	1.0	10	2.5	4	*Do not house heavy breeds on wire.*	
	9–15* weeks	2.0	5	5.0	2		
	15–20 weeks	3.0	4	7.5	1.5		
	21 weeks and up	4.0	3	10.0	1		
Light	1 day–1 week	—	—	0.5	20	25	160
	1–11 weeks	1.0	10	2.5	4	45	290
	12–20 weeks	2.0	5	5.0	2	60	390
	21 weeks and up	3.0	3	7.5	1.5	75	480
Bantam	1 day–1 week	—	—	0.3	30	20	130
	1–11 weeks	0	15	1.5	7	40	260
	12–20 weeks	1.5	7	3.5	3	55	360
	21 weeks and up	2.0	5	5.0	2	70	450

*or age of slaughter

Doors and Windows

A shelter needs both a people-size door and a chicken-size door. You'll use the people-size door to feed and water your chickens, check on their welfare, collect eggs if the nests are inside the shelter, and periodically clean out the bedding and droppings. The people-size door should be of standard size so you don't bump your head every time you go in and out and so you can easily pitch soiled bedding from the shelter into a wheelbarrow or front-end loader outside the door.

In the summer we use a screen door on which we replaced the window screen wire with half-inch hardware cloth. The screen door improves ventilation, and the hardware cloth is tougher than the original screen, keeping chickens and predators each on their own side.

The chicken-size door is called a *pop hole*. Chickens love to sit on the pop-hole ledge, surveying the outside world. When one chicken is occupying that perch, none of the others can get in or out. For that reason a range or pasture shelter often has a wide pop hole so the birds can quickly enter the shelter in case of an aerial attack. But a single-chicken pop hole reduces drafts, keeps out large dogs and small goats, and may easily be set up to close automatically at nightfall and open at dawn.

A pop hole is made by cutting a flap into the wall and hinging the flap at the bottom so it opens downward as a ramp for the birds to get in and out. A good-size single-chicken pop hole is 10 inches wide by 13 inches high (25 by 33 cm); for really large chickens, 12 by 14 inches (30 by 35 cm) would be better. With an automatically closing door, you have to size the opening to fit the mechanism. If you don't have automatic closure, the flap needs a secure latch you can fasten shut to exclude predators after your chickens have gone to roost in the evening.

Our chickens aren't always ready to go in when we do evening chores, and we're not always timely about letting them out in the early morning, so we use pop-hole doors that close and open automatically. You can find suppliers and ideas for homemade variations on the Internet by doing a "chicken door" keyword search. Our system is regulated by sunlight and can be adjusted by the placement of the electronic eye. Since chickens actively forage at dusk, you have to make sure the pop-hole door doesn't close before they are all inside.

Windows let in light and fresh air, and south-side windows capture the sun's warmth. Provide at least 1 square foot of window for each 10 square feet of floor space (0.1 sq m of window per 1 sq m of floor). The windows should slide or tilt open so you can adjust airflow as the weather changes. They should be fitted with screens of ½- or ¾-inch (1.5 or 2 cm) hardware cloth to keep out wild birds, as well as weasels, minks, and raccoons that can tear through standard window screening and poultry netting.

PROBLEM PREVENTION

If your chickens experience growth, reproductive, or behavioral problems, try one of the following approaches:

- Reduce the number of chickens
- Increase the size of their living space
- Provide more roosts

Ventilation

The more time chickens spend indoors, the more important ventilation becomes. Ventilation serves these essential functions:

- Supplies oxygen-laden fresh air
- Removes heat released during breathing
- Removes moisture from the air (released during breathing or evaporated from droppings)
- Removes harmful gases (carbon dioxide released during breathing and ammonia evaporated from droppings)
- Removes dust particles suspended in the air
- Dilutes concentration of disease-causing organisms in the air

Compared to other animals, chickens have a high respiration rate, causing them to use up available oxygen quickly while at the same time releasing large amounts of carbon dioxide, heat, and moisture. As a result, chickens are susceptible to respiratory problems. Stale air inside the shelter makes a bad situation worse, because airborne disease-carrying microorganisms become concentrated more quickly in stale air than in fresh air.

Ventilation holes near the ceiling along the south and north walls give warm, moist air a place to escape. Screens over the holes will keep out wild birds, which may carry parasites or disease. Drop-down covers, hinged at the bottom and latched at the top, let you open or close ventilation holes as the weather dictates. If you're afraid you'll forget, use temperature-sensitive vents with slats that open and close automatically.

During cold weather, not only must you provide good ventilation, but you have to worry about drafts. Close the ventilation holes on the north side, keeping the holes on the south side open unless the weather turns bitter cold.

In warm weather, cross-ventilation keeps chickens cool and removes moisture. The warmer the air, the more moisture it can hold. During the summer, open all the ventilation holes and open windows on the north and south walls.

A screen door, with ½-inch (1.5 cm) hardware cloth secured with screws and washers, provides extra ventilation on hot summer days.

A cupola improves ventilation by letting hot, humid air escape through the roof where temperatures soar during summer.

Fans are another option for ventilation if your shelter has electricity. Poultry-shelter fans come in two styles: ceiling mounted and wall mounted.

A variable-speed ceiling fan keeps the air moving but benefits chickens only if ventilation holes are open, to avoid trapping hot air against the ceiling. Use a ceiling fan only if your ceiling is high enough to keep you from bumping into it and your chickens from flying into it. If you can't bump your head on the fan, chances are it's high enough for them to avoid flying into it, unless it's directly over a perch or other platform from which a chicken conceivably might launch.

A wall-mounted fan sucks stale air out, causing fresh air to be drawn in. The fan, rated in cubic feet per minute, or cfm, should move 5 cubic feet (0.2 cu m) of air per minute per bird. If your flock is housed on litter, place the fan outlet near the floor, where it will more readily suck out dust as well as stale air. Since some dust will stick to the fan, a wall-mounted fan needs frequent cleaning with a vacuum or a pressure air hose.

A fan designed for use in your home won't last long in the dust and humidity generated in the normal chicken shelter. To find fans designed for agricultural use, do a computer keyword search for "barn fan" or "agricultural fan" on the Internet or visit a local farm store or rural-oriented builder's supply.

Temperature Control

The temperature inside a chicken shelter varies throughout the day and with the seasons. Environmental factors that influence indoor temperature include the following:

- The outside temperature and humidity
- The amount of ventilation or draft at chicken level
- The temperature of walls, roosts, nests, feeders, and waterers
- Shade on the shelter thrown by trees or buildings
- Insulation in the ceiling and walls or provided by surrounding shrubs
- The number of chickens living in the shelter, the ages of the chickens, and whether or not they are laying

A chicken's body operates most efficiently at an effective ambient temperature between 70 and 75°F (21 and 24°C). In colder weather, they eat more to obtain the additional energy they need to stay warm. Hot weather is more problematic. For each degree increase, broilers eat 1 percent less, causing a drop in average weight gain. Egg production may rise slightly, but eggs become smaller and have thinner shells. When the temperature exceeds 95°F (35°C), birds may die.

To keep the shelter from getting too hot, treat the roof and walls with insulation, such as 1½-inch (4 cm) Styrofoam sheets, particularly on the south and west sides. Cover the insulation with plywood or other material

QUICK VENT CHECK

Use your nose and eyes to check for proper ventilation. If you smell ammonia fumes and see thick cobwebs, your shelter is not adequately ventilated.

your chickens can't pick to pieces. To reflect heat, use aluminum roofing or light-colored composite roofing, and paint the outside of the coop white or some other light color. Try to maintain grass around the coop, keeping it mowed to a height of no more than 4 inches (10 cm). Plant trees to shade the roof, or install awnings or deep overhangs to shade the walls. An awning or overhang serves the additional purpose of providing a shady, breezy place outdoors for birds to rest.

To enhance heat retention in winter, build the north side of your coop into a hill or stack bales of straw against the north wall. Where cold weather is neither intense nor prolonged, double-walled construction that provides dead-air spaces may be adequate to retain the heat generated by your flock. In really cold weather, you'll need insulation and, to keep moisture from collecting and dripping, a continuous vapor barrier along the walls. Windows on the south wall supply solar heat on sunny days but must be shaded in hot weather. Using a heater for mature chickens won't do them any favors. In a properly constructed shelter, chickens can keep sufficiently warm if they aren't wet or sitting in a draft. In cold weather, chickens stay comfortable by fluffing up their feathers to trap a layer of warm air; a draft removes that warm air, allowing the birds to chill. And a draft during bitter weather can cause combs to freeze. Yet despite the need to avoid drafts in cold weather, good ventilation is still needed to maintain a healthful environment for your chickens.

Electric Wiring

Not all chicken shelters have electric wiring, but you may wish to include lights to extend the laying season or for your own convenience doing evening chores. During foul weather or gloomy days, indoor bright lights encourage normal activities such as eating and dust bathing. But don't fill the entire indoors in bright lights. Install low lights or provide for a darkened area near nests to encourage laying and over perches so birds can rest.

A security light outside the shelter helps deter predators and thieves and is handy during nighttime emergencies. If you use automatic door closers, make sure the security light doesn't go on until after the door is closed. Otherwise the brightly lit yard will encourage chickens to linger outdoors, and they'll get left outside when the door closes.

If electricity isn't already handy to your shelter or you're constructing a portable shelter, 12-volt battery-powered fixtures may be the best option. Use a deep-cycle battery, such as one designed for use in a golf cart, and recharge it as needed. Or use solar power. The same system used to light your chicken shelter can power an electric fence surrounding it.

Since the environment inside a chicken shelter can be corrosive, weather-proof electrical fixtures are more reliable and will last longer than those used in a home. Run all wires through plastic or metal conduits so chickens and rodents can't chew through them and cause a short or fire. Be sure to glue all conduit joints to prevent sagging and keep out insects.

Properly installed wiring reduces the potential for stray voltage, which results in an unpleasant shock when you touch a metal object, light switch, or junction box within the shelter. Improper grounding, as well as undersized or overloaded circuits, increases the potential for stray voltage. Only a properly trained and qualified person should install wiring.

Flooring

The main considerations in deciding what type of flooring to use in your chicken shelter are cost, ease of cleaning, and resistance to predators. You have generally one of four basic options for a chicken-house floor:

Dirt is cheapest and easiest to install, but consider it only if you have sandy soil that ensures adequate drainage. In warm weather, dirt helps keep birds cool, but in cold weather it draws heat away. A shelter with a dirt floor is not easy to clean and cannot be made rodentproof.

Wood offers an economical way to protect birds from rodents but only if the floor is at least 1 foot (0.3 m) off the ground to discourage mice and rats from taking up residence in the space underneath. Wood floors are hard to clean, and the cracks between the boards invariably get packed with filth and bugs.

Droppings boards of sturdy welded wire or closely spaced wooden slats allow manure to fall through where chickens can't pick in it. Not only will the chickens remain healthier, but the droppings will be easier to remove because they won't get trampled and packed down. Even if droppings boards don't cover the entire floor, putting them under roosts (or using them as perches) improves sanitation and simplifies cleaning away the piles of night-deposit droppings.

To build this style floor, construct the perimeter wooden framework and fasten it to either welded wire or 1 by 2-inch lumber, placed on edge for rigidity, with 1-inch (2.5 cm) gaps between boards. Make the floor in removable sections that are small and light enough for you to move easily. When you clean out the droppings underneath, take the sections outdoors and clean them with high-pressure air or a pressure washer. Then let them dry in direct sunlight. Like wood flooring, droppings boards must be high enough off the ground to discourage rodents.

Concrete is the most expensive option but the most impervious to rodents and predators and the easiest to clean and disinfect. As a low-cost alternative, mix one part dry cement with three parts rock-free or sifted soil, and spread 4 to 6 inches (10 to 15 cm) over plain dirt. Level the mixed soil, and use a dirt tamper to pound it smooth. Mist the floor lightly with water, and let it set for several days before turning the chickens in. You'll end up with a firm floor that's easy to maintain.

Bedding

Bedding, scattered over the floor or under droppings boards, offers numerous advantages: it absorbs moisture and manure, cushions the birds' feet, and controls temperature by insulating the ground. Good bedding, also called litter, has these properties:

- Is inexpensive
- Is durable
- Is lightweight
- Is absorbent
- Dries quickly
- Is easy to handle
- Doesn't pack readily
- Has medium-size particles
- Is low in thermal conductivity
- Is free of mustiness and mold
- Has not been treated with toxic chemicals
- Makes good compost and fertilizer

Of all the different kinds of litter I've tried over the years, wood shavings (especially pine) remain my favorite because they're easy to manage — but they're also quite expensive. Chopped straw is a good alternative. Wheat straw is best, followed by rye, oat, and buckwheat, in that order. Chopped straw, especially mixed with shredded corncobs and stalks, makes nice loose, fluffy bedding. On the other hand, straw that hasn't been chopped mats easily and when combined with moist droppings and trampled by the chickens, creates an impenetrable mass that's difficult to clean up. A good alternative to chopped straw is well-dried clippings from a lawn or pasture that wasn't sprayed with toxins — the operative phrases here are "well dried" and "wasn't sprayed."

Dried leaves are sometimes plentiful but pack too readily to make good bedding. Rice hulls and peanut hulls are cheap in some areas but are not absorbent enough to make good litter. Shredded paper is at least as good as rice or peanut hulls and is inexpensive, but tends to retain moisture and to mat, so it must be replaced often. In some areas it's sold by the bale, or you can make your own, given lots of newsprint and a shredder. Coated (shiny) paper is not absorbent enough for this purpose.

Deep litter insulates chickens in the winter and lets them keep cool by burrowing in on hot summer days. Start young birds on bedding a minimum of 4 inches (10 cm) deep and work up to 8 inches (20 cm) by the time they are mature. When litter around the doorway, under roosts, or around feeders becomes packed, break it up with a hoe or rake. Around waterers or doorways, remove wet patches of litter and add fresh dry litter (and fix leaks).

DUST BATHING

Chickens enjoy frequent dust baths to keep their feathers clean and in good condition, which helps prevent injuries and control body temperature. Chickens wallow in dust or litter and work it through their feathers by flapping their wings and kicking their legs. If they're bathing outdoors in the warm sun, sometimes one will lie so still in a wallow that you're tempted to think she's dead. When they're done dusting, they stand up and shake themselves, and you can see the dust billowing out.

Outside, chickens dig holes in the dirt or sand. A dust hole near a gate or doorway is inconvenient and potentially dangerous for an unwary human who steps into it. Paving in these areas prevents dusting, provided the paving extends beyond the shelter's drip line. Otherwise, during wet weather your chickens will dust out a nice big hole for you to trip over in that dry strip of ground between the pavement and the drip line.

Chickens confined indoors, or kept inside during rainy weather, dust in litter. If your coop floor is mostly covered with droppings boards or wire, provide a section of litter where several chickens can dust at the same time. A bright light will encourage them to use the litter for dusting instead of as a community nest for laying eggs.

Under droppings boards, after each cleaning spread at least 2 inches (5 cm) of litter beneath the boards to absorb moisture from manure and make it easier to scoop up. An easy-to-manage combination system is to place droppings boards beneath perches where the majority of droppings accumulate and have open litter everywhere else. Your chickens won't be able to get to the manure piles beneath the droppings boards but will have the open-litter area to dust and scratch in, stirring up the bedding and keeping it light and loose. To encourage scratching, scatter a handful of grain over the litter each day, and let the flock scramble for it.

Roosts

Wild chickens roost in trees. Many of our domestic breeds are too heavy to fly up as high as a tree limb but still like to perch off the ground. You can make a perch from an old ladder or anything else strong enough to hold chickens and rough enough for them to grip, but without being so splintery it injures their feet. If you use new lumber, round off the corners so your chickens can wrap their toes around it. Plastic pipe and metal pipe do not make good roosts; they're too smooth for chickens to grasp firmly. A wood closet rod, on the other hand, is suitable, as is a 2×2 with the edges rounded off.

Allow 8 inches (20 cm) of perching space for each chicken, 10 inches (25 cm) for the larger breeds. If one perch doesn't offer enough roosting space for your number of birds, install additional roosts. Place them 2 feet (0.6 m) above the floor and at least 18 inches (45 cm) from the nearest parallel wall, and space them 18 inches apart. If floor space is limited, install roosts in stair-step fashion 12 inches (30 cm)

apart vertically and horizontally, so chickens can easily hop from lower to higher rungs. Either way, make perches readily removable so the droppings below can be cleared away easily.

Nests

Hens like to lay eggs in dark, out-of-the-way places. Nest boxes encourage them to lay their eggs where you can find them and where the eggs will stay clean and unbroken. You can furnish a large flock with a community nest, offering a minimum of 9 square feet (0.8 sq m) per 100 hens. But any nest large enough to accommodate more than one hen at a time invariably leads to broken eggs. I much prefer individual nests.

Furnish one nest for every four to five hens in your flock. A good size for Leghorn-size layers is 12 inches wide by 14 inches high by 12 inches deep (30 by 35 by 30 cm). For heavier breeds, make nests 14 inches wide by 14 inches high by 12 inches deep (35 by 35 by 30 cm); for bantams, 10 inches wide by 12 inches high by 10 inches deep (25 by 30 by 25 cm).

A rail just below the entrance to the nest gives hens a place to land before entering. Make sure the rail isn't so close to the nest that a chicken roosting on it will fill the nest with droppings. For most chickens, the rail should be no closer than 8 inches (20 cm) from the edge of the nest.

A 4-inch (10 cm) sill along the bottom edge of each nest prevents eggs from rolling out and holds in nesting material. Pad each nest with soft clean litter and change it as often as necessary to keep eggs clean and unbroken. Exactly how often a nest needs cleaning depends on how messy your chickens are: Do they deposit poop while sleeping in the nests at night or hiding in them during the day? Do they track

─── ROOSTING OPTIONS ───

Roosts over Droppings Pit

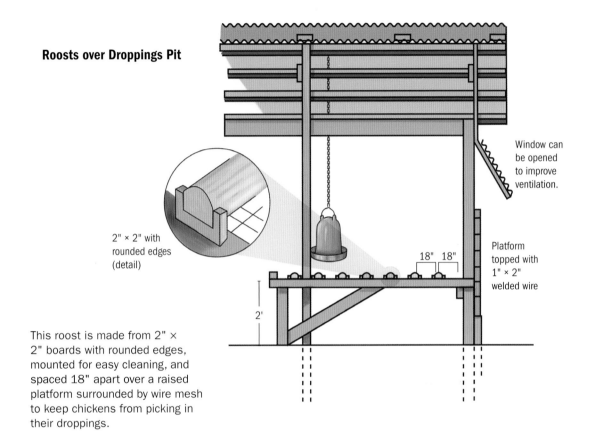

2" × 2" with
rounded edges
(detail)

Window can
be opened
to improve
ventilation.

18" 18"

Platform
topped with
1" × 2"
welded wire

2'

This roost is made from 2" ×
2" boards with rounded edges,
mounted for easy cleaning, and
spaced 18" apart over a raised
platform surrounded by wire mesh
to keep chickens from picking in
their droppings.

Step-Stair Perches

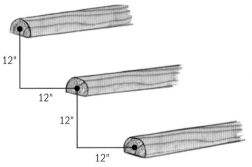

12"

12"

12"

12"

If roosting space is at a premium,
step-stair the perches and space
them 12" apart.

in mud on their feet? Do thin-shelled eggs get broken and mess up the litter? These and other management issues influence how often the litter must be changed.

Place nests on the ground until your pullets get accustomed to using them, then raise the nests 18 to 20 inches (45 to 50 cm) off the ground by setting them on a platform or firmly attaching them to the wall. Raised nests discourage chickens from scratching in them and dirtying or breaking eggs. Further discourage nonlaying activity by placing nests on the darkest wall of your coop or shielding the nests from light.

A 45-degree sloped roof above the nests will keep birds from roosting on top. Better yet, build nests to jut outside the coop, and provide egg-gathering access from outside. The advantages to this situation are that the chickens won't be able to roost on top of the nests, they'll have more floor space indoors, and you'll be able to collect eggs without disturbing your flock.

An alternative design for darkened nests that are easy to clean calls for placing a long bottomless nest box on a shelf. Partition the inside of the box into a series of nesting cubicles, with entrances facing the wall. Allow an 8-inch (20 cm) gap between the wall and the entrances so hens can walk along the shelf at the back; with this plan, chickens won't be inclined to roost overnight with their rear ends hanging into the nests. Slope the top of the box away from the wall to prevent roosting on top, and add a drop panel at the front for egg collection. To clean these nests, first check to make sure no eggs or hens are inside, then pull the box off the shelf and the nesting material will fall out.

Ready-built wooden or metal nests are available from a number of sources. Do an Internet keyword search for "chicken nests."

Even if you don't plan to buy your nests, you'll get lots of ideas for building your own. If your flock consists of a few pet chickens, a small animal carrier makes a handy, and easy-to-clean, ready-built nest.

EXTERIOR NESTS

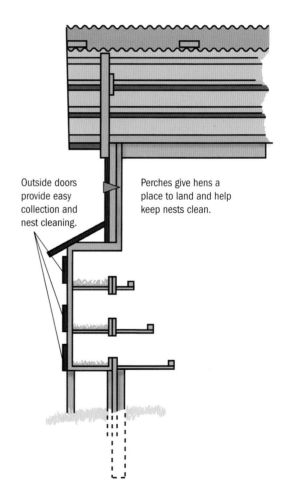

Outside doors provide easy collection and nest cleaning.

Perches give hens a place to land and help keep nests clean.

Exterior nests increase floor space and are easy to maintain from outside the coop.

Cages

Any small enclosure used for confining chickens, strictly speaking, is a cage, and that includes hutches, arks, converted playhouses, and other small coops that keep chickens in close confinement. The industrial practice of caging commercial laying hens has given caged housing a bad name. Commercial laying hens are caged to control their diet and guard them against diseases, predators, and the weather.

NEST BOXES ON SHELVES

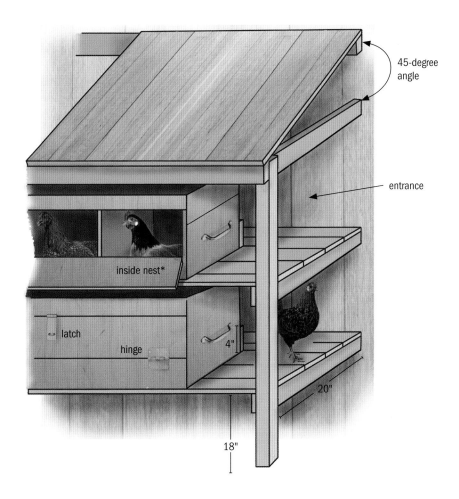

Nest boxes on shelves provide darkened entrances at the back and can easily be cleaned by sliding each box off its shelf. Don't place more than four nests per box.
*Size nests according to the size of your chickens.

Layer cages have no nests but are designed with sloping wire floors so eggs roll to the outside, where they remain clean and easy to collect. The hens are allotted so little space they can scarcely turn around and must have the tips of their beaks cut off so they won't eat each other. Anyone who loves chickens abhors this type of caged confinement.

By contrast, a fellow I once visited housed his chickens by ones, twos, and threes in outdoor hutches. Every day he released birds from one hutch to wander freely for a while, then put them back and released those in the next hutch. Barring bad weather, each bird had at

least half an hour on the ground daily, more on days when the fellow worked in his yard or garden. When each bird's time was up, the fellow easily picked it up and lifted it back into its cage — no frantic chasing and catching involved. His birds were the calmest, most well-adjusted chickens I've ever seen.

Chickens may be caged for any number of reasons. Exhibition birds are caged while being trained and conditioned for the showroom and to control breeding. Breeder cocks may be caged to keep them from fighting with one another or defeathering hens. In such a case, the cocks might be rotated so each has

LEVEL HUTCH

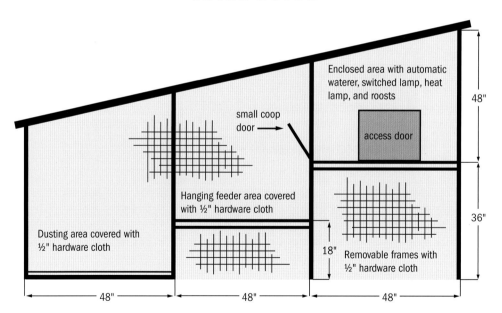

This three-level hutch includes everything a chicken needs to be safe and comfortable.

a turn running with the hens. A defeathered hen might be caged until her feathers grow back. A mother hen might be penned up with her chicks to protect them from cats and other predators. Urban and suburban pet chickens are commonly caged in fancy landscaped hutches to keep them from annoying the neighbors or being eaten by a dog. A pop hole and ladder might be added to let the chickens out while the owner works around the yard.

Cage Styles

Caging chickens can be less expensive than building a chicken shelter from scratch, especially if the cages are enclosed within an existing structure. When I raised exhibition bantams, I scoured the classified section of the local newspaper until I located a rabbitry that was going out of business. For next to nothing I picked up all the cages I needed and modified them to be more spacious. I housed my bantams in pairs and trios in our garage, where I could control their diet for peak health and be sure their valuable eggs would not be hidden, soiled, or stolen by predators. The birds were easily protected from predators, since I closed the garage door at night so raccoons or vagrant dogs couldn't get under the cages and bite off their feet. On nice days, the cages could be moved easily into the yard to let the bantams snack on fresh vegetation through the bottom wire.

Those cages were set up on concrete blocks at each corner, so the droppings fell underneath for easy cleaning. Later we built an outdoor hutch consisting of a row of cages elevated on stout legs and with a shed roof as protection from sun, wind, and rain. From the happy sounds I heard while working in the yard, those chickens were as content as could be.

Cages may be set on sturdy wooden frames, hung from rafters, or attached to the wall. Today I have four all-purpose cages that get frequent temporary use, clipped to the wall with sturdy picture hangers. In the summer, we hang them outdoors in the shade of the barn's eaves; in winter, we hang them inside the warm barn.

Heavy breeds should never be housed on wire, and no chicken should be kept on suspended wire as permanent full-time housing. If you need to keep mature chickens caged, make the floor of batten boards or of solid plywood

CAGE DIMENSIONS

A cage must be roomy enough for a chicken to stretch its neck, wings, and legs without touching the sides or top.

Number of Birds	Width	Depth	Height*
1	30 in (75 cm)	24 in (60 cm)	24 in (60 cm)
2	27 in (68 cm)	32 in (80 cm)	24 in (60 cm)
4	46 in (115 cm)	32 in (80 cm)	24 in (60 cm)

*Add 6 inches (15 cm) in height if you plan to install roosts.

covered with bedding or at least provide resting pads, such as those made for rabbits.

Chickens housed together in a cage must get along, since a chicken that's bullied would have nowhere to get away. Modifications — adding a sunporch, making the unit multileveled, and providing a dust-bathing area — create a variety of environments where birds can get away from each other and avoid boredom or bullying.

Make Your Own Cages

Buying ready-built new cages may be cheaper than making your own, unless you have an inexpensive source for wire. You'll need 1-inch-by-1-inch (2.5 by 2.5 cm) or 1-inch-by-2-inch (2.5 by 5.0 cm) galvanized 12- or 14-gauge welded wire. You'll also need J clips, or ferrules, ferrule-closing pliers, and wire side cutters. If you buy used cages, you may have to redesign them, for which you'll still need the clips, pliers, and cutters.

Use the table on page 83 to determine what size cage you need. Lighter breeds appreciate a roost 6 inches (15 cm) off the cage floor, in which case you may have to make the cages 6 inches higher than you would otherwise so the birds won't rub their combs or topknots against the ceiling. From welded wire cut four sides, a top, and a bottom, and clip them together along the edges. Around the top and bottom edges, add lengths of 10-gauge wire as reinforcement to keep the cage from sagging.

To make a door, cut a 14-inch-square (35 cm) opening in the center of the front wall, 2 inches (5 cm) from the bottom. File the cut ends smooth. From a separate piece of wire, cut a door 14 inches high and 15 inches (38 cm) wide. Using loosely attached ferrules as hinges, attach the door at the bottom, side, or top — the hinge position is strictly a matter of preference, although a door hinged at the bottom so it drops down when it's opened will leave your hands free for working inside.

Latches may be fashioned from all sorts of things, but nothing beats the standard cage-door latch in the illustration. As a safety measure, secure the shut door with a spring clip, hung from a chain attached to the cage to keep the clip from getting lost.

Fences

A stout fence keeps chickens from showing up where they aren't welcome and protects them from predators. Even chickens confined within a range or garden shelter need protection from being harassed by dogs, coyotes, raccoons, and the like or having their shelter rubbed on, climbed on, and bumped by livestock.

Consider the ease of maintenance — spending more time and money now may save you from spending additional time and money later on repairs. Choose a style of fence that blends in with the surroundings, and check for local restrictions that may determine what is and is not allowed. Where an agricultural-style fence or electrified wires are not permitted, a solid wooden fence, a high picket fence, or an attractive rail fence backed with a less visible wire mesh are some possible alternatives.

Chain Link

The ideal chicken fence is made from tightly strung small-mesh woven wire. The best fence I ever had was a 6-foot-high (1.8 m) chain-link fence that came with the first house I owned. In the 11 years I lived there, I lost few birds, most of which were chicks that popped through the fence and got carried off by a neighbor's cat.

CAGE-BUILDING TOOLS

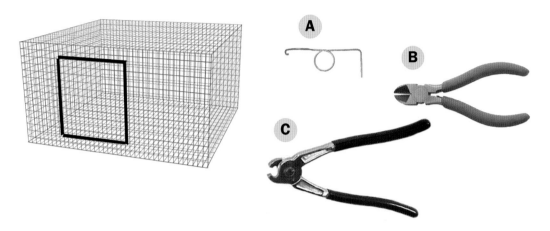

To build or redesign a cage, you will need (A) a latch, (B) wire side cutters, and (C) ferrule-closing pliers, as well as the wire and clips.

I now live on a farm where we enjoy the wildlife as much as we enjoy our poultry. Trouble is, the wildlife have as much interest in poultry as we do. Our large chicken yard (pasture, really) makes the cost of entirely enclosing it with chain link prohibitive. For years we fenced our chickens with the same high-tensile, smooth-wire electric fence that contains our four-legged livestock. It does a good job of keeping out the larger predators but does not keep out the smaller chicken-or-egg eaters and certainly does not keep the chickens in. Occasionally, we lose a bird that wanders into the orchard for lunch and meets a fox with the same idea.

So we have returned to chain link to create yards designed for housing setting hens and growing birds that are more vulnerable to predators than mature birds. Chain link is expensive but virtually maintenance-free. It excludes most predators, provided it is properly constructed. The bottom must be stretched tight with a tension wire or bottom rail, or attached to wide boards turned on edge, to prevent predators from pushing underneath.

The chief disadvantage to chain link is that you'd better be sure exactly where you want your fence before it's installed. If you decide later to make the yard larger or move it over, you'll incur a big expense.

Mesh and Net

If the expense of chain link is prohibitive for you, the next best kind of fence for chickens is wire mesh with openings small enough that neither chickens nor predators can get through. Of the many kinds of wire mesh available, one that works well for chickens and is relatively low on the cost scale is yard-and-garden fence with 1-inch (2.5 cm) spaces toward the bottom and wider spaces toward the top. The small openings at the bottom keep poultry from slipping out and small predators from getting in.

The fence should be at least 4 feet (1.2 m) high, higher if you keep a lightweight breed that likes to fly. Bantams and young chickens of all breeds are especially fond of flying.

To deter raccoons and foxes from burrowing into the poultry yard, dig a trench along the fence line and bury the bottom portion of a fence. Or create an apron by sinking a 12-inch-wide (30 cm) length of wire mesh at the outside bottom. It should be perpendicular to the fence and extend from the bottom of the fence outward, away from the yard. To prevent soil moisture from rapidly rusting the apron, use vinyl-coated wire or brush the apron with roofing tar to slow rusting. Cut and lift the sod along the outside of the fence line, and clip or lash the apron to the bottom of the fence. Spread the apron horizontally along the ground and replace the sod on top. The apron will get matted into the grass roots to create a barrier that discourages digging.

Hexagonal net is a type of wire-mesh fence also called poultry netting, hex net, or hex wire. It consists of thin wire, twisted and woven together into a series of hexagons, giving it a honeycomb appearance. The result is lightweight fencing that keeps chickens in but does not deter determined predators from breaking through with brute strength.

Hex net comes in mesh sizes ranging from 0.5 inch to 2 inches (1.5 to 5 cm). The smaller the mesh, the stronger the fence. The smallest grid, called aviary netting, is made from 22-gauge wire and is used to house chicks and to prevent wild birds from stealing poultry feed. A 12-inch-wide (30 cm) strip of aviary netting securely attached to the bottom of a wire-mesh fence will keep baby chicks from straying out of the yard.

Hexagonal net is included here only because it is so common, but I do not recommend it for permanent fencing. Although the wire is relatively inexpensive, all the necessary support to make it sufficiently tight adds to the cost of erecting the fence. And when all is said and done, the fence has a short useful life, during which it tears easily and requires constant repair.

One-inch (2.5 cm) mesh, woven from 18-gauge wire, is commonly called chicken wire. Rolls range in length from 25 feet to 150 feet (10 to 50 m), in width from 12 to 72 inches (30 to 180 cm) The narrowest wire is used to reinforce the lower portion of a woven wire or rail fence to keep little critters from slipping in or out.

So-called turkey netting, made of 20-gauge wire, has 2-inch (5 cm) mesh and is used for mature birds. Heights range from 18 inches to 72 inches (45 to 180 cm), length from 25 feet to 150 feet (10 to 50 m). Mesh this large is difficult to stretch properly. For a tall fence, therefore, many fencers run two narrow rolls, one above the other, either stapling the butted edges to a rail or fastening them together with cage-making ferrules.

A less common variation, called rabbit netting, has 1-inch (2.5 cm) mesh at the bottom and 2-inch (5 cm) mesh toward the top. It comes in 25-foot (7.5 m) rolls, is 28 inches (70 cm) high, and may be used to pen chicks.

Unless you treat hex wire with great care, don't expect it to last more than five years, if that long. Options in protective coating are galvanizing and vinyl. Some brands are galvanized before being woven, some afterward. The former is cheaper but should be used only under cover, since it rusts rapidly in open weather. Plastic-coated wire is a bit more rust resistant, and some people find the colors more attractive than plain wire.

Hex wire is easy to put up, although it tears readily, and slight tears grow into big holes. Netting also tends to sag. You'll need to erect a stout framework of closely spaced wood posts with a top rail for stapling and a stout baseboard, both for stapling and to deter burrowing; make sure no dips at soil level leave gaps for sneaky critters to slip under. To keep the wire taut on a tall fence, add a rail in the middle as well. Hand-stretch the mesh by pulling on the tension wires — the wires woven in and out at the top and bottom of the netting. Taller netting has additional intermediate tension wires. To avoid snagging skin and clothing, especially around gates, fold the cut ends under before stapling them down.

Secure Gates

No matter how secure your fence is, it's only as secure as your gates. Our commercially installed chain-link poultry run left us to deal with predator-size gaps at the sides and bottoms of the gates.

Even when a gate is initially installed close enough to the ground, traffic from walking, wheelbarrows, mowers, and so forth eventually wears grooves under the gate. Adding a sill solves that problem.

Sink a pressure-treated 4×4 under each walk-through gate and a 6×6 under a drive-through gate, or pour reinforced concrete sills of similar size. This small investment prevents soil compression from creating ruts beneath your gates — helping to keep your birds in and predators out.

A gate without a sill eventually develops ruts underneath that let birds out and predators in.

Electric Fence

Adding offset electrified scare wires to your fence gives you both a physical barrier and a psychological barrier. Should the physical barrier fail (that is, the power goes off), you still have the psychological barrier.

To deter climbers, string an electrified scare wire on offset insulators along the outside bottom of your fence, 10 to 12 inches (25 to 30 cm) above ground level. As insurance, especially if your area experiences winter snow, add a second scare wire 8 to 10 inches (20 to 25 cm) below the top of the fence.

Using a plug-in electric controller to transmit energy to the wires will give them plenty of zap, especially when fast-growing weeds drain away power before you get around to weed whacking. Out in a field, a portable option is a battery-operated or solar-powered energizer of the sort used to control grazing livestock or protect gardens, but you'll have to be more mindful of weed loads.

Electric Net

An all-electric net fence designed specifically for poultry sounds great in principle, but in practice is not ideal. It must be constantly electrified so poultry, pets, and predators won't get tangled in the net; if you live in an area prone to power outages, you must use a battery- or solar-operated energizer and make certain it's always fully functional. Even with the best high-impedance energizer, chickens get tangled in the polywire net and electrocute themselves, in the process tearing the net and thus reducing its useful life.

Other issues — difficulty keeping the net taut; problems getting line posts into rocky soil, drought-ridden clay, or frozen ground; inconvenient guy wires at the corners — are incidental by comparison with finding bullfrogs and other dead animals tangled in the fence. Despite these issues, net fence with a battery-powered controller is popular as a lightweight, movable fence for range rotation. No matter how careful you are not to snag the fence each time you move it, gradually it will fray to the point of requiring replacement.

Chickens aren't as susceptible to getting zapped as other livestock because of their small feet and protective feathers, but they do learn to respect an electric fence. First they have to know their home territory. Whenever you move chickens much outside their previous location, confine them inside their shelter for a day. Then, when you let them into the fenced yard, they shouldn't stray too far from home.

8"

12"

Attach wire mesh fencing to the side of the posts away from the chicken yard. Electrified scare wires toward the top and bottom will discourage four-legged predators.

4

FEED AND WATER

The secret to feeding chickens properly is to imitate as closely as possible their natural diet were they truly free to forage. The more attention you pay to your feeding—and drinking—program, the greater your rewards will be in terms of perky chicks, healthy chickens, abundant eggs, good fertility and hatchability, show awards, and tasty meat.

Water

A chicken drinks often throughout the day, sipping a little each time. A chicken's body contains more than 50 percent water, and an egg is 65 percent water. Therefore, for its body to function properly, a bird needs access to fresh drinking water at all times.

On average, each chicken drinks between 1 and 2 cups (237 and 473 mL) of water each day. Exactly how much a chicken drinks depends on several factors, including these:

- **Age:** Chickens drink more as they get older.
- **If they're laying:** Layers drink twice as much as nonlayers.
- **Temperature:** Chickens drink two to four times more in hot weather.
- **Time of day:** Chickens drink the most at dawn and dusk.

The most important ingredient in your chickens' diet is water. But water quality and availability are not things we normally think much about until a problem arises. Water deprivation is a serious matter. Pound for pound, a meat bird requires about 1½ times as much water as feed, and hot weather increases the amount of water needed. If broilers don't get enough water, they won't eat well and therefore won't grow well. Similarly, laying hens that don't get enough to drink won't lay well. Hens deprived of water for 24 hours may take another 24 hours to recover. Deprived of water for 36 hours, they may go into a molt, followed by a long period of poor laying from which they may never recover.

Deprivation easily occurs when water needs go up during warm weather but the amount of water furnished remains the same. Chickens prefer water at a temperature between 50 and 55°F (10 and 13°C). The warmer the water, the less they'll drink. In summer, put out extra waterers and keep them in the shade or frequently furnish your flock with fresh, cool water. Electrolytes in the water stimulate drinking and replenish electrolytes lost due to heat stress.

Providing feed on a free-choice basis lets your chickens eat whenever they feel peckish.

WATER HEATERS

Heating Pan

Immersion Heater

To keep water from freezing, set a metal drinker on a thermostatically controlled heating pan or drop a sinking immersion heater into a trough.

Water deprivation occurs in winter when the water supply freezes. In cold weather, bring your chickens warm (not steaming hot) water at least twice a day. Or keep waterers from freezing with water-warming devices, available from a farm store or livestock-supply catalog. You might choose to use an immersion heater in a water trough, place a metal bell fountain on a pan heater, or wrap heating coils around automatic watering pipes. For a small number of chickens, a plug-in heated water bowl makes a handy option. You can save electricity that would otherwise be lost by unnecessarily heating water on the warmer days by plugging each water-heating device into a thermostatically controlled outlet adapter (Thermo Cube TC-3).

Water Quality

Even when they have plenty of water, chickens may be deprived if they don't like the taste. Factors affecting water quality include the following:

- **Color:** Water should be colorless.
- **Odor:** Water should be odorless.
- **Opacity:** Water should be clear.
- **Bacteria:** Water should be free of bacteria.

Medications in the water can cause chickens not to drink. Do not medicate water when chickens are under extreme stress, such as during hot weather or at a show.

Large amounts of dissolved minerals can make water taste unpleasant to chickens. To

MILK-FED CHICKENS

Milk is 87 percent water, and the remainder is loaded with protein, carbohydrate, fat, vitamins, and minerals. Best of all, chickens love it. If you raise dairy animals and have excess milk, modest amounts make a terrific supplement for your chickens. Even better is whey left over from cheese making. Whey contains most of the whole milk's protein but little of the fat.

Offer the milk in a container that can be scrubbed easily between fillings, and place it where milk splatter won't stick to buildings and fences — a chicken that gets milk on its comb will shake its head and send milk droplets flying in all directions.

Milk helps put weight on meat birds, but layers that drink too much milk get fat, and fat hens don't lay well. A good rule of thumb is to feed no more than 10 pounds (4.5 kg) of liquid milk per 50 pounds (22.5 kg) of rations consumed.

find out if your water supply contains a high concentration of minerals, have the water tested. If total dissolved solids exceed 1,000 parts per million (ppm), find an alternative source of water for your flock.

Chickens should not have to get their drinking water from puddles or other stagnant, unhealthy sources. Instead, provide fresh, clean water in suitable containers. Never expect your chickens to drink water you wouldn't drink yourself.

Waterers

Waterers, also called drinkers, come in many different styles. The best waterer has these features:

- Keeps water clean and free of droppings and other chicken-generated debris
- Doesn't leak, drip, or tip over easily
- Furnishes enough water to last all day
- Is easy to clean

Automatic, or piped-in, water is the best kind because, at least in theory, it ensures the chickens never run out. But piped-in water isn't without disadvantages. Aside from the expense of running plumbing to the shelter, water pipes can leak if not properly installed and can freeze in winter, unless buried below the frost line or wrapped in electric heating tape. And if you're on a well and the power goes out, so does the water.

Silt and other particles in the water may clog the automatic mechanism, so drinkers must be checked at least daily to make sure they remain functional. If your water system lacks a built-in filter, you can minimize clogging by putting an inexpensive fine-screen filter, of the sort used for drip irrigation systems, in the connection between the hose and drinker, and clean it regularly. When shopping for an automatic waterer, look first at how difficult it is to clean. If it can't be emptied and swabbed out easily or quickly disconnected for a good scrub, pass it up. Another feature to watch for is how easy the device is to bend or break. A chicken attempting to roost on the device or chickens running into it while chasing each other may provide enough impact to tip the waterer (causing a leak) or break it off (causing a minor flood). And if larger animals, such as turkeys or four-legged livestock, have access to the same area as your chickens, the drinker must be sturdy enough to withstand their activities as well.

Automatic drinkers that connect to your main water line may need a pressure-reduction valve, which is either built into the device or installed in the line separately. An automatic waterer designed for pets is an example of a device with a built-in pressure regulator and offers an inexpensive option for use with a few chickens. An example of a separate reducer is a float-controlled toilet tank installed between the main line and the drinker; when a chicken drinks, water from the tank refills the drinker and water from the main line refills the tank.

When installing a water line specifically to supply automatic drinkers, include a check valve to avoid the possibility that syphoning will cause contaminated water to back up into the main line. If, as is commonly done, you connect an automatic device to a faucet by means of a hose, use only a standard outdoor hose, as the hose will be under pressure all the time. Any other kind of hose will leak.

A chicken-size water trough has a float valve to regulate water depth. The trough may be placed directly on the ground but is less apt to collect shelter debris if mounted on a stand or bracketed to the wall — provided, of course, the trough may be easily removed for cleaning. A rotating reel on top will prevent chickens from roosting on the edge and fouling the water.

Instead of hooking it up to a water line, you could set up an automatic drinker on a gravity flow system, which is handy where running a water line would be expensive or inconvenient. Water is stored in a tank of some sort, which may be filled periodically with a hose or by hauling water in buckets, or may be arranged to collect rainwater. We affixed an agricultural water tank to a small trailer that may be pulled by a riding mower to a water outlet for filling, and then pulled back to location. We put a standard valve (faucet) on the tank, making setup exactly like attaching to a regular faucet or a hose. A few pieces of scrap copper pipe tossed into the tank helps control algae.

In addition to single drinkers, automatic systems are also available with multiple drinking stations. These devices come in two basic styles, nipples and cups. To drink from a nipple, the chicken taps its beak against a metal pin to obtain water by the drop. By contrast, a cup holds a small amount of water and automatically refills each time a bird drinks. Larger cups are available for use in open housing, but they tend to collect henhouse debris and are not easy to clean.

HOMEMADE NIPPLE WATERER

To drink from a nipple, the chicken taps its beak against a metal pin to obtain water by the drop.

Water nipples may be used to make a drinker from just about any type of water container ranging from an empty milk jug to a 5-gallon bucket, a 55-gallon drum, or a PVC pipe connected to the main water line. Installing nipples is easy — just drill a hole of suitable size and insert the nipple so it hangs down vertically.

Properly installed nipples are virtually trouble-free. They keep bedding and poop out of the drinking water, and they don't need to be frequently scrubbed out. Just check each nipple daily to see that it isn't leaking or clogged (tap it with a finger to make sure water comes out).

Recommendations made by nipple manufacturers range widely from one nipple per three birds to one nipple per ten. In practice, you should have at least two nipples so two chickens can drink at once. With a larger flock, furnish about one nipple per six chickens.

As with any drinker, the height must be adjusted to accommodate the size of the chickens; keep the nipple at, or just above, eye level. An Internet keyword search for "chicken water nipples" will net you lots of drinker design ideas, as well as information on how to establish appropriate water pressure for in-line nipples.

For a small number of chickens, hand-filled and hand-carried bell-shaped waterers are available in a variety of sizes and made of plastic or galvanized metal. Inexpensive plastic 1-gallon (4 L) drinkers work okay for young birds but hold only enough water for a few chickens and are easily knocked over by rambunctious birds. In addition, the plastic cracks after a time, especially if exposed to direct sunlight or freezing weather, and the cost of replacing those "inexpensive" waterers adds up fast.

Metal waterers are sturdier than those made of plastic and come in larger sizes, holding 2 gallons (8 L), 3 gallons (12 L), 5 gallons (20 L), or greater. As in all things,

TYPES OF WATERERS

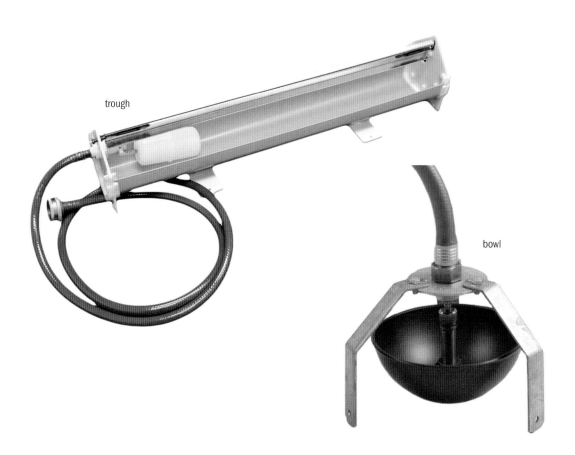

trough

bowl

Two styles of automatic watering devices for multiple chickens are the trough and the bowl.

you get what you pay for — a cheap metal waterer will rust through faster than a quality galvanized drinker.

Drinking Stations

Provide enough waterers so at least one-third of your birds can drink at the same time. Even if you have so few chickens that one drinking station appears to be adequate, furnish at least two, spaced well apart, to ensure that all chickens can get a drink without fighting or being chased away.

In a large shelter, provide one drinking station at least every 8 yards (7 m). Additional waterers outside the shelter encourage chickens to spend more time outdoors and are absolutely necessary during hot weather.

To make drinking easy and minimize contamination, position waterers so the top edge is the height of the birds' backs. Make sure the surface is level so water won't drip out.

Bell waterers tend to get tipped over, the rim gets filled with debris, and the bottoms eventually rust through from constant ground-contact moisture. Hanging solves these problems for the metal 2-gallon (8 L) size. Since manufacturers don't provide a handy way to hang them, you have to devise your own method. An easy way is to fashion a hanger out of a piece of 10-gauge solid/rigid wire with the ends wrapped around the drinker handle and the middle bent to hang from a hook or spring clip at the end of a chain coming down from the rafters.

For larger, heavier waterers and those used outdoors, place each drinking station over a miniature droppings pit. Build a wooden frame of ½-inch-by-12-inch-by-42-inch boards. Staple strong wire mesh to one side, and set the frame, wire side up, on a bed of sand or gravel. Place the waterer on the wire so chickens have to hop up onto the mesh to get a drink. Any spills will fall out of reach through the wire.

No less than once a week, clean waterers with soap and water, and disinfect them with either vinegar or a solution of chlorine bleach — one part bleach to nine parts water. A grout brush, designed for cleaning kitchen and bathroom tiles, fits perfectly into the rim of a bell waterer for easy cleaning.

A 2-gallon (8 L) bell waterer may be hung from a chain so it can't tip over or fill with debris.

The Natural Chicken

Deciding what to feed your chickens is easier when you know what a chicken would normally eat, given a choice. Knowing how a chicken digests what it eats is also helpful, since a chicken's digestive system is a tad different from that of a human. Recognizing normal feeding behavior is not essential to developing a decent diet for your flock, but knowing what behavior to expect will enhance your chicken-watching experience.

A Diet of Their Choosing

A chicken free to choose what it eats is an opportunistic omnivore. Truly free-ranging chickens snack on a smorgasbord of grains and other seeds, tender greens, succulent fruits and vegetables, worms and slugs, crawling and flying insects, frogs and lizards, rodents, and small birds. They'll scratch in the manure of other animals to glean insects, larvae, and undigested grains. They'll eat eggs and fresh meat, including other chickens and dead animals of other species. Given access, they'll devour milk, yogurt, and other dairy products.

What chickens are not is vegetarian. They also are not primarily grazers, although they do like fresh greens in their diet the same way we humans enjoy a fresh salad with our meals. Chickens evolved in the jungle, an environment offering a broad variety of tasty things to eat. Barring a jungle environment, the next nearest thing is forest understory, which offers a tempting mini smorgasbord but is fraught with danger.

When we developed our present garden, we initially incorporated chickens as part of the setup. One day, while we were working in our as-yet-unfenced garden, we decided to let the chickens out of their yard. Instead of foraging close by, they headed straight for the surrounding woods to scratch for insects in the leaves. Before long we heard a single loud squawk and knew we'd lost a chicken. Moments later we heard a second squawk. When the flock returned from their outing in the woods, sure enough — we were short two hens, apparently carried off by a pair of foxes.

We eventually gave up the idea of combining chickens and garden but only because our garden is so close to our bedroom window that we got tired of the roosters' early-morning wake-up call. Although a garden offers quite a variety of plant and animal life, indiscriminately turning chickens loose in a growing garden is a decidedly bad idea. They will chow down on or scratch up newly emerging seedlings, peck at red ripe tomatoes and strawberries, and demolish cucumbers and squash. But a well-managed system of combining chickens with a garden can benefit both the chickens and the garden, although in most cases it won't come close to furnishing complete nutrition.

Digestion

To stay ahead of predators, chickens evolved with the ability to eat a lot at one time, then move to a safe place where they can digest what they've eaten. Consequently, they forage actively at dusk, and digestion continues while they roost overnight — resulting in piles of droppings beneath the perch.

The expression "as scarce as hen's teeth" — in reference to something extremely rare or nonexistent — came about because chickens have no teeth. Their digestion results entirely from a combination of chemical and mechanical actions.

CHICKENS AND GARDENS

Schemes for safely combining chickens with gardening are probably as old as chicken-keeping itself. Here are but a few of the many possibilities for gardening with chickens:

- Put the chicken house next to the garden, where you can easily toss plant refuse to the flock and collect nitrogen-rich manure for the compost pile.

- Surround the garden with a double-fenced chicken yard, or "moat," creating a bug-free, weed-free zone that discourages entry by garden plant marauders, including deer, rabbits, and groundhogs.

- Let chickens into your garden late in the day, giving them an hour or so to glean bugs and nip leaves but not enough time to do serious damage before they're ready to go to roost (keep them out while tomatoes are on the vine, though, as birds invariably make a bee-line for ripe tomatoes).

- Choose a breed with heavy leg feathering, since they tend to scratch less than others and will therefore do less damage to your seedlings.

- Build a portable shelter to fit over raised beds so you can rotate the birds along with your veggies. Variations on this plan are discussed at length by Andy Lee and Pat Foreman in their book, *Chicken Tractor*.

- Divide the garden area in two with the chicken house in the middle. Garden on one side and confine the chickens to the other, alternating these uses annually.

Whatever system you use, keep chickens away from crops you plan to eat to avoid contamination with droppings that may carry salmonella, *E. coli*, or other pathogens harmful to humans. Pathogens may absorb into a plant's cells, where they cannot be washed off. To be on the safe side, for root crops and any other crop in which the edible portion touches soil, keep chickens away for 120 days; for crops without soil contact, 90 days is sufficient.

Saliva starts breaking down feed as soon as it enters a bird's mouth. The chicken's tongue pushes the feed toward the back to slide down the throat (the *esophagus*) into an expandable *crop*, where it is temporarily stored. A full crop bulges in a manner similar to a chipmunk's cheeks.

From the crop, feed trickles into the *true stomach* (the *proventriculus*), where enzymes break it down further. It then passes into the gizzard (the *ventriculus*), or mechanical stomach. The *gizzard* consists of strong muscles surrounding a tough pouch filled with small stones or grit that the chicken has swallowed for grinding up grains and other hard feedstuffs. Chickens that eat only processed foods — such as chick starter or layer pellets — don't need grit. But chickens that eat grains and other hard substances do need grit to grind up their food and make it digestible. Chickens that have a place to scratch in dirt generally pick up all the natural grit their bodies need;

DIGESTIVE SYSTEM

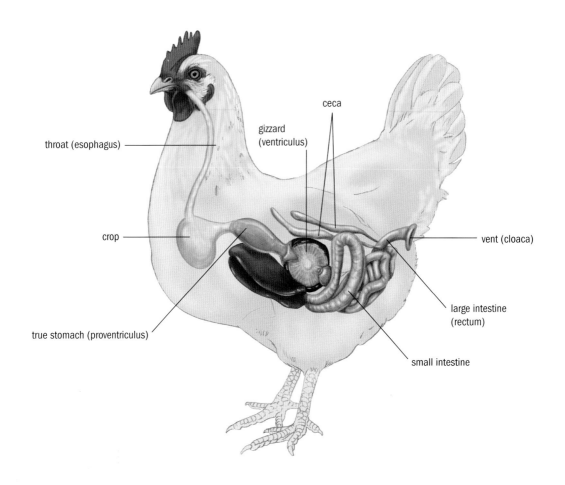

throat (esophagus)

gizzard
(ventriculus)

ceca

crop

vent (cloaca)

large intestine
(rectum)

true stomach (proventriculus)

small intestine

chickens in confinement must be fed grit as a dietary supplement.

From the gizzard, feed passes into the *small intestine*, where nutrients are absorbed. Between the small and large intestine are two blind pouches, called *ceca*, that have no known function — they may harbor beneficial microbes that aid digestion, in addition to the microbes populating the small and large intestine. The *large intestine*, or *rectum*, is relatively short and absorbs most of the water from the digested feedstuffs.

The large intestine ends at the *vent*, or *cloaca*, where a final bit of moisture is absorbed before wastes leave the body, expelled in the form of droppings. In a healthy chicken, feed passes through the entire digestive system within three to four hours of consumption.

A healthy chicken has a two-part dropping. The firm brownish part on the bottom is feces. On top is a smaller white part, which corresponds to a human's urine. A healthy chicken doesn't excrete urine but expels blood wastes in the form of semisolid uric acid, called *urine salts* or *urates*, that appear as white pasty caps on top of droppings. In addition to these frequent droppings, the ceca empty their contents two or three times a day, producing smelly, pasty droppings without the white caps.

Feeding Behavior

A chicken's normal feeding behavior is to peck and swallow. A chicken that gets hold of something tasty that's too big to swallow may pick it up in its beak and start running. Other chickens will give chase. One of three things happens next, reminiscent of a football game: the first chicken drops the piece, and another bird grabs it and runs; one of the others takes the piece away from that chicken and keeps running to stay ahead of the pack; another of the chickens giving chase catches up with the lead chicken and they engage in a tug-of-war.

Food running is so instinctive that baby chicks do it. A chick that finds a worm, for instance, may grab it and start running, thus attracting the others to give chase. Toss a cooked spaghetti noodle into the brooder, and the chase is on. The purpose of food running is supposedly to break up something large into pieces small enough to swallow. I've seen lone chickens find something to eat, grab it, and run. Sometimes they stop and shake it apart, but other times the act of running attracts far-away chickens to come give chase. In the case of, say, a chunk of apple or a piece of bread, the chicken that first grabs it is so busy trying to get away from the others that it may never get so much as a taste. The bird would have been far better off to attract less attention by quietly pecking off pieces small enough to swallow.

But let's say an unfortunate mouse attempts to cross the chicken yard and is grabbed by a quick-thinking chicken. In the course of food running, the mouse is torn apart and several chickens share the bounty. So food running does serve the purpose of breaking up too-large food items for sharing, provided the chickens are careful in selecting what to run with.

Tidbitting is a mother hen's act of breaking up a food item into pieces small enough for her chicks to swallow, but in contrast to food running, the hen remains at the spot where she found the food item and sounds the food call to gather her chicks. She'll then repeatedly pick up and drop the large piece, which might cause small pieces to break off, and also draws attention to the piece to encourage the chicks to peck at it. She might even hold the piece in her beak while the chicks peck at it.

A cock tidbits as a ploy to gather hens around him. He may or may not have actually found something tasty to eat. If not, as soon as a hen comes running he'll switch tactics and begin his courtship dance.

Head shaking and beak beating are other ways a chicken breaks up food. The bird picks up the food item in its beak and either shakes it or rubs it on the ground to break off pieces.

Ground scratching stirs up the soil to bring up seeds, insects, worms, and bits of grit. Baby chicks scratch instinctively. When I start newly hatched chicks, I initially put their feed in a shallow tray where they can find it easily. As soon as they scratch in the feed, scattering it out of the tray, it's time to switch to a chick feeder they can get their heads into but not their feet.

Feeding behavior includes cleaning particles of food from the head and beak, which a chicken does by scratching its head and beak with a claw. It may also clean its beak by wiping it on the ground. Beak wiping serves the additional purposes of keeping the beak sharpened for pecking and worn down to prevent the ends from growing so long or out of balance the bird can't peck properly.

Feed Choices

A chicken living under natural conditions — meaning in the jungle — enjoys a widely varied diet that furnishes an array of vitamins, minerals, protein, energy, and everything else a chicken needs to remain healthy. Confined chickens still have the same basic nutritional needs, which may be provided by a judicious combination of commercially available or home-mixed rations, grains, sprouts, table scraps, greens, and various supplements.

Commercial Rations

Early formulations of rations developed for chickens in confinement fell short of providing all the necessary nutrients. The formulas were initially tweaked to resolve nutritional issues and then to push for better performance: faster growth of meat birds and better egg production from layers. At the same time, methods were sought to reduce feed costs, and one way to do that is to incorporate agricultural wastes. Stick your head in a freshly opened bag of commercial chicken feed, and you might not like the odor.

In most rations, corn furnishes energy and soybean meal supplies protein. Soy meal has come under increasing scrutiny for reasons having to do with such things as solvent residues left after the oil has been extracted and inherent properties of soy that cause it to inhibit the absorption of other nutrients. You'd be hard pressed, however, to find commercial rations that don't include soy meal as the main source of protein.

Rations for chicks contain a high amount of protein. As birds grow, they gradually need less protein and more energy. Commercial rations, formulated according to age, include chick starter ration, grower ration, developer ration, and lay ration. Meat birds have their own formulations for starter/grower ration and finisher ration intended to induce rapid growth. Layer ration may be available with different levels of protein, the higher levels used during hot weather when hens tend to eat less and also to improve the hatchability of eggs collected for incubation.

The variety of choice you have in rations will depend on where you live. In many areas developer, finisher, and breeder rations are not available, and lay ration comes in only one protein level. You'll have more choices if you live in an area where chickens are numerous. If you're really lucky, you'll have a mill close by offering superior and freshly mixed rations. Commercial rations come in three basic forms: mash, pellets, and crumbles.

Mash is feed that has been ground to various degrees of coarseness but is still recognizable so chickens can pick out what they like. It is most commonly available at a mill that does not have the necessary equipment to extrude pellets and is also the typical form of home-mixed rations.

Pellets are made from mash that has been compressed; each pellet has an identical nutritional value so the chickens can't pick and choose. The other advantage to pellets is that if they get dropped on the ground (as they tend to do around feeding stations), they are likely to be picked up and eaten. Their disadvantage is that pellets quickly satisfy chickens' nutritional needs and then they have nothing to do, and bored chickens pick on one another.

Crumbles are crushed pellets. They are fed to chicks that aren't yet big enough to swallow whole pellets and to mature chickens, so they take longer to eat and therefore are less likely to become bored. The main disadvantage of crumbles is that any spilled or dropped on the ground is usually wasted.

Home-Mixed Rations

Exclusive of housing, feed accounts for 70 percent of the cost of keeping chickens, so it's only natural to look for ways to keep the cost down by mixing your own. Doing so also gives you better control over the quality of your chickens' diet.

The chief disadvantage to a home-mixed ration is that the chickens can pick out what they like. Furthermore, formulating rations is the most complex aspect of poultry management and isn't something to take lightly if you're just starting out. Ration formulation requires the following:

- Availability of appropriate feedstuffs
- Analysis of feedstuff composition
- Knowledge of the nutritional needs of chickens
- Ability to mix feed in a quantity your flock will use within four weeks

If you decide to mix your own, you might purchase all of the feedstuffs separately or grow some of them yourself. You'll need to include carbohydrates for energy (corn and other grains), protein (a combination of ingredients that together furnish all the amino acids

needed to create a complete protein), and vitamins and minerals (some of which will be in carbohydrate and protein ingredients; others will require additional feedstuffs).

Of these ingredients, protein offers the greatest challenge. Soybeans are the most common source of protein in chicken feed, but they first must be heated by roasting or other means to deactivate soy's natural protein inhibitor. Roasted soybeans are high in fat, which provides energy as well as protein. Chickens are most commonly fed soybean meal, which is what's left after soy oil has been removed from the beans. Alternative sources of protein include other oilseed meals (such as peanut, safflower, and sunflower), grain legumes (alfalfa, beans, and field peas), and animal sources (fish meal, meat and bone meal, and dried whey).

The exact nutritional analysis of each ingredient varies with its source, the time of year it was grown, the place where it was grown, and the method by which it was grown and harvested. Purchased feedstuffs may be labeled. Average values are suggested in *Nutrient Requirements of Poultry* (see Recommended Reading on page 402) and similar source books. Information also may be available through your local Extension office.

Ask local feed outlets and your state Extension poultry specialist to help you find other chicken keepers who mix their own poultry rations, and solicit their help in ironing out the wrinkles based on the availability of local feedstuffs. Purchasing ingredients in bulk to share with others who mix their own is a good way to keep costs down while maintaining freshness.

You'll need a way to mix the ingredients together. Depending on the volume of feed you need, you may find a local mill willing to mix for you. Most mills want to mix at least 1,000 pounds (500 kg), if not a ton (1,000 kg), which may be more than you can use within about four weeks. You might get a mixer that operates off the power take-off (PTO) of a farm or garden tractor. For small volumes, a manually operated compost tumbler works fine, and for really small volumes simply combine and stir.

You'll also need a grinder, otherwise known as a feed mill or heavy-duty flour mill, that produces a coarse grind to improve the digestibility of corn, beans, and large grains. Mills come in hand-crank or motor-driven

FORMULATING HOME-MIXED RATIONS

This table offers an easy way to formulate your own rations. It lists a variety of feedstuffs from which you can choose to mix 100 pounds (45 kg) of starter, grower, or layer ration. For the most nutritious blend, select a combination of ingredients from each line that adds up to the total weight for that line.

Although you needn't limit yourself to ingredients listed, substitute only ingredients of similar nutritional value. As an example, instead of alfalfa meal you may use alfalfa pellets or alfalfa hay fines (the bits of vegetation that collect at the bottom of a livestock hay feeder) or provide pasture where your chickens can forage for fresh tender greens.

If you have a goat or a cow that produces more milk than your household uses, you may substitute fresh milk for the milk powder. Two pounds (0.9 kg) of milk powder is equivalent to about

Ingredient
Coarsely ground grain (corn, milo, oats, wheat, rice, etc.)
Wheat bran, rice bran, mill feed, etc.
Soybean meal, peanut meal, cottonseed meal (low gossypol), sunflower meal, sesame meal, etc.
Meat meal, fish meal, soybean meal
Alfalfa meal (not needed for range-fed birds)
Bone meal, defluorinated dicalcium phosphate
Vitamin supplement (supplying 200,000 IU vitamin A, 80,000 ICU vitamin D_3, 100 mg riboflavin)
Yeast, milk powder (not needed if vitamin supplement is balanced)
Ground limestone, marble, oyster shell, aragonite
Trace mineral salt or iodized salt (supplemented with 0.5 ounce manganese sulfate and 0.5 ounce zinc oxide)
TOTAL

From "Feeding Chickens," *Suburban Rancher,* Leaflet #2919, University of California

versions. The hand-crank style is fine for a few chickens, but if you have a sizable flock, you'll be happier with a quicker and less taxing motor-driven mill.

A good, comprehensive resource for ration formulation is *Feeding Poultry* by G. F. Heuser (see Recommended Reading on page 402).

Although it includes some nutrient tables, its greater value is in its extensive discussions of nutrient sources and feeding methods, as well as its numerous recipes for complete starter/grower rations, layer rations, and breeder rations.

20 pounds (9 kg) of liquid milk needed for every 100 pounds (45 kg) of feed consumed. Offer the milk in a separate container; mixing it into the dry ration invites spoilage.

Whatever you have available that's suitable for feeding chickens, determine where it fits in the table and substitute like amounts. If the feedstuff doesn't come with a nutritional label, you can determine approximate nutritional equivalents by consulting *Nutrient Requirements of Poultry* or a similar source; most of them contain page after page of tables listing the average nutritional contents of various feedstuffs.

Imperial Measurements			Metric Measurements		
Starter	**Grower**	**Layer**	**Starter**	**Grower**	**Layer**
46 lb.	50 lb.	53.5 lb.	20.5 kg	22.5 kg	24 kg
10 lb.	18 lb.	17 lb.	4.5 kg	8 kg	7.5 kg
29.5 lb.	16.5 lb.	15 lb.	13 kg	7 kg	6.7 kg
5 lb.	5 lb.	3 lb.	2.3 kg	2.3 kg	1.3 kg
4 lb.	4 lb.	4 lb.	1.8 kg	1.8 kg	1.8 kg
2 lb.	2 lb.	2 lb.	1 kg	1 kg	1 kg
+	+	+	+	+	+
2 lb.	2 lb.	2 lb.	1 kg	1 kg	1 kg
1 lb.	2 lb.	3 lb.	0.5 kg	1 kg	1.3 kg
0.5 lb.	0.5 lb.	0.5 lb.	200 g	200 g	200 g
100 lb.	**100 lb.**	**100 lb.**	**45 kg**	**45 kg**	**45 kg**

Meeting Protein Needs

Chickens of different ages and levels of production have different protein needs. Broilers require the greatest amount of protein (20 to 24 percent), while mature cocks outside breeding season require the least (9 percent). Chickens eat to meet their energy needs and need more energy to stay warm in cold weather. Since energy is less expensive than protein, you can save money by increasing their ration's carbohydrates (by reducing the protein) during winter.

You can easily adjust the protein level of any ration by combining it with a supplemental ration. You'll need to know how much protein you want to have and also the protein content of both the ration and the supplement. To raise protein, choose a supplement that's higher in protein than the ration; to reduce protein, choose a supplement that's lower in protein than the ration.

Using a method called *Pearson's square*, you can easily determine how much ration and how much supplement you must combine to get the protein content you want. Begin by drawing a square on a piece of paper. In the upper left corner, write the percentage of protein of the regular ration. In the lower left corner, write the percentage of protein contained in the supplement. At the center of the square, write the percentage of protein you want to end up with.

Moving from the upper left toward the lower right (following the arrow in the illustration), subtract the smaller number from the larger number. Write the answer in the lower right corner. Moving from the lower left toward the upper right (again following the arrow), subtract the smaller number from the larger number. Write the answer in the upper right corner. The number in the upper right corner tells you how many pounds of ration, and the number in the lower right corner tells you how many pounds of supplement you need to mix together to achieve the desired amount of protein.

The illustration below shows two typical examples. In the first case, 16 percent lay ration is combined with 8 percent scratch to create a 9 percent cock maintenance diet. Note that since we want to reduce the protein content, the number in the lower left corner

PROTEIN REQUIREMENTS

Type	Age	Ration	% Protein
Broilers	0–3 weeks	broiler starter	20–24
	3 weeks–butcher	broiler finisher	16–20
Layers	0–6 weeks	pullet starter	18–20
	6–14 weeks	pullet grower	16–18
	14–20 weeks	pullet developer	14–16
	20+ weeks*	layer	16–18
Cocks	maintenance	layer + scratch	9
Breeders	20+ weeks*	breeder	18–20

*Layer or breeder ration should not be fed to pullets until they start laying at 18–20 weeks for Leghorn-type hens, 22–24 weeks for other breeds.

PEARSON'S SQUARE

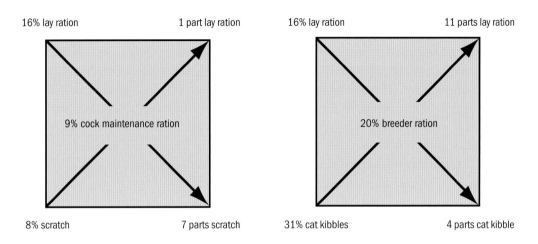

16% lay ration	1 part lay ration
9% cock maintenance ration	
8% scratch	7 parts scratch

16% lay ration	11 parts lay ration
20% breeder ration	
31% cat kibbles	4 parts cat kibble

Using Pearson's square, you can determine how much ration and how much supplement to combine to get the desired protein content.

must be less than the number in the upper left corner. Pearson's square shows that we need to combine 1 pound (0.5 kg) of lay ration with 7 pounds (3.2 kg) of scratch to get a 9 percent cock maintenance ration.

In the second example, 16 percent layer ration is combined with 31 percent cat kibble to create a breeder-flock ration containing 20 percent protein. Since we now want to increase the protein content, the number in the lower left corner is greater than the number in the upper left corner. Pearson's square shows that we should mix 11 pounds (5 kg) of layer ration with 4 pounds (1.8 kg) of kibble to get a 20 percent breeder ration.

Different feedstuffs have different weights, so you won't get an accurate mix if you measure by volume (bucketfuls) instead of by weight (pounds or kilograms). Use a spring scale or weigh yourself on a bathroom scale holding an empty bucket. Add feed to the bucket until you increase the total weight by the amount you need. In combining feedstuffs, use rations of similar consistency. If, for example, you use soybean meal to boost the protein in pelleted ration, the meal will filter out and fall to the bottom of the trough. You'd do better to combine soybean meal with crumbles, perhaps moistening the result at feeding time to keep the meal from sifting out.

Whenever you adjust the protein in your flock's diet by more than a percentage point or two, make the change gradually. Too rapid a change can cause intestinal upset and diarrhea.

Scratch

In an effort to avoid feeding commercially formulated rations, some chicken keepers rely on scratch grains. Chickens love *scratch* — a mixture containing at least two kinds of grain, one of which is usually cracked corn. A common mixture consists of one-third each cracked corn, wheat, and oats. Avoid mixtures containing barley — even if your chickens eat it, they'll find it hard to digest.

It's called scratch because when it's scattered on the ground, the chickens scratch for it, perhaps thinking they'll turn up more grain kernels. Like people, individual chickens have their preferences. When fed scratch, some may eat it all, others may pick out only the corn. When I fed scratch containing milo (sorghum grain), some of my chickens refused to eat the milo; others ate only the milo.

Scratch is high in energy and low in vitamins, minerals, and protein. Too much scratch in the diet radically reduces total protein intake. In growing birds, insufficient protein leads to feather picking — the chicks eat protein-rich feathers in an attempt to obtain enough of that important nutrient. Scratch should never be fed to chicks younger than 8 weeks of age. In the diet of laying hens, insufficient protein reduces egg production and the hatchability of incubated eggs and makes hens fat and unhealthy. Scratch should therefore be fed sparingly, if at all.

Consider scratch to be in the same food group as candy, and you will be unlikely to overfeed it. Scratch does have its uses, aside from being offered as an occasional treat:

- It can be a training device: throw down a handful while crooning "Here chick, chick, chick"; your chickens will learn to come when you call.
- It may be used to gather your chickens into their shelter so you can close them in for the night before they'd otherwise be ready to go inside.
- It may be used to trick chickens into stirring up their bedding to keep it loose and dry: toss a handful over the litter once a day, traditionally late in the afternoon when birds are thinking of going to roost.
- It gives active range-fed layers an extra energy boost and raises plump, tasty corn-fed broilers.
- It may be used to reduce protein in the diet of cocks outside breeding season, thus minimizing the cost of feeding them.
- If fed at dusk in winter, it gives chickens energy to stay warm overnight.

In the summertime, when energy needs go down, reduce the scratch or switch to whole oats. Including whole oats as part of the summer diet minimizes heat stress and improves egg production in hot weather.

Sprouts

Sprouting grains before you feed them improves their vitamin, mineral, and protein content, and provides an excellent source of green feed during winter and spring when naturally growing greens are not plentiful. Any edible grain, seed, or legume may be sprouted. Be sure to use fresh seed that has not been treated with a fungicide or any other chemical (as farm crop and garden seeds often are). Most scratch grains contain cracked corn, which of course won't sprout, but any whole grain intended for livestock should sprout quite nicely. Do a keyword search for "seed sprouting" on the Internet, and you will find all sorts of sprouting seeds and bulk sprouting trays and barrels, as well as instructions on how to sprout various kinds of grains and seeds. Oats are a particular favorite for chickens. One pound (0.5 kg) of oats makes about 3½ pounds (1.5 kg) of sprouts.

If you sprout any or all of the grain portion of your chickens' diet, feed the sprouts separately. Combining them with dry ingredients in a ration mix can cause the dry feeds to go moldy if they are not eaten up right away. Chances are your chickens will gobble down the sprouts first, so no need to worry about them going moldy.

Table Scraps

Feeding your chickens table scraps helps increase the variety in their diet. Just about anything you eat is suitable for feeding your chickens, with a few important caveats:

- Never feed raw potato peels, which chickens can't digest easily — cook potato peels or avoid them.
- Don't feed strong-tasting foods like onions, garlic, or fish, which impart an unpleasant flavor to poultry meat and eggs.
- Don't feed avocados or guacamole unless you are absolutely certain it includes none of the brown seed cover, which contains the toxic compound *persin* that can be deadly to chickens.
- Never feed anything spoiled or rotten, which can make chickens sick.
- Avoid feeding fried foods, which are difficult to digest and unhealthy.
- Never feed anything that contains caffeine or alcohol.
- Avoid feeding anything that is high in fat, sugar, or sugar substitute.

Above all, don't overdo any single item. The resulting nutritional imbalance can cause slow growth, reduced laying, and poor health.

GRAIN SPROUTS

To make sprouts you'll need a large jar or small bucket with a cover of wire screen, plastic screen, or cheesecloth to let in air. Since sprouting causes grain to expand eightfold, start with no more grain than fills one-eighth of the container.

Thoroughly wash the grain, cover it with fresh warm water, and soak it for 5 to 12 hours at room temperature. The length of soak depends on the size of the seeds; soak small seeds for 5 hours, medium seeds for 8 hours, and grains for 10 to 12 hours.

Rinse the soaked grain in warm (not hot) water and drain. At this point, the container should be about one-fourth filled with plump grain. Turn the container on its side so the grain distributes evenly along its length, and place it in a warm, darkened place. Keep the grain moist by rinsing it with warm water three times a day, taking care to drain well each time so the grain won't sour or ferment.

Depending on the temperature and how well you keep the grain moistened, it will sprout in two or three days. Sprouted grains have the best flavor, tenderness, and nutritional value when tiny green leaves just begin to appear. If you have more than you wish to feed your chickens immediately, you can store the remainder in the refrigerator for a day or two to keep it fresh.

Supplements

Depending on how your chickens are managed, they may need supplemental grit, calcium, phosphorus, or salt. Flaxseed is an additional supplement used to boost the omega-3 content of eggs.

Grit, in the form of small pebbles and large grains of sand pecked up and swallowed by a chicken, lodges in the bird's gizzard. When grains and fibrous vegetation pass through the gizzard, muscular action breaks them up by grinding them together with the grit. In short, grit serves as a chicken's teeth.

Chickens that eat only commercially prepared rations need no grit, since the rations are sufficiently softened by the bird's saliva. Range-fed chickens need grit to grind up plant matter. Any chicken that eats whole grains and seeds needs grit. And since grit eventually gets ground up and passed through the digestive system, it needs to be constantly replenished.

Yarded or pastured birds pick up natural grit from the soil but may not get enough. Granite grit, available from any farm store carrying poultry rations, should be offered in a separate hopper and available at all times.

Laying hens need calcium to keep eggshells strong. The amount of calcium a hen needs varies with her age, diet, and state of health; older hens, for instance, need more calcium than younger hens. Hens on pasture obtain some amount of calcium naturally, but illness may cause a calcium imbalance. In warm weather, when all chickens eat less, the calcium in a hen's ration may not be enough to meet her needs, and a hen that gets too little calcium lays thin-shelled eggs. On the other hand, a hen that eats extra rations in

SUPPLEMENT STATION

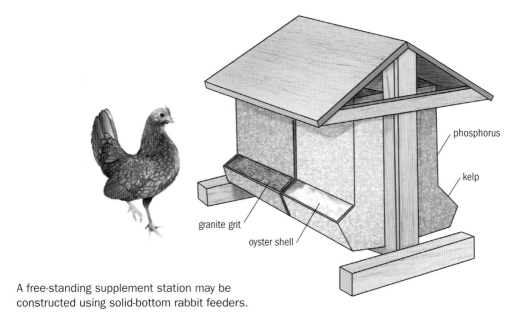

A free-standing supplement station may be constructed using solid-bottom rabbit feeders.

an attempt to replenish calcium gets fat and becomes a poor layer.

Eggshells consist primarily of calcium carbonate, the same material found in oyster shells, aragonite, and limestone. All laying hens should have access to a separate hopper full of crushed oyster shells, ground aragonite, or chipped limestone (not dolomitic limestone, which can be detrimental to egg production).

Phosphorus and calcium are interrelated — a hen's body needs one to metabolize the other. Range-fed hens obtain some phosphorus and calcium by eating beetles and other hard-shelled bugs, but they may not get enough. To balance the calcium supplement, offer phosphorus in the form of defluorinated rock phosphate or phosphorus-16 in a separate hopper and make it available at all times.

All chickens need salt, but only in tiny amounts. Commercially prepared rations contain all the salt a flock needs. Range-fed chickens that eat primarily plants and grain may need a salt supplement. Salt deficiency causes hens to lay fewer, smaller eggs and causes any chicken to become cannibalistic.

SALT CAUTION

Chickens that do not have access to water at all times may be poisoned by even a normal amount of salt. In warm weather, when chickens need more water than usual, make sure they never run out; in winter do whatever it takes to keep drinking water from freezing. If for any reason your chickens could possibly lose access to water at any time during the day, remove the salt hopper until the problem is corrected.

Loose salt (not rock salt) should always be available to range-fed chickens in a separate hopper. Iodized salt is suitable, although either a trace-mineral salt mix or kelp will supply your chickens with many other necessary minerals in addition to salt.

Feeding Routines

How often you feed your chickens is yet another matter of choice. Some chicken keepers like big feeders that needn't be filled often. I prefer smaller feeders that hold too little for the feed to get stale. Since I also collect eggs at least twice a day, checking the feed and water while I'm there anyway is no big deal.

I feed in the morning after breakfast and in the evening after dinner. Other chicken keepers feed first thing on rising and last thing before retiring. Still others feed once a day. How often you feed is not as important as making sure the feed and water never run out.

However, you don't want feed to sit long enough to get stale, as it loses nutrients and becomes unpalatable. Additionally, feed left in the open overnight attracts rodents, raccoons, and other pilfering critters, thereby increasing your feed bill.

Feed left where rain can get in eventually will become moldy — a definite health hazard to chickens. On the other hand, most chickens love moistened feed once they get used to it. Moistening the ration is a good way to encourage eating on hot summer days, and crumbles moistened with warm milk or water is a real treat on cold winter days. But moist feed should be eaten up in half a day or less, and the feeder must be scrupulously cleaned to prevent moldering feed from lurking in the corners.

How Much to Feed

The amount a chicken eats varies with the season and temperature, as well as with the bird's age, size, weight, and rate of lay. Since chickens eat to meet their energy needs, the amount of a particular ration a bird will eat also depends on the ration's energy density. A chicken fed the same ration year-round will eat more during cold weather, because it needs more energy to keep its body warm. As some extremely general guidelines, expect to feed:

- Each mature bantam about ½ pound (0.25 kg) of feed per week
- Each mature light-breed chicken about 2 pounds (1 kg) of feed per week
- Each mature midweight dual-purpose chicken about 3 pounds (1.5 kg) of feed per week
- Each mature heavy-breed chicken about 4 pounds (2 kg) of feed per week
- Each meat bird about 10 pounds (4.5 kg) of total feed to reach mature butchering age

A chicken that doesn't get enough to eat won't grow or lay well. A chicken may eat too little if it goes through a partial or hard molt, if it's low in the peck order, if the weather turns hot, or if it finds its ration unpalatable. Birds can be particularly fussy about texture. They don't like dusty or powdery mash. They also don't like (and shouldn't be fed) moldy or musty rations.

If your chickens don't seem to be eating enough, perk up their appetites. Begin by feeding more frequently, even though the trough may already be full. Offer variety — chickens are particularly fond of milk, cottage cheese, yogurt, tomatoes, salad greens, and sprouts. Moistening the feed may also increase appetite: for each dozen birds, stir a little water into ¼ pound (0.1 kg) of ration fed daily. If your appetite-stimulating attempts fail, the issue may not be the ration but poor health.

Feed that disappears too fast is a sure sign something is wrong. Your chickens may be infested with worms. Take a sample of droppings to your vet for a fecal test, and worm your chickens as necessary. In winter, rapidly disappearing feed may mean your chickens are too cold. Eliminate indoor drafts, and increase the carbohydrates in their ration. Disappearing feed may not be your chickens' fault at all — make sure rodents, opossums, wild birds, and other creatures are not dipping their snouts into the trough.

Feeding Methods

Two different methods are used to feed rations to chickens — free-choice feeding and restricted feeding. In choosing which method you prefer, decide whether the advantages outweigh the disadvantages.

Free-choice feeding involves leaving rations out at all times so chickens can eat whenever they wish. The obvious advantages of free-choice feeding are that it saves time and ensures no chicken goes hungry. On the downside, feed is always available where rodents, wild birds, and other livestock (notably goats) may gobble it down; feed can get dirty, wet, or moldy between feedings; and some breeds, especially the heavier ones, tend to get fat when fed free choice.

Restricted feeding entails feeding chickens often but giving them only small amounts at a time. Show birds may be kept on a restricted regime so they'll look forward to human visits, in which case they're usually fed as much as they'll eat in a 15-minute period twice daily.

Cornish-cross broilers are often restricted so they don't grow so fast their little legs can't carry them. Pullets of the heavier breeds may be fed limited amounts to keep them from maturing too rapidly, because early laying results in fewer eggs of smaller size. Older, lightweight hens and breeders in the dual-purpose or meat categories may be put on a restricted diet to keep them from getting fat and lazy. A restricted-feeding program is time-consuming and chickens lowest in the peck order may not get enough to eat. This type of feeding program works only if you have enough feeders to allow all birds to eat at the same time. Because birds eat quickly and then have plenty of time to get bored, restricted feeding may lead to cannibalism.

Range Feeding

Chickens housed on pasture may be fed their ration either as free choice or restricted, in addition to having access to natural forage. How much of their diet they obtain by foraging depends on the quality of the pasture and the number of chickens competing for it. Chickens on pasture eat seeds and insects, in addition to tender greens, but under normal conditions whatever nutrition they derive from the pasture is unlikely to replace a significant percentage of their ration.

The chief benefits of pasturing are to increase dietary variety, provide healthful exercise and fresh air, and reduce the concentration of parasitic worms in the diet.

Unlike cows and sheep, chickens are not primarily grazers and cannot digest large amounts of tough fiber. Short pasture perennials are therefore more suitable for chickens than taller plants.

Among warm-season greens, alfalfa is a good choice where adequate rainfall or irrigation is available. Lespedeza has a similar nutritional value and grows well in southern regions, although in the colder north it must be seeded as an annual. Ladino and alsike clover are other popular warm-season choices. Add a little plantain, both narrow and broad leaf, as well as chicory, and the chickens will love you for it. And don't worry about dandelions; the chickens will take care of them for you.

Orchard grass is a cool-season pasture grass with a broad leaf that chickens like, and it gives them an early start on spring greens. A mixture of grasses extends the season and might include perennial ryegrass, fescue, Kentucky bluegrass, Canada bluegrass, and timothy.

Any of the cereal grasses make good cool-season pasture. Oats seeded in spring grow quickly; seeded in midsummer, they

MIXED GREENS

If you don't have pasture but do have at least a little garden space, a mixture of greens grown just for your chickens will keep them healthy, as well as ensure richer-tasting deep yellow yolks. You can either cut and feed the greens or turn the chickens into the plot for a limited time each day to peck and scratch. Any mixture of lettuces, spinach, and other greens will delight your chickens.

Nichols Garden Nursery offers a Chicken Greens mix that regrows as a cut-and-cut-again crop. The Sand Hill Preservation Center offers two blends of Brooder Yard Greens, one for spring growth and the other for fall.

offer late-summer and fall forage. Rye, wheat, and barley seeded in the fall may be grazed all winter into spring. Since chickens pluck rather than graze, the pruning action causes these plants to tiller, or grow more stems, and consequently more grain if you let the cereal grasses go to seed. Alternatively, plant them as a garden cover crop for your chickens to graze, and turn them under before they go to seed. Spelt makes another good winter cover crop that furnishes greens in early spring when little else is growing.

Plants that are in the vegetative, or growing, stage are more nutritious than tough, stemmy plants, which chickens won't eat anyway except if they're half starved. Unless your flock follows some other kind of livestock in a grazing rotation, during times of rapid vegetative growth — when plants grow faster than the chickens can eat them — you will have to get out the mower or Bush Hog to keep the pasture mowed down. Cutting plants short not only keeps them growing but also lets in sunlight to help minimize the buildup of infectious organisms.

MOVING A RANGE SHELTER

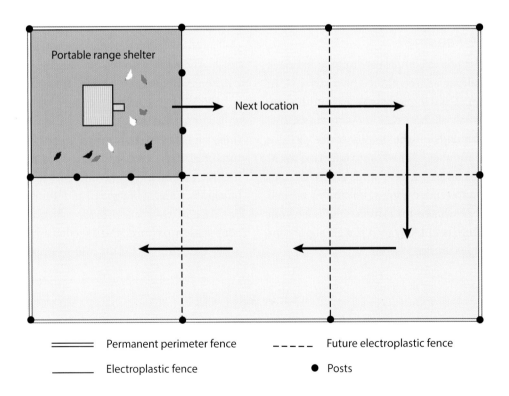

Permanent perimeter fence	Future electroplastic fence
Electroplastic fence	● Posts

For a fenced-range flock, move the shelter when pasture has been grazed down to 1 inch (2.5 cm) or when bare spots appear.

Chickens tend to stay close to their shelter and can quickly overgraze an area, trample the pasture, and destroy plants by digging holes for dust baths. They are more inclined to forage widely where trees give them a sense of security, not to mention shade. Another way to encourage birds to venture forth is to space watering stations some distance from their shelter.

You might also scatter scratch grains on the ground, choosing a different place each day so the chickens don't keep scratching up one area. Since foraging causes chickens to burn off extra energy, you can safely feed ranged chickens up to ¾ pound (0.3 kg) of scratch per two dozen birds.

When the area has been grazed down, or if bare spots appear, move the chickens to new ground. You'll be doing nothing more than imitating the natural conditions under which plants evolved and under which they grow best. In nature, a flock moves together to avoid predation, quickly grazes down an area (during which it scratches up the ground and deposits large amounts of manure), then moves on.

How long a flock takes to graze down a given area depends on a number of factors, including the size of the flock, the kind and condition of the pasture, temperature, and rainfall. When sun, rain, and warm weather combine to help plants grow quickly, a small flock might graze a given area for two weeks or more. In cool, hot, or dry weather, when plants grow slowly, the same flock may graze down the same area to nothing in a matter of days. Chickens that are confined within a range house need to be moved to new ground daily. Exactly how small an area you can confine your flock to, and how long you can keep them there, may be determined only through watchful experimentation within these parameters:

- Let chickens in when plants are no more than 5 inches (12.5 cm) tall.
- Move the chickens when the plants have been grazed down to 1 inch (2.5 cm) or when bare spots appear.
- Do not return chickens to the same piece of ground twice within the same year.
- The longer a range shelter stays in one place, the more time the pasture requires for restoration once the shelter is moved.

Over the years, pasture soil will increase in acidity. When soil pH drops below 5.5 as determined by a soil test, spread lime at the rate of 2 tons per acre (2 metric tons per 0.5 hectare). Then let the pasture rest to give plants time to rejuvenate and to break the cycle of parasitic worms and infectious diseases.

Toxic Plants

Some weeds found in pasture may be toxic but should not be a problem if your flock has plenty of other food choices. Most toxic plants don't taste good and therefore are not tempting to eat, except to a starving bird. Birds nibble here and there to get a variety in their diet, but a bite or two of a toxic leaf or seed is unlikely to create a problem. (The greater danger is to a house chicken, which may be tempted to eat toxic house plants if they are the only available greens.)

Whether or not a specific plant is toxic may vary with its stage of maturity, growing conditions (such as drought), and other environmental factors. Should a chicken get a potentially toxic dose, the effect also will depend on the bird's age and state of health. The accompanying table lists common plants that could pose a danger. Some mushrooms are toxic as well, but mushrooms would have a hard time getting a foothold where poultry are active.

Common Name	Botanical Name
Black locust	*Robinia pseudoacacia*
Black nightshade	*Solanum nigrum*
Bladderpod, bagpod	*Sesbania vesicaria*
Castor bean	*Ricinus communis*
Corn cockle	*Agrostemma githago*
Crown vetch	*Coronilla varia*
Death camas	*Zygadenus* spp
Jimsonweed, thorn apple	*Datura stramonium*
Milkweed	*Asclepias* spp
Oleander	*Nerium oleander*
Poison hemlock	*Conium maculatum*
Pokeberry	*Phytolacca americana*
Potato	*Solanum tuberosum*
Rattlebox	*Daubentonia punicea*
Vetch	*Vicia* spp
Water hemlock, cowbane	*Cicuta* spp
Yew	*Taxus* spp

Feeders

Feeders come in many different styles, the two most common being a long trough and a hanging tube. When looking at designs, consider that chickens are notorious feed wasters. Feeders that encourage wastage are narrow or shallow and lack a lip that prevents chickens from *billing out* — using their beaks to scoop feed onto the ground. Chickens will spend minutes on end scooping rations out of a feeder. Why they do it is anybody's guess. Maybe they're hoping to find something more interesting, or maybe they just need something to do.

At any rate, a feeder with a rolled or bent-in edge reduces billing out. To further discourage this behavior, keep feeders at the height of the chickens' backs — which means, if you raise chicks, changing the height as the chicks grow to maturity.

Regardless of its design, a good feeder has these important features:

- Discourages billing out
- Prevents contamination with droppings
- Is easy to clean

Trough Feeders

Never fill a trough feeder more than two-thirds full. Chickens waste approximately 30 percent of the feed in a full trough, 10 percent in a two-thirds-full trough, 3 percent in a half-full trough, and approximately 1 percent in a trough that's only one-third full. Obviously, you'll save a lot of money by using more troughs so you can put less feed in each one.

Since you fill a trough from the top and chickens eat from the top, trough feeders tend to collect stale or wet feed at the bottom. Never add fresh feed on top of feed already in the trough. Instead, rake or push the old feed to

FEEDERS

This trough has an anti-roosting reel that rotates and dumps any bird that tries to hop on. Allow 4 inches (10 cm) of trough space for each bird, counting both sides if birds can eat from either side.

Hang a tube feeder to the height of the chickens' backs. Allow 1½ inches of circumference (3.75 cm) for each bird.

one side, and at least once a week empty and scrub the trough.

A good feeder discourages chickens from roosting on top and contaminating feed with droppings. A trough mounted on a wall allows little room for roosting. A free-standing trough may be fitted with an anti-roosting device that turns and dumps any chicken trying to perch on it. In my experience, somehow chickens still manage to get their droppings into the feed.

Tube Feeders

After having used trough feeders for years, I've found I much prefer tube feeders. Since you pour feed into the top and chickens eat from the bottom, feed doesn't sit around getting stale. A tube feeder is fine for pellets or crumbles but works well for mash only if you fill it no more than two-thirds full. Otherwise the mash may pack and bridge, or remain suspended in the tube instead of dropping down. The best way to adjust a tube feeder to the right height as a flock grows is to hang it from the rafters by a chain and hook.

Chickens, especially young ones, like to roost on the edge of the feeder top, or hop inside for a private snack, thereby fouling the feed. A tube feeder, therefore, should be fitted with a cover. These days most hanging feeders come without a lid, though some, but not all, manufacturers offer a fitting lid as a separate option. You can recycle the lid from a plastic bucket, if you find a size that fits your feeder, by notching out opposite sides to fit under the feeder handle. Chickens will still roost on top, so when you lift the lid to refill the feeder, you'll have to brush off droppings to prevent them from falling into the feed.

A better (though admittedly odd-looking) option is to cut the bottom from an empty plastic gallon jug (bleach, vinegar, or anything else nontoxic) and hang it over the top of the feeder, upside down, by a string through a hole poked into the center. Chickens don't like the dangling thing that jiggles when they jump up onto the feeder rim, and roosting on top of the feeder instantly comes to an end.

The base cut from a plastic jug and hung over the top of a feeder makes a strange-looking contraption, but successfully keeps chickens from roosting on the rim.

Feeding Stations

If you feed free choice, put out enough feeders so at least one-third of your chickens can eat at the same time. If you feed on a restricted basis, you'll need enough feeders so the whole flock can eat at once. As a general rule, allow each mature chicken at least 1.5 inches (3.75 cm) of space around a tube feeder or 1 inch (2.5 cm) of space along a trough feeder. If the trough is accessible from both sides, count both sides in your calculation; for example, an 18-inch (50 cm) trough has 36 inches (100 cm) of feeder space.

Even if one feeder would be enough for your flock, furnish at least two to ensure weaker birds don't get chased away by the bullies. If you have more than one rooster, furnish at least one feeding station per rooster. Each cock will gather his hens around a feeder, and fighting will be reduced.

Placing feeders inside the shelter keeps feed from getting wet, but encourages chickens to spend more time indoors. If you have to keep feeders indoors, practice good litter management by moving them every two or three days to prevent concentrated activity in one area.

Placing feeders under a covered outdoor area keeps feed out of the sun and rain and encourages the flock to spend more time in fresh air. But outdoor feeders attract wild birds, especially sparrows and starlings, and leaving a full feeder in the open overnight invites pilfering by opossums and other wildlife.

Feed Storage

Storing extra feed so you won't run out is a good idea, but don't stock up too far ahead.

From the moment it's mixed, feed starts losing nutritional value through oxidation and other aging processes. Any prepared feed should be used within about four weeks of being milled. Allowing a week or two for transport and storage at the farm store, buy only as much as you can use in two to three weeks.

Store feed off the floor on pallets or scrap lumber, away from moisture. After opening a bag, pour the feed into a clean plastic trash container with a tight-fitting lid and keep it in a cool, dry place, out of the sun.

For feed storage, a plastic container is preferable to a galvanized can, since metal sweats in warm weather, causing feed to get wet and turn moldy. A closed container slows the rate at which feed goes stale and keeps out rodents. To calculate an appropriate-size container for your flock, about 15 pounds of feed fits into a 5-gallon bucket (7 kg of feed fits into a 20 L bucket), and feed is packaged in 25- and 50-pound bags (in metric countries 5-, 10-, 12.5-, 20-, 25- and 50-kg bags).

Take care when you fill the container not to spill any feed on the ground. Spilled feed attracts rodents, which may carry diseases that could infect your birds. And the rodents won't stop at eating spilled feed but will chew holes through your stored sacks and eat up incredible quantities of costly rations.

Use up all the feed in the container before opening another bag, and never pour fresh feed on top of old feed. If you have a little feed left from a previous batch, pour it into the container's lid, pour the fresh bag into the container, and put the older feed on top where you'll use it first.

5

ROUTINE MANAGEMENT

The two biggest management issues for most chicken keepers are preventing chickens from pecking each other, which causes feather loss and open wounds, and keeping at bay predators that maim or kill chickens for fun or food. Routine management also includes protecting your chickens from weather extremes, managing litter, dealing with manure, keeping beaks and toenails trimmed, and properly handling and transporting birds to minimize stress.

Cannibalism

Cannibalism is the nasty habit chickens have of pecking at one another. It usually starts with one bird and spreads to others. Depending on the cause, multiple birds may peck at each other's bodies, leading to feather loss and eventually bleeding wounds, or the whole flock may gang up on one unfortunate individual and peck it to death. Identifying and removing the cannibalistic offenders may stop the problem before it gets too far along, but preventing it in the first place is far easier than trying to stop it once it starts.

Cannibalistic pecking is entirely different from peck-order fighting to establish dominance, although frequent fighting to adjust the peck order may lead to bloody injuries, which in turn lead to pecking and cannibalism. Constantly disrupting the peck order by introducing new birds to the flock also causes stress that may lead to feather pulling, vent picking, and other forms of cannibalism.

Chicks that lack opportunities for pecking may peck their own toes or the toes of other chicks nearby.

No matter whether you live in the country, the suburbs, or in town, guarding your flock against predators is an important part of flock management.

Causes

No one is exactly sure what causes cannibalism. Since one of the three main activities of a chicken is eating (the other two are avoiding being eaten and making more chickens) and a chicken's mode of eating is to peck and swallow, cannibalism quite likely stems from the chicken's need to peck. If a bird is not kept busy pecking to satisfy hunger, it will keep itself busy pecking to make mischief.

Early forms of cannibalism among chicks include toe picking (a chick's toes look remarkably like little worms) and feather picking (chickens love tasty red treats like strawberries, ripe tomatoes, and newly emerging blood-filled feather quills). Hens may peck each other if they have too few nests to lay in. Stressful situations that can lead to irritability and pecking include overcrowding, boredom, lack of exercise, bright lights, excessive heat without proper ventilation, too little perching space, too few feeding or watering stations, and feeders or drinkers too close together.

If the chickens don't like a change in their ration, they may peck each other in seeking alternative sources of food. High-calorie, low-fiber rations or a nutritional imbalance (too little salt or protein) can lead to pecking as well. Pellet feed is more likely to lead to pecking than other forms of ration, because the pelleting process cooks the feed and thereby increases its digestible calories; pellet-fed chickens quickly satisfy their nutritional needs and then have little else to do but pick on one another. Likewise, a restricted-feeding program causes birds to eat quickly and then have plenty of time to get bored.

An infestation of external parasites can cause a pecking frenzy. Lice and mites irritate the skin and feathers, causing chickens to peck at themselves in trying to relieve the itch, and injured skin invites pecking by other chickens. Likewise, any bleeding injury encourages pecking. Chickens are omnivorous eaters, and chicken meat tastes as good to them as any other fresh food.

Control

Pecking is much easier to prevent than to control. However, certain measures sometimes prove useful once the pecking starts. The first thing to do is identify and remove any instigators and any birds that have been pecked to the point of having bare patches or bloody wounds. Pecked areas invite more pecking.

Switch to red lights that make blood more difficult to see, or change bright lights to dim lights. A good guideline for dimming the lights is to provide just enough light so you can barely read a newspaper. If the temperature inside the housing is hot, open windows or turn on a fan to stir the air and reduce the temperature. Like people, chickens get irritable when they're too hot.

FORMS OF CANNIBALISM

Form	Likely Group
Toe picking	Chicks
Tail pulling, feather picking	Growing birds
Vent picking	Pullets
Head picking	Cocks; any birds in adjoining cages
Egg eating	Hens

Alleviate boredom by letting chickens run outside, where they can spend time exploring and finding things to peck besides each other. At times when they must be confined indoors, provide toys such as shiny aluminum pie tins attached to the wall at head height for them to peck at or swinging perches for them to play on. Straw bales give chickens something interesting to explore and also attract insects for them to peck at. Feeding a portion of the ration as grains, table scraps, or garden greens scattered over the litter or across the yard gives them something to hunt for and peck.

Since salt deficiency causes chickens to crave blood and feathers, try adding one tablespoon of salt per gallon of water in the drinker for one morning, then repeat the salt treatment three days later. At all other times provide plenty of fresh, unsalted water.

Pine tar and various other preparations have been devised or recommended for smearing on wounds to make them less palatable for picking. Some of them work for a short time, especially when picking first starts, but none works well after picking has become serious.

Prevention

Cannibalism is a management problem. Prevent it among chicks by avoiding crowding and by reducing brooding temperature and increasing ventilation as the chicks grow. Chicks raised on wire are more likely to become cannibalistic than chicks raised on litter; perhaps they peck each other as a substitute for pecking the ground.

Furnishing adequate perch space helps prevent pecking by giving chickens more places to get away from one another. To prevent pecking from below, make sure the bottom perch is no closer to the floor than about 18 inches (45 cm). Providing suitable areas for dust bathing both helps minimize external parasites and gives chickens something to do besides peck each other. In general, establishing a rich environment that encourages normal exploring, nesting, and foraging behavior helps reduce pecking.

Patrol your chicken house and yard regularly, and repair broken wire or protruding nails that may cause bleeding wounds. Until it heals, isolate and treat any chicken that's been injured or has reddened skin irritated by a parasite infestation. Similarly, remove any chicken from the flock that is lame, ill, or not growing well, as chickens tend to peck at weaker birds. While chicks are growing, do not combine different age groups, as the stronger, older birds will peck the weaker, younger chicks.

Chickens are less likely to become cannibalistic when their diet is high in sources of insoluble fiber, such as grains, bran, seeds, and dark leafy greens (including pasture forage). Feeding crumbles rather than pellets helps deter cannibalism, because crumbles take longer to eat. Feeding mash is even better, because chickens spend time searching through the ground particles for their favorite tidbits, thus taking longer to eat and having less time to peck each other. Protein deficiency can lead to cannibalism, and protein requirements change with age and season, so take care to adjust rations as needed to ensure your chickens are getting adequate protein.

Do not use bright lights to push pullets to mature early, which increases the risk of vent picking. When an undeveloped pullet starts laying before the age of about 20 weeks, a too-large egg may cause her vent to protrude, and the contrasting color attracts other chickens to peck. Sometimes the pecking continues to the point of killing the pullet by pulling out her intestines. Darkening the nesting area helps prevent vent picking.

Egg eating is a form of cannibalism that usually starts when eggs get broken in the nest, either because too few nests cause hens to crowd together or because a nutritional deficiency creates thin shells. Once chickens find out how good eggs taste, they break them on purpose to eat them. The only way to stop

egg eating and keep it from spreading is to remove the culprit early. Identify instigators by checking for egg yolk smeared on beaks or by catching the eaters in the act.

In commercial operations chickens have the tips of their beaks removed so they may be crowded together without eating each other. Blinders, specs, or so-called peepers are sometimes used to prevent or control cannibalism by keeping chickens from seeing directly ahead to aim a peck, but they may lead to eye disorders. A better option is better management.

Because some breeds and strains are more likely than others to peck at each other, an excellent way to avoid cannibalistic behavior is to select a breed or strain that is not genetically prone to engage in it. Leghorns and other light, high-strung breeds are more likely to peck each other than the heavier, more sedate of the American and Asiatic breeds. I rarely had a cannibalism problem when I raised New Hampshires or barred Plymouth Rocks but had to be vigilant when I raised breeds that are more oriented toward egg production. If you have a problem breed or strain, avoid perpetuating the problem by not hatching chicks from birds that behave aggressively, pick each other's feathers, or otherwise engage in undesirable behavior.

Predators

"What are you building there, a bunker?" My visiting uncle was referring to the concrete foundation of an under-construction chicken

Once a chicken discovers how tasty eggs are, keeping her from eating them will be nearly impossible.

house on our new farm. Looking at it through his eyes, maybe it was overkill. On the other hand, a neighbor had told us, "Chickens don't live long out here," and nothing we could do would stop predation.

Well, as long as we kept our flock in that bunker, we never lost a chicken, except for two that disappeared one day when we let them out to forage and they wandered into the woods to scratch in the dry leaves. That first (and last) time we allowed the chickens to roam from their bunkered yard, two hens fell victim to a pair of foxes with hungry kits.

Later we moved the hen house to one end of our barn, some distance from the house, and soon learned that our chicken bunker had lulled us into complacency about the local predator population. Plenty of critters out there enjoy dining on home-grown poultry as much as we do.

Identifying Predators

The first step in deterring a predator is to identify it. Each critter leaves its own calling card that lets you know which animal you're dealing with. Having raised chickens for some 40 years, I've seen quite a few of these signs, but every now and then I get stumped, largely because the predators haven't read the books and don't always conform to their own standard operating procedure.

One sure sign, of course, is tracks. If you're having trouble finding tracks, spread sand on the ground where the predator will likely step, smooth out the sand, and confine your chickens until you have a chance to look for tracks in the sand after the prowler may have visited. This method requires persistence if you're dealing with a predator that comes around irregularly.

In an active poultry yard, tracks quickly get obliterated, so you can't count on tracks alone. Your best guide is to examine where, how, and when birds turn up dead or missing.

Missing chickens were likely carried off by a fox, coyote, dog, bobcat, owl, or hawk. One time I was working in my yard and could only watch helplessly as a hawk swooped down and carried away a full-grown bantam hen that had been happily scratching in the orchard. Although a hawk rarely carries off a full-grown bird, we take great care to protect chicks, as small birds are particularly attractive and easy prey for hawks and other predators.

Although an owl may carry away a small bird, its calling card is more likely to be a dead bird with just the head and neck missing. Neither hawks nor owls are shy about marching right into the poultry house. One icy winter morning we entered our chicken house to find a young owl snugged in among the roosting chickens, and none too eager to leave. One warm spring day, we found a hawk inside our chicken coop killing a cockerel, with a second dead cockerel nearby.

If you live near water, a mink may be doing the dirty deed. Raccoons, too, will carry off a chicken and may raid the poultry yard as a cooperative venture, then squabble over their kill. You may find the carcass some distance from the house, the insides eaten and feathers scattered around.

A snake will eat chicks without leaving a trace. We once found a black snake in our brooder after he had gulped down a couple of chicks, then (being too fat to slip back out through the wire) curled up under the heat lamp to sleep off his fine meal.

Domestic and feral house cats will make chicks disappear but they leave the wings and feathers of growing birds. On rare occasions a cat will kill a mature chicken, eating the meatier parts and leaving the skin and feathers, and sometimes other parts, scattered around. On my first chicken ranchette, I lost nearly every chick hatched by my hens until I live-trapped a cat someone had turned loose to fend for itself.

I accidentally learned the best way to train a cat to leave chickens alone when my new kitten followed me to the chicken yard. She took an interest in some baby chicks, and the mother hen puffed up to twice her normal size and chased the kitten. For the rest of her life, whenever any chicken happened by, that cat laid her ears back and skulked away.

Rats will carry off baby chicks without a trace. A rat will pull a chick down into its tunnel, but a too-large bird may get stuck and

ANIMAL TRACKS

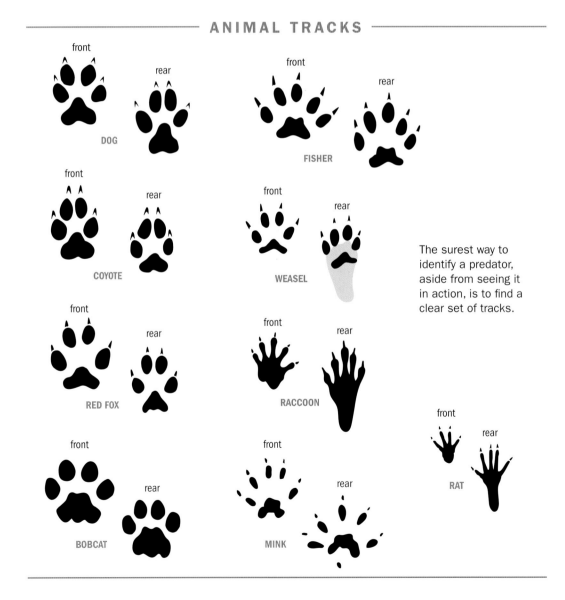

DOG — front, rear

COYOTE — front, rear

RED FOX — front, rear

BOBCAT — front, rear

FISHER — front, rear

WEASEL — front, rear

RACCOON — front, rear

MINK — front, rear

RAT — front, rear

The surest way to identify a predator, aside from seeing it in action, is to find a clear set of tracks.

you'll find its body, usually head first, at the tunnel opening. Older birds are more likely to be found chewed up by rats, usually around the bone. A rat leaves droppings near feeders and may get protein by pulling off and eating feathers from roosting birds. Rat tunnels and holes gnawed in walls provide entry for other predators.

Chickens found dead in the yard, but without any missing parts, were likely attacked by a dog. Dogs kill for sport. When the bird stops moving, the dog loses interest, which is why you often find the victim of a canine attack near where it was killed. I once found a full dozen of my fryers dead and lined up neatly on the walkway. I was trying to guess what kind of predator could have done such a thing when my new puppy came bounding up with yet another fryer to add to his collection.

Like dogs, weasels and their relations (ferrets, fishers, martens, mink, and so forth) also kill for sport. If you find bloodied bodies surrounded by scattered feathers, you were likely visited by one of them. Fishers and martens may kill and stash chickens, then return and eat later. Weasels can sneak into housing through openings as small as 1 inch (2.5 cm) and sometimes run in family packs that can do significant damage in an amazingly short time.

If you find dead birds that have been flattened, the only thing you know is that some kind of predator frightened them. In trying to get away, they piled in a corner or against a wall and the ones on the bottom suffocated.

Parts missing from a dead bird can help you identify the culprit. A chicken found next to a fence or in a pen with its head missing is likely the victim of a raccoon that reached in, grabbed the bird, and pulled its head through the wire. Or a bird of prey could have frightened your birds into fluttering against the wire, and any that got stuck in the wire lost their heads.

When you find a bird dead inside an enclosure with its head and crop missing, your visitor was a raccoon. If the head and back of the neck are missing, suspect a weasel or mink. If the head and neck are missing, and feathers are scattered near a fence post, the likely perpetrator was a great horned owl.

Just as a raccoon will reach into a pen and pull off a chicken's head, so will it also pull off a leg, if that's what it gets hold of first. Dogs, too, may prowl underneath a raised pen, bite at protruding feet, and pull off legs.

Bitten birds, either dead or wounded, may have been attacked by a dog. If they are young birds and the bites are around the hock, suspect a rat. If the bites are on the leg or breast, the biter is likely an opossum. 'Possums like tender growing birds and will sneak up to the roost and bite a chunk out of the breast or thigh of a sleeping bird. On the rare occasion a 'possum kills a chicken, it usually eats it on the spot.

Birds bitten around the rear end and that have their intestines pulled out have been attacked by a weasel or one of its relatives. A hen that has prolapsed may look similar, as the protruding red tissue attracts other chickens to peck, and if they peck long enough and hard enough before you intervene, they will eventually pull out her intestines. Other signs of cannibalism that may be mistaken for signs of predation are missing toes and wounds around the top of the tail of growing chickens. A hen with slice wounds along the sides of her back got them after being repeatedly mated by a sharp-clawed rooster.

Missing eggs could have been eaten by rats, skunks, snakes, opossums, raccoons, dogs, crows, or jays. Rats, skunks, and snakes make off with the entire egg, rats and skunks by rolling them away. A skunk that has been pilfering eggs may leave a faint odor; however, if you think you smell skunk but find pieces of shell in or around the nest, the raider is more likely an old boar raccoon, which emits a similar odor.

A snake eats the egg right out of the nest. During the summer, a 4-foot king snake lives in the hay-storage area of our barn. We're happy to have him clear out the rodents. We call him The Terminator (Mr. T for short) and don't mind that he pilfers the occasional egg laid by a hen that has wandered into the barn, but we have to take great care that he can't get to our baby chicks.

Jays, crows, 'possums, raccoons, dogs, and occasionally skunks leave telltale shells. Jays and crows may carry empty shells quite a distance from where they found the eggs, while a 'possum or 'coon leaves empty shells in or near the nest. Sometimes after cleaning out a nest, a bold 'possum will curl up in the nest and take a nap.

Predator Control

The two ways to control predation are to eliminate the predator's point of entry or eliminate the predator. The easiest way to eliminate the point of entry is to confine chickens inside, if not all the time, then at least at night. An electronically operated chicken-size door that closes when the sun goes down and opens at sunup is helpful if you aren't always Johnny-on-the-spot to close up your chickens at dusk and let them out at dawn. If your predator is likely a fox or coyote, however, keep your chickens in until late morning, by which time these critters have ceased marauding.

Cover the shelter's ventilation openings with ½-inch (1.5 cm) hardware cloth, not chicken wire that a fox, dog, or raccoon can rip through. A deep concrete foundation, as

A snake pilfers eggs without leaving a trace and may eat several at one time.

our poultry bunker had, discourages predators from digging. A bright security light deters some predators and helps you see when you detect a problem at night.

Some of the most difficult predators to control are those that fly down from the sky. If your yard is small enough, covering the top with wire mesh is ideal. A larger yard might be crisscrossed with wires strung 7 feet (2 m) off the ground or at least high enough not to snag whatever equipment you use to maintain the area. From these wires hang old CDs or DVDs, which will twirl in the slightest breeze. In turning, the discs twinkle as they catch sunlight or the beam of a nighttime security light. Any raptor thinking of landing in the yard will stir up enough breeze with its wings to spin the CDs and scare itself away. Persistently windy weather may work the CDs loose, so you have to be diligent about putting back up any that come down.

For pastured poultry, moving the housing every couple of days confuses predators or at least makes them suspicious. Anchor portable housing with skirting that's tight and close to the ground; each time you move the shelter, double check for dips where weasels can weasel in. Keep shelters away from trees and fence posts where raptors may land before swooping down to pick off a chicken. Many four-legged marauders don't like to expose themselves by crossing an open field, so keep grass, weeds, and brush mowed.

A good close-mesh fence will keep out most prowlers. Burying the bottom of the mesh with the lowest 6 to 12 inches (15 to 30 cm) bent outward (away from the poultry yard) helps deter diggers. Using electrified scare wires, or an all-electric fence, increases the barrier's effectiveness.

Electric-fence baiting is a trick used to teach predators, including delinquent dogs, to respect the fence. If you find a dead chicken, hang it from a hot wire so the predator will get a shock when it comes back for another bite. If you don't use electric fencing, hang the chicken from a wire attached to an inexpensive fence controller. The bait should be nose height to the predator in a normal ambling stance, or about 6 inches (15 cm) above ground for weasels and other small creatures to 3 feet (1 m) for big dogs.

For a sizable flock, especially pastured, a guardian animal such as a dog or burro makes a good investment. If you opt for a canine, be sure he's reliable. A poultry breeder once remarked that her thriving repeat business was largely due to dogs — not just neighborhood dogs and roaming packs that preyed on chickens, but also pets owned by the same people who kept her busy hatching chicks to replace all the chickens that were eaten.

If you have a problem with a predator that comes back repeatedly, you might call your local wildlife or animal control agency to see if they'll send out a trapper. Another option is to set a trap yourself. If you use a live trap with the intent of releasing the predator in some far-off location, be aware that many animals are territorial and eventually find their way back home. Others come in families, so catching one won't necessarily solve your problem. And if your marauders are a family of 'possums, think twice about eliminating them — you may end up with a rat problem instead (see page 132).

A predator-control option favored by many rural folks is to stand guard and shoot. If the marauder is your neighbor's dog, be sure to check local laws regarding your obligation to notify the neighbor about your intentions. And if you're dealing with a wild animal that's protected by law, your best option is to eliminate the point of entry.

One perfectly legal method that works on every predator prowling between dusk and full daylight is a solar-powered blinking red light called Nite Guard. The pulse of light it emits once every second between dusk and daylight gives a wary predator the impression another animal is watching, causing the prowler to skulk away. It even works on two-legged thieves, who believe it's a security system. We mounted one next to the chicken-size door of our henhouse and have had no more problems with early-morning pilfering of feed, eggs, or chickens.

PREDATOR DETECTIVE KEY

Clues	Likely Time	Predator
One or Two Birds Killed		
Entire chicken eaten on site	dusk or dawn	hawk
Bites in breast or thigh, abdomen eaten; entire bird eaten on site	night	opossum
Deep marks on head and neck, or head and neck eaten, maybe feathers around fence post	night	owl
Entire chicken eaten or missing, maybe scattered feathers	early morning	coyote
One bird gone, maybe scattered feathers	dusk or dawn	fox
Chicks sometimes pulled into fence, wings and feet not eaten	nightly	domestic cat
Chicks killed, abdomen eaten (but not muscles and skin), maybe lingering smell	night	skunk
Head bitten off, claw marks on neck, back, and sides; body partially covered with litter	night	bobcat (rare)
Bruises and bites on legs; partially eaten chick with head down tunnel	night	rat
Backs bitten, heads missing, necks and breasts torn, breasts and entrails eaten; bird pulled into fence and partially eaten; body found away from housing, maybe scattered feathers	every 5–7 nights	raccoon
Several Birds Killed		
Birds mauled but not eaten; fence or building torn into; feet pulled through cage bottom and bitten off	anytime	dog
Bodies neatly piled, killed by small bites on neck and body, back of head and neck eaten	night	mink
Birds killed by small bites on neck and body, bruises on head and under wings, back of head and neck eaten, bodies neatly piled; faint skunklike odor	night	weasel
Rear end bitten, intestines pulled out	night	fisher, marten
Chicks dead, maybe faint lingering odor	night	skunk
Heads and crops eaten	every 5–7 nights	raccoon

My first flock of chickens came with a 1-acre (0.4 ha) ranchette in a rapidly suburbanizing area, where the chief threats were dogs, rats, and ever-tightening zoning regulations. Because of the latter, we now raise chickens on a rural farm at the end of a dirt road, where a steady and varied parade of industrious wildlife attempts to share our birds. Because the wild animals delight us as much as they attempt to frustrate our poultry-keeping efforts, and because we are encroaching on their territory, we do our best to identify the source of any predation and take appropriate defensive measures to protect our flocks while letting the wildlife be.

PREDATOR DETECTIVE KEY

Clues	Likely Time	Predator
One Bird Missing		
Ranged bird missing, feathers scattered, or no clues	dusk or dawn	fox
A few scattered feathers or no clues	dusk or dawn	hawk
Fence or building torn into, feathers scattered	anytime	dog
Ranged bird missing, feathers scattered or no clues	dusk or dawn	cougar* (rare)
A few scattered feathers or no clues	night	owl
Small bird missing, lingering musky odor	night	mink
Ranged bird missing, no clues	night	bobcat (rare)
Several Birds Missing		
No clues	anytime	human
Ranged birds missing, feathers scattered, or no clues	dusk or dawn	fox
Ranged birds missing, no clues	early morning	coyote
Ranged birds missing, no clues	day	hawk
Chicks missing, no clues	day	snake
Small birds missing, bits of coarse fur at shelter openings	night	raccoon
Chicks or young birds missing	night	rat, cat
Eggs Missing from Nest		
No clues	day	snake
Empty shells in and around nests	anytime	dog
Empty shells in nest or near housing	day	jay, crow
No clues	night	rat
No clues or empty shells in and around nests, maybe faint lingering odor	night	skunk
Empty shells in and around nests	night	raccoon, mink
Empty shells in and around nests	nightly	opossum
Eggs Missing under Broody Hen		
No clues or faint lingering odor	night	skunk

*A cougar is also known as a catamount, mountain lion, panther, or puma.

RODENTS

Rats and mice are a particularly insidious type of predator. They're everywhere, they breed rapidly, and they can't take a hint. They invade any time of year but get worse during fall and winter, when they move indoors seeking food and shelter. Rats eat eggs and chicks, and rats and mice both eat copious quantities of feed and, by moving from one place to another, can spread various diseases. To add insult to injury, rodents gnaw holes in housing and burrow underneath, providing entry for other predators. Whether or not you find evidence, you can safely assume you have a rodent problem.

Discourage rodents by eliminating their hideouts, including piles of unused equipment and other scrap. Store feed in containers with tight lids, and avoid or sweep up spills immediately. Aggressive measures include getting a cat or a Jack Russell terrier, and, if you see rats and you're experienced with a gun, shooting 'em. Don't bother with techie solutions like ultrasound black boxes and electromagnetic radiation, which are as ineffective as they are expensive.

Poisoning is a *last* resort, because you never know whether you might poison a pet, a child, or harmless wildlife. Besides, bait stations work only if the rodents can find no other source of feed, which is pretty unlikely in your average backyard poultry situation. Traps of all sorts are invariably messy, no matter whether they kill or trap live rodents, but they are an option when all else fails.

Weather Considerations

Under temperate conditions, a chicken doesn't need central heat and air to remain comfortable year-round. In cold weather, a chicken's body warms itself by producing approximately 35 BTUs per hour through physical activity and metabolic processes sustained by feed. In moderately warm weather, a chicken's body cools itself by transporting internal heat to the external environment with the aid of its circulatory, respiratory, and excretory systems. In extremely hot or cold conditions, however, your chickens' temperature-regulating systems may need a little help.

Body Heat

During long periods of extreme cold or heat, laying hens stop production and all chickens suffer stress. Chickens generally suffer less in cold weather than in hot weather, as long as their drinking water doesn't freeze and their housing is neither damp nor drafty. When temperatures reach 104°F (40°C) or above, chickens can't lose excess heat fast enough to maintain the proper body temperature and may die. A chicken's body controls temperature by transferring heat between itself and its environment in the four ways described below.

Radiation involves heat transfer between a chicken and nearby objects. If the environment is warmer than the bird's body, nearby objects warm the bird; if the environment is cooler, heat radiates from the bird to the environment. Examples of radiation control include putting a reflective roof on the coop to prevent radiant heat gain from the sun in summer and insulating the roof to prevent radiant heat loss from birds in winter. At low temperatures, most of the heat a bird's body loses is through radiation and convection.

Convection is the transfer of heat between a bird and the surrounding air. Drafts and breezes cause warm air close to a bird's body to be replaced by cooler air. In winter, convection causes a bird to chill. At low temperatures, you can reduce convection by eliminating drafts that carry away warm air trapped in a bird's ruffled feathers.

In summer, convection cools a bird but only as long as the surrounding air is cooler than the bird's body temperature of 103°F (39.5°C). At air temperatures above 70°F (21°C), you can improve convection by opening doors and windows and, if necessary, by installing a fan. At 85 to 90°F (29 to 32°C), a chicken exposes more of its body to moving air by holding out its wings.

Conduction is heat transfer between a chicken and objects its body contacts — floors, litter, nests, and so forth. As you might expect, warm objects warm the bird, cool objects cool it. The parts of a bird's body having the most contact with external objects are its feet, but since the feet are quite small, their conductive influence is minimal. Holes for dust baths are also sources of conduction: cool soil or fresh litter can be a significant factor in keeping a bird cool in summer; warm, composting litter provides warmth in winter.

Excretory heat transfer is a type of conduction that occurs when a bird drinks cool water, warms the water within its body, and eliminates the warm water in its droppings. At high temperatures, chickens increase the rate of heat loss by drinking more than usual, which causes their droppings to become loose and watery. Off-color droppings during the heat of summer may be a sign the birds aren't getting enough to drink. At high temperatures, excretion and evaporation account for most of a chicken's body-heat loss.

A hot chicken tries to stay cool by holding its wings away from its body and panting.

Evaporation is the loss of latent body heat that occurs when the environmental temperature approaches a chicken's body temperature, and its body heat vaporizes liquid on the body's surface. Evaporation is an effective cooling method only when the relative humidity is low. Each 17°F (9.5°C) increase in air temperature doubles the air's capacity to carry moisture, up to a point — air at any temperature can accept only so much moisture, and if the air is already saturated, it can't hold any more.

Respiratory heat transfer is a type of evaporation occurring when a bird inhales air that's cooler than the bird's body, and exhales moisture-laden warm air. The moist-air passages in the bird's extensive respiratory system—which include not only lungs but also air sacs among its organs, and air spaces in some of its bones — help a bird lose internal body heat. In warm weather, a bird increases the rate of heat loss by panting. Since coops tend to be high in humidity due to moisture produced by respiration and excretion, good ventilation and proper litter management are important temperature-control measures.

Low humidity on summer days lets you take advantage of evaporation to cool birds by frequently hosing down the coop's outside walls and roof, and occasionally misting adult chickens, when the following conditions prevail:

- Air temperature is above 95°F (35°C).
- Air humidity is below 75 percent.
- Air circulation (convection) is good.

In winter, humidity in the air chills chickens by drawing their body heat to the surface to vaporize. Temperatures low enough to freeze moisture in the air also can cause frostbite to combs, wattles, and toes. Frostbite is therefore more likely to occur in damp housing than in dry housing.

At any given time, a combination of all four forms of heat transfer determine whether a bird gains or loses body heat. A well-feathered bird, for example, will be comfortable at a temperature of 50°F (10°C) if the air is still and the sun is shining, while the same bird will be miserable at 68°F (20°C), if it's out in the wind and rain.

Preventing Frostbite

Preventing frostbite involves a combination of management measures. To prevent frostbite

- Reduce humidity by improving ventilation and removing damp patches of litter around doorways and waterers.
- Rake or pitchfork litter regularly — loose litter is dryer and is a better insulator than compact litter.
- Eliminate drafts by filling cracks and crevices in walls.
- Install perches in the least drafty part of the housing.
- Use perches that are wide enough (at least 2 by 2 inches [5 by 5 cm]) to allow birds to cover their toes with breast feathers at night.
- If the ceiling is not within 2 feet (0.6 m) of perches to keep body heat close to the birds, install a heat lamp over the roost and plug it into a thermostat set to turn on the heat when the temperature drops below 35°F (2°C). Enclose the lamp in a sturdy wire guard so it can't be damaged in the event of a collision with an airborne chicken. In a small coop, a few well-placed electric lightbulbs should supply sufficient heat.
- Feed a small amount of scratch in the morning to kindle body warmth until birds are warmed by radiant heat from the sun.
- During cold days, stimulate appetites with a little mash moistened with warm milk or water.
- Increase interest in eating by feeding often or by frequently stirring rations.

FROSTBITTEN COMBS AND WATTLES

Frozen combs and wattles look pale. If you discover the condition while the part is still frozen, apply a damp, warm cloth (105°F [40.5°C]) to the frozen part for 15 minutes or until it thaws. Do not rub. After the part has thawed, gently apply an antiseptic ointment such as Neosporin. Isolate the bird from other chickens to prevent further injury, and keep an eye on it to see that the comb and wattles heal properly.

Frozen combs and wattles discovered after they have thawed are red, hot, swollen, and painful. The bird doesn't want to move or eat. If the part has thawed, warming is not necessary. Gently coat the part with Neosporin, and isolate the bird. After the swelling goes down, the part may peel, itch, turn scabby, develop pus, and eventually fall off.

Seriously frozen combs or wattles shrivel and eventually die back. If a part turns black, the tissue has died and gangrene has set in — the comb or wattle must be surgically removed to avoid septicemia.

Pale gray or white tips on the comb are an early sign of frostbite. Roosters, with their larger combs, are typically more susceptible than hens.

- Feed a little scratch at nightfall to increase body warmth during nighttime perching.
- Coat combs and wattles with petroleum jelly as insulation against frozen moisture in the air.

Reducing Heat Stress

As with preventing frostbite, preventing heat stress of your flock also involves a combination of management measures. Heat stress may be avoided by following these simple precautions:

- As water consumption goes up, increase the number of watering stations.
- Frequently fill waterers with cool water.
- Keep water cool by placing waterers in the shade.
- Add electrolytes to water to stimulate drinking and to replenish electrolytes depleted due to the heat.
- Otherwise avoid medicating the water — if birds don't like the taste, they'll drink less.
- Since hot weather causes birds to eat less and rations to go stale faster, ensure freshness by purchasing feed more often and in smaller quantities.
- Distribute feeders so birds don't have to travel far to eat.

- Encourage eating by feeding early in the morning and by turning on lights during cool morning and evening hours.
- Open windows and doors and/or install a ceiling fan to increase inside air movement.
- Eliminate crowded conditions by removing some birds or expanding their current housing.
- Do not confine chickens to hot spaces such as trapnests (see page 244) or cages in direct sunlight or where ventilation is poor.
- Provide plenty of shade where the chickens can rest. If necessary, put up an awning or tarp.
- Do not disturb your chickens during the heat of the day.
- Hose down the coop roof and outside walls several times a day.
- Lightly mist adult birds (never chicks; they can easily chill) when the temperature is high and humidity is low.

Weather Preferences

Mature chickens can adapt to temperature extremes through gradual exposure. A slow, steady shift in temperature therefore causes much less stress than a sudden change. When a chicken becomes acclimated to warm temperatures, it pants less readily and is less likely to die at what might otherwise be a lethal temperature. Likewise, a bird that's used to warm weather is less tolerant of a sudden shift to cold weather. Breed also plays a role in weather tolerance.

Hot-weather intolerant. Breeds that are loosely feathered, like Orpingtons, and those that are heavily feathered, like the Asiatics (Brahma, Cochin, Langshan) and Americans (Plymouth Rock, New Hampshire, Rhode

Island Red) suffer more in hot weather than lightly feathered breeds, and hens in lay suffer more than those not in production.

Cold-weather intolerant. Low temperatures affect lightly feathered breeds such as Hamburgs, Naked Necks, and the Mediterraneans (Buttercups, Leghorns, Minorcas); sparsely feathered breeds such as Shamo; and breeds with short, close feathers, such as Cornish and Modern Game. Not only are the latter two inadequately insulated against cold weather, but neither breed does well in winter confinement.

Crested varieties (Polish, Houdan, Crevecoeur) are vulnerable to freezing if their copious head feathers get wet. Single-comb breeds suffer more in cold weather than rose-comb breeds; large-comb breeds like Dorkings suffer more than birds with smaller combs. Cocks suffer more than hens, since they have larger combs and, unlike hens, usually don't sleep with their heads tucked under a wing.

Cold-weather tolerant. The Chantecler, created in Canada as a dual-purpose breed with small comb and wattles, lays well in winter and can withstand cold weather. The table called *Large Breed Groups* in chapter 1 indicates each breed's general weather adaptability.

Coop Cleanup

"Cleanliness is next to godliness, filthiness is next to death," said the hand-lettered sign tacked inside the henhouse of an old-timer I once visited. It was his daily reminder that safeguarding a flock's health is dependent on good sanitation. Sloppy sanitation is the most common cause of failure when raising chickens.

Good sanitation includes frequently cleaning feeders and waterers, disinfecting reused housing or equipment, and regularly cleaning the house and yard. Unless a flock has experienced a health problem, a properly designed and maintained coop needs a completely thorough cleaning no more often than once a year. Even in the healthiest environment, disease-causing organisms build up over time; a thorough annual cleaning will remove 95 percent of the contamination.

To minimize disease-causing organisms that flourish in the warm months of summer, schedule your major cleanup for spring. Choose a warm, dry, sunny day, and wear a dust mask so you won't breathe the fine dust you'll stir up, which otherwise may cause nasal irritation and coldlike symptoms. Lightly mist the walls and equipment to keep dust out of the air.

Remove all movable feeders, waterers, perches, and nests, and clear out the old bedding. My favorite tool for moving loose litter is a snow shovel, but a coal shovel works well, too. A pitchfork is more suitable for caked or packed areas. With a hoe, scrape manure from perches, walls, and nests.

Remove dust and cobwebs especially from corners and cracks, light fixtures, and window screens. Where electricity isn't handy, an old broom makes a serviceable dust mop. I like a shop vac, because it captures more dust, but I lost several vacuums before learning the fine dust has to be frequently cleared from the filter to avoid burning out the vacuum; clean the filter whenever the vacuum stops drawing well or feels hot.

Mix 1 tablespoon (15 mL) of chlorine bleach per gallon (3.75 L) of boiling water and use the solution to disinfect all troughs, perches, and nests, and then scrub the inside of the coop. Leave the doors and windows open to hasten drying. While you're waiting for the coop to dry, rake the yard and pick up any junk that might be lying around. Piles of scrap wood or discarded equipment keep out the sun's healing rays and attract insects, snakes, and rodents.

Dealing with Manure

Chicken manure is made up largely of undigested feed. For each 100 pounds (45 kg) of ration your chickens eat, expect 45 pounds (20 kg) of droppings, dry weight, which amounts to about 1 pound (0.45 kg) of manure per week for a large breed. Fresh droppings contain, in addition to feed residue, intestinal bacteria, digestive juices, mineral by-products from metabolic processes, and water. The white pasty stuff on top of a dropping is the chicken's equivalent of urine, consisting mostly of nitrogenous waste and water. Water makes up about 85 percent of the total weight of fresh chicken droppings, and its evaporation contributes to henhouse humidity, not to mention odor.

A management decision you must make is whether to do one massive annual cleanup of manure or break that task up into smaller, more frequent cleanings. Manure left in one place over a period of time attracts both fly predators and fly parasites. When you remove the manure, you also remove the natural fly predators. With too few predators left to destroy them, fly eggs in the smallest clump of manure that might be left behind will hatch into hundreds of flies. If you remove manure during the summer months, be prepared to either continue cleaning out the manure and bedding thoroughly at least once a week until fly season ends or institute some other fly-control measure.

Pasturing chickens in portable housing is one way to avoid a manure problem — simply move the shelter often enough to keep manure from accumulating. In stationary housing, litter absorbs much of manure's moisture, minimizing both humidity and odor. Frequent manure removal also minimizes humidity and odor but introduces new problems: it's labor intensive (who has time?); it requires a year-round system for dealing with the manure; it uses up more bedding; it leads to fly problems in warm weather.

Composting Litter

Old-time farmers, notorious for being behind in their work, tossed a layer of fresh bedding on top of the old litter whenever the place seemed a little messy. When they needed spring fertilizer, they spread all the used litter on their fields and started over with fresh bedding.

During World War II, when feed for livestock became scarce and expensive, the Ohio State Agricultural Experimental Station began seeking ways to reduce the need for animal protein in chicken feed. They discovered that

CHICKEN DIAPERS

If you bring a chicken into your house, you'll obviously need a different approach to manure management. Because chickens are difficult to housebreak, reusable plastic-lined chicken diapers have been developed in various sizes, styles, and colors for use with chickens kept as house pets, convalescing from an injury or illness, or being primped for show. An Internet keyword search for "chicken diaper" will reveal numerous sources, as well as detailed instructions on how to make your own.

decomposing litter is rich in vitamin B_{12}, a vitamin found only in animal protein and a promoter of health, growth, and reproduction. They learned that chickens housed on naturally composting litter don't need this otherwise expensive nutrient in their rations because they got plenty of it scratching and pecking in the compost.

Any good bedding is suitable for composting litter. Lay down 4 inches (10 cm) of whatever clean bedding you've chosen in the spring, after you've done your annual cleaning, or when you first start out with an empty coop. Whenever the surface gets packed or matted, break it up and stir in a little fresh bedding, enough to absorb prevailing moisture. The goal is to have the bedding 10 inches (25 cm) deep by the start of winter.

Keep adding fresh litter as needed to absorb the amount of manure your flock deposits. If your birds are not crowded, 12 to 15 inches (30 to 38 cm) of bedding should strike the right balance. Decomposing litter reduces in volume like a compost pile. Once the litter reaches enough volume to start actively composting, the amount of volume reduction will roughly equal the amount of new litter added.

Rather than becoming filthy, as you might expect, properly managed built-up litter gradually ferments, and after about six months the resulting compost develops sanitizing properties. Furthermore, the heat produced by fermentation keeps a flock warm during the cooler months, and flies are less of a problem in warmer months because accumulated dry manure attracts natural fly predators and parasites.

Managing composting litter involves raking or stirring the bedding as often as necessary to keep the surface from crusting over.

Manure accumulating beneath perches may need to be removed or, better yet, collected in a droppings pit.

Adequate ventilation is needed to ensure the litter retains the right amount of moisture for good fermentation. To test litter moisture, pick up a handful and squeeze. If the moisture level is just right, the litter will stick slightly to your hand but will break up when you let go. If it's too dry, it won't stick to your hand; if it's too wet, it will ball up and not break apart easily when you drop it.

If the bedding is either too damp or too dry, the environment becomes unpleasant and unhealthful for chickens and humans alike. Excessively dry litter not only fails to ferment properly (and therefore is not self-sanitizing) but also creates a dust problem. Dampen dry litter with an occasional light sprinkling of water, followed by stirring.

Excessive moisture is more often a problem than excessive dryness. Wet litter with a high manure content — perhaps the result of housing too many chickens in too small a space — can cause painful burns to hocks and footpads, which then become infected and cause lameness. Damp litter favors the growth of disease-causing molds and bacteria, and promotes the survival of viruses, parasitic worms, and protozoa (such as those causing coccidiosis). Damp litter also releases ammonia fumes that irritate avian (and human) eyes and respiratory tracts, opening the way to disease.

When litter is damp enough to emit ammonia, the first thing you'll notice is the odor. If your eyes burn and your nose runs, the ammonia level has become high enough to increase your birds' susceptibility to respiratory disease. If the ammonia concentration gets so strong that birds' eyes become inflamed and watery and the chickens develop jerky head movements, ammonia blindness may soon follow. To keep litter from getting too moist do the following:

- Remove any damp spots that develop around doorways and drinkers.
- Adjust doorways, and fix dripping drinkers to eliminate persistent damp patches.
- Repair roof leaks and correct drainage problems that cause indoor puddling.
- Insulate the ceiling, if necessary, to prevent winter condensation from dripping on litter.
- Add fresh litter more often or decrease the number of birds housed.
- Aerate the litter more often by loosening and turning it.
- Provide good ventilation to remove excess moisture from the air.

MINIMIZING NITROGEN LOSS

The smell of ammonia coming from chicken manure in either the henhouse or the compost pile means nitrogen is evaporating. To reduce nitrogen losses, periodically apply ground rock phosphate or ground dolomitic limestone. Either substance combines with nitrogen to keep it from evaporating and to improve manure's fertilizer value. As a side benefit, it also helps keep litter dry. For an average-size backyard coop, apply 1 pound (0.45 kg) per week stirred into litter, or 2 pounds (0.9 kg) per week scattered over the droppings pit.

You can keep a flock on the same composting litter for years, provided the bedding doesn't get damp and remains warm, and no serious disease breaks out. On the other hand, if you need the nitrogen-rich bedding to fertilize your garden or you live where summers are quite warm, you may wish to clean the house each spring and begin the summer with fresh, cool litter.

If frequent disease outbreaks in your area make the reuse of litter unwise or if you raise successive batches of birds for short production periods, instead of composting litter, use deep litter: When you bring in a batch of new birds, spread the cleaned floor with 6 inches (15 cm) of fresh bedding. During the production period, stir the litter as necessary to prevent surface matting. At the end of each production period, remove the litter and thoroughly clean the coop.

Using Manure as Fertilizer

Smart poultry keepers don't look at manure as a nuisance waste product but as a valuable commodity that may be used or sold for gardening. The combination of aged manure and litter has average nitrogen (N), phosphate (P), and potash (K) values of 1.8, 1.4, and 0.8, respectively (as percentages of total weight), outranking most other barnyard droppings.

A good minimum yearly application is 45 pounds — approximately the amount produced by one hen each year — per 100 square feet of garden. (In approximate metric equivalents, a minimum application on a 10 sq m garden is 23 kg.) A generous application is 90 pounds per 100 square feet. (In metric equivalents, for a 10 sq m area, a generous application is 45 kg, or about a 4.5 cm layer.) In practical terms, if you spread a 1-inch (2.5 cm) layer over 100 square feet (9.3 sq m) of soil, you'll have applied roughly 55 pounds (25 kg); a 1½-inch (3.8 cm) layer will give you a little less than 90 pounds (40 kg).

Never feed growing plants fresh chicken manure — its high nitrogen content is hot enough to burn them. Furthermore, the excessive amount of nitrogen in fresh droppings encourages unbalanced plant growth, such as weak plant stalks and forked carrot roots.

Row-crop farmers generally prefer to spread fresh manure over bare soil and turn it under at the end of the growing season in autumn, giving the manure time to age in the ground over winter. This practice, called "sheet composting," requires keeping chickens off the cropland for at least a year to prevent the spread of disease and parasites.

Many gardeners prefer to compost the manure for spring application to growing plants. Composting not only transforms nitrogen into a form that's readily usable without damaging plants but also destroys bacteria, viruses, coccidia, and parasitic worm eggs. Built-up litter composts naturally, stabilizing or fixing both nitrogen and potash in the process.

Raw manure that's composted without being mixed with some carbon-containing substance (shavings, straw, weeds, grass clippings, and the like) will overheat and dry into a powdery white nutrient-poor ashy substance called *fire fang*. If you're not an experienced composter, you can find all the information you need in any good book on gardening. My favorite is *Gardening When It Counts*, by Steve Solomon.

Trimming Procedures

The claws and beaks of chickens are made of keratin, the same substance as your fingernails and toenails, and like your nails, they continually grow. Chickens evolved in an environment in which their claws and beaks naturally remained in balance by wearing down as they grew. But in backyard confinement, sometimes they grow too long and need to be trimmed. A cock's spurs, too, can grow too long for the bird's comfort.

Claw Trimming

A chicken uses its claws to scratch the ground for food and to scratch an itch. When a chicken doesn't have hard surfaces to scratch its claws against, they continue to grow until they curl, and then the chicken can't walk properly.

Nails that don't wear down naturally need to be periodically trimmed. Cocks have their claws trimmed to prevent injury to hens during breeding, and chickens groomed for show must have their nails neatly trimmed.

How often claws need trimming depends on how fast they grow. And the rate of growth depends on the environment and the time of year, so keep an eye on the claws and trim them as often as necessary.

Use a pair of canine toenail clippers or heavy shears, and finish by filing away sharp edges. Trim away small amounts at a time to avoid snipping a nail too short. If you should accidentally draw blood, stop the bleeding by applying an astringent such as witch hazel, styptic powder, or alum or encourage rapid clotting with a little flour or cornstarch.

FERTILIZER VALUE OF CHICKEN MANURE
The fertilizer value of chicken manure varies with its age and also with the nutritional intake of birds at various stages of growth.

	% Nitrogen (N)	% Phosphate (P)	% Potash (K)
Chicken Manure, by Degree of Freshness			
Wet, sticky, caked	1.5	1.0	0.5
Moist, crumbly to sticky	2.0	2.0	1.0
Crumbly but not dusty	3.0	2.8	1.5
Dry, dusty	5.0	3.5	1.8
Fresh Manure, by Age of Birds			
Baby chick	1.7	1.3	0.7
Growing chick	1.6	0.9	0.6
Hen	1.1	0.8	0.5
Comparison, Barnyard Averages (Fresh)			
Cow	0.6	0.2	0.5
Horse	0.7	0.3	0.6
Steer	0.7	0.3	0.4
Duck	1.1	1.4	0.5
Goat	1.3	1.5	0.4
Turkey	2.0	1.4	0.6
Rabbit	2.4	1.4	0.6

Sources: USDA and The Pennsylvania State University

Beak Trimming

A chicken uses its beak for gathering food and for exploring and manipulating objects in the environment, for preening, nesting, and engaging in social interactions. A beak that grows out of balance interferes with the chicken's ability to eat and enjoy other activities that are necessary for its well-being.

In a natural setting, the beak wears down as fast as it grows. The chicken wipes its beak on the ground to clean it, at the same time sharpening the beak for pecking and keeping it from growing too long. The upper half of the beak is naturally a tad longer than the lower half, but when a chicken lacks opportunities to keep it worn down to a proper length, the upper half grows so long it interferes with eating and other activities.

When the upper half just begins to overlap the lower half, you can trim it back with a fingernail file. Once it has passed the filing stage, use toenail clippers or the same canine clippers used on claws. If you don't let the upper beak grow too far, the part that needs to be trimmed away will be lighter in color than the rest of the beak. When in doubt, look inside the chicken's mouth and you can see where the live tissue ends.

Trim a little at a time to make sure you don't get into live tissue and cause pain and bleeding. In most cases only the upper half needs trimming. On rare occasions the lower half may need a little reshaping, especially if a too-long upper half pushes the lower half in the opposite direction.

When the upper and lower halves grow in opposite directions, the beak may be trimmed to let the bird peck properly. But in most cases a crossed beak is a genetic defect, and such a chicken should not be used for breeding. You will know it's genetic if the crossed beak appears almost from the time a chick hatches and keeps growing crossed no matter how often you trim.

Beak trimming is not the same procedure as debeaking — although the commercial industry now euphemistically calls debeaking "beak trimming" or "beak conditioning" — which is cutting a beak so it remains permanently

Crossed beak (left) is a genetic deformity that sometimes can be corrected with frequent trimming so the chicken can eat properly. A top half of a beak that grows longer than the bottom half (right) also must be trimmed.

short to prevent cannibalism. Birds in a properly managed backyard flock should not need permanent debeaking.

Temporary debeaking, however, may be the lesser of two evils when chicks persistently peck each other and cannot be stopped. Use nail clippers to remove one-fifth of the upper portion, which should grow back in about six weeks.

Trimming Spurs

Cocks use their spurs as weapons for fighting each other and fighting off predators. Cock fighting is the oldest known spectator sport. Some cocks retain the natural fighting spirit and will attack anything or anyone. Others are more selective and may attack only women; or only men; or only a person wearing floppy pants, a yellow slicker, or dangling boot laces; and so on. The only way to cope with a cock that attacks without provocation is to get rid of the bird, period.

Most hens have little rudimentary knobs instead of spurs, although some have real spurs that can grow quite long. And some hens get pretty feisty, although you'd be hard pressed to find a hen as lethal as a man-fighting cock.

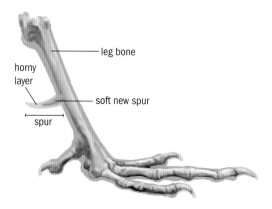

The spur is an extension of the leg bone that curves and becomes covered with a tough, pointy spur case as a cock matures.

Long spurs may affect a rooster's ability to walk and to breed and are dangerous to other chickens and to humans. Spurs are trimmed to prevent injury to the cock's handlers, prevent the wounding of hens during breeding, minimize injury in peck-order fights, and groom old cocks for show. A spur that curls back into a cock's leg must be trimmed to prevent lameness.

The *spur* is an extension of the leg bone, covered with the same tough keratinous material that makes up claws and beaks. The spur starts out as a little bony bump. As the cock matures, it gets longer, curves, hardens, and develops a sharp pointed tip. The tip of a mature spur may be trimmed with a Dremel tool or clipped with a pair of toenail or canine clippers and the edges smoothed with a file. Take care not to clip off too much at once, or you may damage the soft new tissue underneath and cause pain and bleeding.

An old hardened spur casing that has grown long and dangerous may be twisted off to reveal a fresh short spur underneath. Exhibitors typically spruce up old show cocks by twisting off their spurs every couple of years.

To twist off a spur, take the following four steps. Don't be in a hurry, and never break off a spur or you may damage the tender new growth and cause serious bleeding.

Soften the juncture between the spur case and the leg by liberally working in some vegetable oil.

Gently twist the spur back and forth on its axis, continuing to apply oil until the case feels loose.

Continue twisting back and forth until eventually the hard case pops off.

Since the freshly exposed soft tissue is vulnerable to injury, carefully apply a liberal amount of antibacterial ointment such

Clipping the flight feathers keeps a chicken from being able to fly until the feathers grow back.

as Neosporin, and confine the cock alone to a roomy area for a few days while the new spurs harden.

Wing Clipping

When a chicken repeatedly flies out of its enclosure, its wing feathers may be trimmed to protect the chicken from predators, to protect vegetable and flower beds from the chicken, or to keep the bird from getting into the wrong breeding pen. Breeds that are best known as flyers include Leghorn, Hamburg, Old England Game, and nearly any bantam. Heavier breeds may fly while they are young but rarely after they fully mature. For breeds that don't fly high, a tall fence may be adequate. If the chickens are kept in a small yard, lightweight netting secured over the top should keep them in.

Brailing is sometimes used to control a young bird that will eventually grow too heavy to fly and also works for a show bird when you don't want to mar its appearance by feather clipping. Brailing involves wrapping one wing with a soft cord so the wing can't be opened for flight. The cord, or *brail*, must be removed occasionally so it doesn't become tight enough to cause damage and so the bird can exercise its wing muscles.

Clipping involves the use of sharp shears to shorten the first 10 flight feathers of one wing. Clipping the flight feathers causes a bird to lack the balance needed for flight. It lasts only until new feathers grow during the next molt, which may be a few months in young birds or up to one year for older chickens. The clipped feathers may not readily fall out during the molt, requiring your assistance before new feathers can grow in. Clipping, therefore, should be considered a last resort after all other methods of confinement have failed.

Handling Chickens

Whenever you work around your chickens, move in a way that avoids causing them stress. Sing or talk softly as you approach your flock, so they'll hear you coming. Walk slowly but deliberately while moving among them. Ask children not to run and scream in the chicken yard, and if you bring your dog into the yard, train it not to bark, run, or chase chickens. When picking up an individual chicken, always handle it gently.

Catching a Chicken

Occasionally, you'll need to catch your chickens to perform routine procedures, including examining and treating them for parasites, moving them, grooming them for show, and so forth. If you spend a lot of time with your birds, catching one is as simple as bending down and picking it up. Chickens that are handled regularly will gather around your feet expectantly.

The easiest way to catch a chicken that isn't tame is to pick it off its perch at night. If you have to catch a wily chicken during the day, the job will be easier with the aid of a net, catching hook, catching crate, or a helper.

With a helper, slowly herd the bird into the corner of a mesh fence or, better yet, inside a building with the entries closed so the chicken can't get out. The chicken will most likely try to escape along one side; two people can guard both sides of the corner, as well as the middle space between you. Move deliberately, taking great care not to panic herded birds or cause them injury by making them pile into each other or against equipment. When you and your helper get close enough, one of you should either clamp down quickly on the bird's two wings or reach beneath it and grasp both legs.

A catching crate helps you avoid exciting or injuring your birds. You can make one either from scratch or by modifying an existing cage or bird carrier. Despite its fancy-sounding name, a catching crate is nothing more than a lightweight but strong wooden or wire box with an opening at one end designed to fit against the chicken-size door to the chicken house or pen. The opening has a slide-down door you close after herding one or more birds from the pen into the crate. A trapdoor at the top lets you handily reach in to remove the

bird you want. When rounding up chickens at pasture, use a pair of panels to guide them into the catching crate's open end.

A catching hook, also called a *fowl catcher,* consists of a shaft of heavy-gauge wire bent at one end into a hook, with a plastic or wooden handle at the other end. With this hook you snare a running chicken by one

HOW TO CATCH A CHICKEN

To catch a chicken with a hook, slip the hook around one leg and pull toward you.

leg, pull it toward you, and quickly pick it up while it's off balance. Using a catching hook requires quick thinking and physical agility.

A chicken-catching net consists of a fishnet pocket attached to a metal hoop with a handle. It looks just like a butterfly net. When you get close enough to the chicken, toss the net over the bird, then quickly retrieve the bird from inside the net. A really wild chicken might slip out of the net, so clamp the hoop against the ground and hold it tight until you are in a position to get the chicken out. Some athletic and well-coordinated souls can catch a chicken in midair and quickly retrieve it from the net before it has a chance to escape.

Whether using a hook or a net, if you scatter some grain on the ground, the chickens will gather around, and if you're stealthy enough, you can sneak up and nab the one you're after while it's busy pecking. Once you start to catch an untame chicken, don't give up. If you let it get away, you will have a more difficult time trying to catch it next time.

Carrying a Chicken

To carry a tame chicken, reach beneath the bird and grasp its legs with one hand, then bring your other arm around to cradle it. Be sure to get hold of both legs to keep the chicken from scratching you. Once you have the chicken safely cradled, take a moment to stroke its neck and wattles. By giving the bird a pleasurable experience, you'll have an easier time catching it in the future.

A frightened chicken will try to get free by flapping its wings and paddling its feet and may slice you with a claw. To hold a frightened chicken, carry its weight on the arm with the hand holding the legs, and with the other hand hold the wings over its back. Do not carry a chicken by its wings alone, as doing so may cause damage to the bird as well as to you (from flailing claws and spurs). Covering a bird's eyes by laying a handkerchief over its head usually helps calm it down.

Chickens are less stressed when carried one at a time and held with the head upward. But in the event you have to carry more than one bird at a time, turn them upside down and hold onto both legs. Most of the time a chicken held by both legs with its head downward will stop struggling. Be sure to hold onto both legs; otherwise the chicken may churn in an attempt to get away, possibly injuring itself or you. Carry only as many chickens in one hand as you can comfortably hang onto — never more than four at a time.

Transporting Chickens

A chicken may be safely transported in anything from a paper sack to a pet carrier, provided the container satisfies these criteria:

- The chicken can't get out.
- The container is not so big the bird can hurt itself flying in an attempt to get out.
- The container is not so small the bird can't stand up and move.
- The container has no sharp edges or other injurious protrusions.
- The chicken has access to drinking water; in long-distance transit at least provide water during occasional stops.
- The container is sufficiently ventilated to allow the chicken to breathe.
- The chicken is protected from drafts, cold, and rain.

HOW TO CARRY A CHICKEN

To carry a tame chicken, hold its legs underneath with one hand while cradling it with the other arm.

To carry a frightened chicken, hold its legs with one hand and let its weight rest on that arm, while restraining the wings with your other hand — do not carry the chicken by its wings alone.

A car trunk is not suitable for transporting chickens, as it may accumulate lethal carbon monoxide fumes. A stock rack or wire cage on an open pickup bed is not suitable, either, unless some form of wind protection is provided. A pickup with a topper may be suitable if the topper is not airtight and the truck is never parked where the chickens are left suffering in the hot sun. Chickens will be safe and comfortable traveling inside a car or truck that humans are comfortable riding in, but don't forget to protect the floor and seats from droppings with a tarp, some feed sacks, or several layers of newspaper.

Enterprise Integration

Chickens combine well with many other rural enterprises, to the benefit of both the chickens and the other enterprises. Some combinations, however, need extra management or should be avoided all together.

Chickens work really well with most pastured livestock and benefit both from the pasture and from scratching in livestock manure. By working through livestock manure, they help reduce livestock parasites. If you are considering integrating your chickens with other livestock, keep these considerations in mind. Chickens have difficulty remaining healthy in wet areas, which includes habitats that are ideal for ducks, geese, and other waterfowl. Chickens and pigs don't make a good long-term combination because both are susceptible to avian tuberculosis and bird flu. And chickens kept together with dairy goats will foul the bedding that milkers lie in and, if nesting in hay mangers, will

Chickens make good company for pastured sheep.

foul the goats' fodder with muddy feet, manure droppings, and the occasional broken egg.

Chickens work well in an orchard, where they dine on the windfall fruit that might otherwise harbor insects that cause damage to trees and fruit. And on hot days, chickens enjoy the shade provided by the trees. But keep in mind that over time the chickens will create a system of dust holes that cause difficult footing for the humans who maintain the trees and pick the fruit.

Properly managed, chickens combine well with a garden and with small-fruit and berry production. Proper management includes ensuring that the chickens neither eat more than their fair share nor scratch up new plantings. Also, for the sake of food safety for humans, chickens should be kept away from crops within 90 days of harvest, 120 days if the crop's edible portion has contact with the soil.

Having chickens next to the garden where they can be fed the refuse (weeds, surplus vegetables, beetles, and so on) and letting them scratch in fallow areas works fine. Letting them discharge waste on harvestable crops is not a good idea. Our garden usually has some area where chickens can scratch, such as where we'll plant the corn or beans before the soil is warm enough, or where we've pulled the early peas after they are harvested. The concept of the so-called chicken tractor (some people alternatively use portable electroplastic netting) is to confine them to such areas and not let them loose in the entire garden (except before planting and after harvesting).

DIVERSIFICATION WITH CHICKENS

Excellent with Chickens	Reasons
Cattle, sheep, meat goats	Chickens obtain nutrients from undigested feed in livestock manure, keep down flies, and reduce livestock parasites.
Horses	Chickens obtain nutrients from horse manure and make the horses less likely to spook.
Orchard	Chickens get exercise and fresh air while eliminating bugs and windfall fruit.
Pasture	Chickens benefit from sunlight, fresh air, and green feed while fertilizing the forage.
Good with Chickens	**Reasons**
Garden	Chickens benefit from green feed and contribute fertilizer but will damage plants.
Small fruit	Chickens benefit from exercise and will keep down bugs but will eat fruit.
Forest	Chickens benefit from fresh air and exercise but attract woodlot predators.
Poor with Chickens	**Reasons**
Dairy goats	Chickens foul goat bedding and hay, and goats eat chicken rations.
Pigs	Chickens and pigs share susceptibility to avian tuberculosis.
Waterfowl	Damp conditions created by waterfowl are unhealthful for chickens.
Wetlands	Damp conditions are unhealthful for chickens, and water bugs transmit disease and parasites.

6

HEALTH CARE

Owners of backyard flocks report few health problems, according to a thorough assessment of poultry flocks released by the United States Department of Agriculture. That's great news, because a chicken can get sick pretty fast, and by the time you notice, you're usually too late to do anything about it. Even if you're perceptive enough to recognize early signs of illness, treatment with medications is often harmful to the environment and to the future health of your flock.

Chickens maintained in a healthful environment can live a dozen years or more. A few well-kept pets have survived into their 20s. Starting out with healthy chickens and employing good management practices go a long way toward keeping them healthy.

Biosecurity

Biosecurity technically means taking precautions to protect your flock from infectious diseases but is often used as a catchall word to encompass any measure you take to protect your chickens from harm. The closer you adhere to the following practices, the more likely your chickens are to remain free of disease.

Day-to-Day Health Management

Commonsense routine management goes a long way toward preserving the health of your chickens and protecting them from harm. Good management starts before you bring home your first chickens.

Acquire only healthy stock. When possible, buy directly from a breeder, rather than through an auction or other live-bird market where chickens from various sources are brought together and may exchange diseases. Before bringing home any bird, examine it for the typical signs of good health, including bright eyes; smooth, shiny feathers; smooth, clean legs; and full, bright comb.

Feed a balanced ration. To be active, healthy, and productive, chickens need a variety of vitamins, minerals, protein, and energy as described in chapter 4.

Water, water everywhere. Ensure a continuing and plentiful supply of drinking water that is clean and free of chemicals.

Provide a sound environment. Chickens need a dry, well-ventilated shelter that keeps them safe from predators, as described in chapter 3.

Give them enough space. Unsanitary, unhealthy conditions are often caused by crowding, which forces chickens to live perpetually in filth.

Practice good sanitation. Remove manure piles and damp litter as they accumulate. Scrub waterers at least once a week. Regularly clean feeders to prevent the accumulation of moldy feed.

Burn or deeply bury dead birds and other animals. Chickens can get sick from picking at dead, diseased birds or botulism-tainted animals (that perhaps wandered into the yard and died after having been hit by a car). Exactly how you dispose of bodies will depend on your local regulations.

Keep an accurate flock history. Should a problem arise, details that might come in handy when determining what might be affecting your flock include the birds' ages, familial strains, and vaccinations; past diseases in your flock or on your place; and contact with other flocks (by means of moved birds or human visitors). You'll also want to note any suspicious signs of illness you observe, even if nothing immediate develops, and the circumstances surrounding any death in your flock.

Reduce Outside Exposure

Diseases can come to your chickens from a variety of sources. Although you needn't be overly worried (unless an infectious avian disease breaks out in your area), following a few simple measures will help reduce your flock's exposure to disease-causing organisms.

Keep your chickens away from other chickens. Many diseases are spread by carriers — birds that transmit disease without themselves showing signs. Anytime you introduce new chickens, you run the risk of introducing a disease. If you acquire a new bird or bring one back from a show, isolate it for at least two weeks, until you're certain the bird is healthy. Better yet, put one of your own birds in isolation with the new bird, as the new chicken may appear healthy while carrying a communicable disease that your sacrificial bird will catch.

Stay away from other flocks, and keep other poultry people away from yours. This biosecurity measure is one of the hardest to observe, since those of us who enjoy chickens like to visit other people with the same interest. To avoid spreading disease, cover your shoes with disposable plastic bags before going into someone else's yard, and keep bags handy for use by visitors to your yard.

Disinfect equipment used by other flocks. Anytime you purchase used equipment or reuse equipment from a previous flock, scrub it well, disinfect it, and dry it in the sun before putting it to immediate use. Equipment that will not be reused right away may be made adequately sanitary without the use of harsh chemicals. Clean it thoroughly of manure and other organic debris, scrub it well with hot water, dry in the sun, and store in a clean, dry environment until you need it.

Keep wild birds away. Wild birds are especially likely to carry diseases — on their feet as they fly from flock to flock pilfering grain, in parasites on their bodies, or as illnesses they have in common with chickens. Screen windows, and if possible, place netting over your chicken run to keep out wild birds.

Control insects and rodents. Other four- and six-legged critters living in your chickens' midst may bring in diseases. Keep your yard free of weeds and piled debris that attract vermin. Manage manure to minimize flies.

When vistiting another chicken yard, cover your shoes with disposable plastic bags or bootees, and ask guests to do the same when visiting yours.

Minimize Harm

Some biosecurity measures are easier said than done. Those of us who raise poultry often keep multiple species and multiple ages together at one time, and we love to visit other local flocks. However, the more biosecurity measures you observe, the better chance your flock has of remaining healthy and safe.

Breed for resistance. Hatch chicks only from 100 percent healthy breeders. If you make it a practice to breed birds that are at least 2 years old, you'll have plenty of time to weed out any that are susceptible to disease.

Tend to chicks first. When brooding baby chicks, take care of the chicks before you tend to the older chickens in your flock. Chicks need time to acquire natural immunity through gradual exposure to the pathogens in their environment; if you work with your older birds first, you could expose chicks to too much, too soon.

Avoid mixing birds of different ages. *Salmonella*, *E. coli*, and other bacteria may cycle back through the flock from older birds to younger ones. As these bacteria cycle through the flock, they can become more lethal. The influenza virus can also cycle forward from young birds to older ones.

Avoid mixing birds of different species. Germs that are relatively harmless in one species may have a devastating effect on another.

Confine your flock. As picturesque and down-home as roaming chickens may be, getting hit by a car, being eaten by a dog, or being poisoned by lawn spray is bad for a bird's health.

Medicate only when necessary. Medicating your flock every time a bird looks droopy can cause stronger germs to develop that eventually won't respond to drugs at all.

Reduce stress. Chickens are always under stress in one form or another. To avoid introducing more stress, move gently among your chickens, handle them with respect, and avoid making more than one major change at a time. When stress is unavoidable, such as before and after a move or during unpleasant weather, boost your flock's immunity with a vitamin supplement.

Parasites

A *parasite* is any living thing that lives off another living thing (the *host*) without providing any benefit in return. Technically, all infections are caused by parasites of one sort or another. But most of us think of a parasite

Severity	Stressor
COMMON STRESSORS	
Usually minor	Too much time between hatching and first food or water
	Nutritional deficiency in chicks due to inadequate breeder-flock diet
	Cold, damp floor
	Debeaking
	Rough handling
	Low-grade infection
	Eating spoiled feed
	Unusual noises or other disturbances
	Extremely high egg production
Moderate	Chilling or overheating during first weeks of life
	Extremely rapid growth
	Chilling or overheating during a move
	Sudden exposure to cold
	Extreme variations in weather or temperature
	Unsanitary feeders, drinkers, or litter
	Internal or external parasites
	Insufficient ventilation or draftiness
	Vaccination
	Competition between sexes or individuals
	Medication (severity depends on drug used)
Serious	Overcrowding
	Nutritional imbalance
	Insufficient drinking water
	Combining chickens of various ages
Severe	Suffocation caused by piling (see page 305)
	Lengthy periods without feed or water
	Inadequate number of feeders
	Poorly placed feeders causing starvation
	Onset of any disease

Adapted from: *Farm Flock Management Guide*, Floyd W. Hicks, Pennsylvania State University

as an animal form (such as a worm or a mite) that lives on or within another animal form (such as a chicken).

Parasites are spread by wild birds, rodents, and chickens. They can be brought into a flock on used feeders, waterers, nests, and other equipment that hasn't been thoroughly cleaned before being reused. Exactly which parasites are most likely to infect your flock depends in part on your management style and in part on the area of the country in which you live. Your local veterinarian or state Extension poultry specialist can tell you which parasites to watch out for.

Animal-form parasites are common among chickens, but not all pose a serious threat. A heavy infestation of even the most benign parasite, however, causes stress that increases a chicken's susceptibility to other infections. Internal parasites live inside a chicken's body, usually in some part of its digestive tract; the most common internal parasites that invade chickens are worms and protozoa. External parasites that live on or attack the outside of a chicken's body include mites; lice; and a host of fleas, flies, and other minor pests.

Worms and Worm Control

Worms are most likely to become a problem where chickens are kept on the same ground year after year. Two categories of worm infect chickens: roundworms and flatworms.

Roundworms (nematodes) are the most significant parasitic worm; several species afflict chickens, and they can do a great deal of damage. The most likely nematodes to infect a chicken are the following:

- *Cecal worms* — short worms that invade the ceca (two blind pouches attached to the intestine), causing either no signs or weight loss and weakness. Although the most common nematode, they rarely cause a serious problem.
- *Large roundworms (ascarids)* — long yellowish-white worms that invade the intestine, causing pale heads, droopiness, weight loss, diarrhea, and death. Most chickens become resistant to ascarids by 3 months of age.
- *Capillary worms* — hairlike worms that invade the crop and upper intestine, causing droopiness, weight loss, diarrhea, and sometimes death. When chickens sit around with their heads drawn in, capillary worms are the likely culprits.
- *Gapeworms* — red, fork-shaped worms that invade the windpipe, causing gasping, coughing, and head shaking (in an attempt to dislodge the worm). These parasites are quite serious in young birds, as they can cause death through strangulation.

Flatworms are far less common in chickens than are roundworms. Flatworms fall into two categories:

- *Tapeworms (cestodes)* — long, ribbonlike segmented white worms that invade the intestine, causing weakness, slow growth or weight loss, and sometimes death.
- *Flukes (trematodes)* — broad leaf-shaped worms that attach themselves either inside the body or beneath the skin. Flukes are a problem primarily in swampy areas and where sanitation is abysmal.

Life Cycles

To control parasites effectively, you have to know something about their life cycles. Some have a direct life cycle; others have an indirect life cycle.

In a direct life cycle, a female parasite inside a chicken's body sheds eggs that are expelled in the chicken's droppings. The infective parasite egg may then be eaten by the same chicken or by a different chicken. Assuming the parasite is a guest of the chicken it invades, the chicken becomes its host. A direct-cycle parasite goes directly from one host to another.

In an indirect life cycle, parasite eggs expelled in a chicken's droppings are eaten by some other creature, such as an ant, a grasshopper, or an earthworm. A chicken cannot become infected (or reinfected) by eating a parasite egg but rather becomes infected (or

DIRECT LIFE CYCLE OF ROUNDWORM

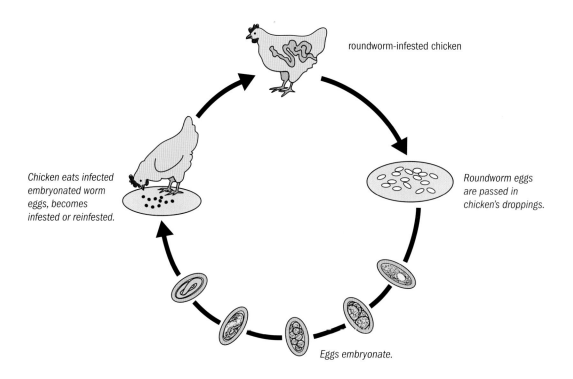

roundworm-infested chicken

Roundworm eggs are passed in chicken's droppings.

Eggs embryonate.

Chicken eats infected embryonated worm eggs, becomes infested or reinfested.

Worms with a direct life cycle live in a chicken's body and shed eggs that pass out with the chicken's droppings and are eaten by the same or another chicken to begin a new cycle.

reinfected) by eating a creature containing a parasite egg. Because the parasite goes from a chicken to some other host and back to a chicken, the other creature is called an intermediate host. Parasites requiring an intermediate host are said to have an indirect life cycle because they cannot infect one chicken directly after leaving another. Most roundworms and all tapeworms are indirect-cycle parasites.

Pastured chickens are most likely to be infected by indirect-cycle parasitic worms whose life cycles involve ants, earthworms, slugs, or snails. Litter-raised flocks are likely to be infected either by direct-cycle parasites or by indirect-cycle parasites involving beetles, cockroaches, or earthworms. Chickens housed in cages are likely to be infected by parasites whose life cycles involve insects that fly.

INDIRECT LIFE CYCLE OF ROUNDWORM OR TAPEWORM

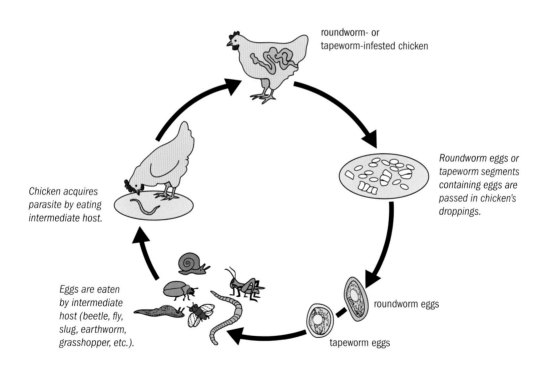

roundworm- or tapeworm-infested chicken

Roundworm eggs or tapeworm segments containing eggs are passed in chicken's droppings.

roundworm eggs

tapeworm eggs

Eggs are eaten by intermediate host (beetle, fly, slug, earthworm, grasshopper, etc.).

Chicken acquires parasite by eating intermediate host.

Worms with an indirect life cycle shed eggs that pass out with the chicken's droppings and are eaten by an intermediate host, which is then eaten by the same or another chicken to begin a new cycle.

PARASITIC WORMS AND THEIR HOSTS

Parasite	Intermediate Host
Ascarid	None (direct cycle)
Capillary	None or earthworm
Cecal	None or beetle, earwig, grasshopper
Fluke	Dragonfly, mayfly
Gapeworm	None or earthworm, slug, snail
Tapeworm	Ant, beetle, earthworm, slug, snail, termite

Worm Control

Under proper management, including good sanitation, chickens gradually develop resistance to parasitic worms. A bird that has the opportunity to acquire resistance through gradual exposure to the worms in its environment gets an unhealthy load only if it is seriously stressed, such as by crowding, unsanitary conditions, or the presence of some other disease. Treating your chickens without knowing whether or not they have worms is a waste of money and may actually be harmful because

- It interferes with the chickens' development of resistance to parasites.
- It can cause parasite populations to become drug resistant.

In contrast to deworming indiscriminately based on the biased advice of dealers who make a living selling poultry drugs, a more sensible approach involves submitting fecal samples periodically for examination, then developing a deworming schedule based on your flock's need. A veterinarian will do a fecal test for a few dollars. By taking fecal samples to your vet on a regular basis (perhaps every three months for a year), you can find out whether or not your chickens have a parasite problem and whether the problem varies in severity with the season. For example, most intermediate hosts proliferate during warm weather — becoming dormant in winter in northern areas — so you may need to deworm only once a year, in autumn. In a warm, humid climate, intermediate hosts thrive year-round, requiring more aggressive measures for controlling the intermediate hosts or deworming the chickens.

If you prefer to avoid drug use, sooner or later someone will tell you that the best way to keep chickens free of worms is to feed them diatomaceous earth (DE), sometimes called fossil flour, consisting of diatom fossils ground into an abrasive powder that supposedly shreds delicate worm bodies. But when diatomaceous earth gets wet (as it does when combined with a chicken's saliva), it softens and loses its cutting edge, and once it passes through the gizzard it's too ground up to retain any sharp edges. But feeding your chickens DE won't hurt, and it contains a large number of beneficial trace minerals, so go ahead if it makes you feel better.

Coccidiosis

Protozoa are simple and tiny creatures; some are harmless while others cause serious illness. The protozoa most likely to infect chickens are *coccidia*. *Coccidiosis* is the most common cause of death in young birds.

Although many different animals may be infected by coccidia, the species that infect chickens do not affect other kinds of livestock. The opposite is also true — chickens cannot get coccidiosis from other animals.

Coccidia have a direct life cycle. For each egg that hatches in a chicken's intestines, millions are later released in the bird's droppings. All ground-fed birds are exposed to coccidia throughout their lives, but a properly maintained pastured flock is less likely to become infected than birds living in crowded conditions, housed on damp litter, or allowed to drink water fouled with droppings.

Gradual exposure to coccidia allows a chicken to develop immunity. In young birds that are not yet immune, illness or death occurs when poor sanitation exposes them to too many coccidia too rapidly. Chicks raised in a wire cage and later moved to a coop with litter have had little exposure and therefore have no immunity and can become seriously infected. Even mature birds can become infected in hot, humid weather when coccidia proliferate more rapidly.

Nine different forms of coccidia are caused by nine separate species of protozoa, each invading a different part of the chicken's intestine. A chicken may be infected by more than one species at a time. Birds become immune in two ways: through gradual exposure or by surviving the illness. But they become immune only to the species occurring in their environment. Healthy chickens brought together from different sources may not all be immune to the same forms and may therefore transmit the disease to one another — with devastating consequences for the flock.

In young chickens the main signs of coccidiosis are slow growth and loose, watery, or off-color droppings. If blood appears in the droppings, the illness is serious — birds may survive but are unlikely to thrive. The disease may develop gradually, or bloody diarrhea and death may come on fast. In mature birds, the chief sign is a decrease in laying. Infected older birds that appear healthy shed billions of eggs that readily infect younger birds.

If you suspect coccidiosis, take a sample of fresh droppings to your veterinarian and ask for a fecal test to find out what kind of coccidia are involved and which medication to use. Not all anticoccidial medications work against all types of coccidia, and using the wrong drugs can do more harm than good.

Chicks reared early in the year while the weather is cool are in the best position to develop gradual immunity as the weather warms. If you raise chicks late in the season or in a year-round warm climate where coccidiosis is difficult to control, you may need to use a *coccidiostat* to prevent the condition, either by feeding a medicated starter ration or by lacing the drinking water with the coccidiostat. But medicate water with caution; chicks drink more in warm weather and can ingest a toxic dose.

SIGNS OF COMMON EXTERNAL PARASITES

Parasite	Signs
Red mites	Red or black specks crawling on skin at night, hiding in nests and woodwork cracks during the day; pale comb and wattles, weight loss, death among young birds and setting hens
Northern fowl mites	Red or black specks around vent or on eggs in nests, insect eggs along feather shafts; dirty-looking vent and tail, weight loss, drop in laying
Scaly leg mites	Swollen-looking shanks and toes with raised, crusty scales; mites are too small to see with the naked eye
Lice	Pale insects scurrying on skin, white eggs clumped at base of feathers; dirty-looking vent and tail area, weight loss, reduced laying and fertility

Mites

Several species of mite survive by consuming the skin, feathers, or blood of a chicken. Mites cause irritation, feather damage, increased appetite, low egg production, reduced fertility, retarded growth, and sometimes death.

Red mites, also known as chicken mites, are active in warm climates during summer and are especially attracted to broody hens. They invade chickens primarily at night, when they appear as tiny red or black specks on a bird's body. After feeding, they crawl along roosts to find a place to hide during the day. Adult mites can live for many months in unoccupied nests or housing. Control red mites by thoroughly cleaning the coop and dusting every crack and crevice with an insecticide approved for poultry.

Northern fowl mites are active in cool climates during winter, causing scabby skin and darkened feathers around the vent. You'll know your chickens have this mite if you see tiny specks crawling on eggs in nests or on birds during the day. They may be repelled from nests by using cedar chips as nesting material, especially for vulnerable setting hens. Since these mites increase rapidly, if you spot them, act fast by dusting birds and nests with an approved insecticide. Northern fowl mites cannot live long in an unoccupied chicken house, which will be free of them after about three weeks.

Scaly leg mites burrow under the scales of a chicken's legs, making the scales stick out and causing the miserable bird to walk around stiff legged. Since leg mites travel slowly from one bird to another, they may be controlled easily by brushing perches monthly with a mixture of one part kerosene to two parts linseed oil (not motor oil) and at the same time coating each chicken's legs with petroleum jelly (Vaseline).

Older birds are more likely to be infected than younger birds and are difficult to treat because the mites burrow deeply. Soak the bird's legs in warm soapy water, and gently scrape or rub off the softened dead scales, then smother the mites by coating the legs with petroleum jelly daily for two weeks or until the mites are gone.

Lice

A chicken with a properly shaped beak can minimize lice on its body by grooming; a chicken that has been debeaked or has an overgrown beak is more likely to have lice because it cannot groom properly. Lice chew on a chicken, causing the bird to break off or pull out its feathers trying to stop the irritation. The resulting damage makes the plumage look dull or rough. Louse-infested chickens don't lay well and have reduced fertility.

Lice spread through contact with an infested bird or its feathers and live their entire lives on a bird's body. You can easily see the straw-colored pests scurrying around on a chicken's skin, the

Northern fowl mites feed 24/7, leaving blackish debris and scabby skin around the vent.

scabby dirty areas they create around the vent and tail, and louse eggs (called nits) clumped in masses around the feather shafts.

If you spot signs of lice on one of your chickens, chances are they all have lice. An effective way to reduce the louse population is to diligently rake up and remove nit-laden feathers from the house and yard. Treatment involves applying a delousing product approved for poultry to the shelter walls, roosts, nest boxes, floors, and chickens. This treatment won't kill nits, so it must be repeated two more times at seven-day intervals to zap lice that hatch between times.

Flies

Although flies are not technically parasites, their larvae can be. Flies that bother chickens

EXTERNAL PARASITE ZAPPERS

A number of options are available for dealing with external parasites. Some of them are applied to individual chickens; others are applied to the housing. Some measures are environmentally friendly; others are not. When external parasites get out of control, sometimes a combination of methods is needed to solve the problem.

Dust baths in dry soil or fine road dust, and the preening that follows, help a chicken rid its feathers of irritating parasites. For further control, old-time poultry keepers lace dustbins with wood ashes, diatomaceous earth (DE), or lime-and-sulfur garden powder. But chickens are highly suscep-tible to respiratory problems, and breathing in these foreign materials can make matters worse. Adding DE to the dust bath as a general practice is not a good idea, but for chickens that are already infested, the benefits of DE (if you're inclined to use it) may well outweigh the dangers.

Pet shampoo and flea dip make excellent parasite-control options, especially if you wash your chickens for exhibition anyway. On the rare occasion when I acquire a new adult bird, I bathe it with pet shampoo before turning it out.

Linseed oil or another natural oil applied to cleaned roosts, nests, and cracks in walls or floors is a messy but effective way to rid housing of parasites that spend part of their time off a bird's body. Take care when using oil, as it can cause a fire hazard in a wooden building.

Systemic inhibitors permeate a chicken's entire system, making it unappealing to external as well as internal parasites. Ivermectin (trade name, Ivomec) is an over-the-counter cattle dewormer that is not approved for poultry but is sometimes used anyway — dropped or squirted into a chicken's mouth at the rate of 0.25 cc per adult chicken, 5 to 7 drops per mature bantam — but should never be used in a flock raised for meat or eggs. Overuse can result in resistant parasites, and an excessive dose is toxic to chickens.

Insecticide may be your only recourse for a serious parasite infestation. The list of insecti-cides approved for poultry is short and changes often, so check with your Extension poultry spe-cialist or veterinarian for the latest information. Never use a nonapproved product on chickens raised for meat or eggs. Even an approved insecticide is toxic and must be handled with care, so read labels and follow all precautions.

fall into two categories: biting flies and filth flies. Filth flies don't actually attack chickens the way biting flies do, but both spread disease.

Biting flies are found primarily around bodies of water. The main ones that bother chickens are blackflies (also known as buffalo gnats or turkey gnats) and biting gnats (also known as midges, no-see-ums, punkies, and sand flies), the bites of which cause irritation. These flies are difficult to control, so your best bet is to keep chickens away from streams and stagnant water.

Filth flies, including the common housefly, don't bite, but they do transmit tapeworms (when eaten by a chicken) and spread diseases on their feet. Flies breed in damp litter and manure, so control involves keeping litter dry — fix leaky waterers and roofs, regrade to prevent runoff seepage around the foundation, and improve ventilation. Chickens help control flies by eating the flies and their larvae. If flies get out of control, avoid using insecticides, or you will end up with a resistant fly population. Instead, set out fly traps or a good-quality flypaper such as Sticky Roll, or introduce natural fly predators. Properly managed litter and manure accumulated over the summer will develop a natural population of fly predators.

Health, Disease, and Disease Resistance

Most backyard flocks kept for family meat and eggs, or just for fun, rarely experience serious illnesses. Your flock may not be so lucky if

USING FLY PREDATORS

Fly predators are tiny parasitic wasps that live on or near manure and other decaying matter, where they attack flies but do not sting or bite humans, chickens, or other animals. The female wasp seeks out fly pupae and deposits from one to a dozen eggs (depending on the wasp species) into each pupa. Moving from pupa to pupa, she lives just long enough to deposit all of her 50 to 100 eggs.

The eggs develop into wasp larvae that feed inside the host fly pupa, thereby killing it. In 14 to 28 days the mature wasps emerge, and the females begin searching out new host pupae in which to lay eggs.

Under favorable conditions, parasitic wasps will populate on their own, but you also may purchase them from biological pest control suppliers. Because different species of predating wasp favor different species of fly, some suppliers offer combinations of wasps — including *Muscidifurax raptor, M. zaraptor, Spalangia cameroni, S. endius, S. nigra,* and *S. nigroaenea* — that together attack more than one species of fly, typically the house fly (*Musca domestica*) and the barn fly (*Stomoxys calcitrans*).

The best time to begin releasing parasitic wasps is in mid- to late May. Periodic releases until mid-August are recommended so flies won't overwhelm the wasps' ability to control them. Based on your location and setup, many suppliers offer guidelines on how many wasps you need and how often to release them. To assess their effectiveness, collect fly pupae from different spots, put them in a jar, and wait to see whether flies or wasps emerge.

other flocks are housed nearby, if you habitually buy or trade chickens, if you regularly show your chickens, or if you run a small-scale commercial operation requiring a regular turnover in your poultry population. If any of these situations pertain to you, you'd be wise to get a comprehensive health-care guide such as *The Chicken Health Handbook*, which covers the large number of diseases — some mild, some devastating — to which chickens are susceptible.

As a chicken ages, its state of health changes in two opposing ways:

- It develops resistance to some diseases in its environment.
- It becomes susceptible to other diseases that take significant time to develop.

From a health perspective, the older a breeding chicken gets, the more valuable it becomes, since by its longevity it demonstrates a certain hardiness that's likely to be inherited by its offspring. From a production standpoint, however, the older a chicken gets the less economical it becomes, with the profit curve dropping ever more sharply as long-term diseases take their toll. For the latter reason, many poultry owners won't keep a chicken more than one or two years, three at the most.

Because commercial producers can't afford the economic loss resulting from a serious disease, and because mixing birds of different ages increases the likelihood of disease, commercial folks follow the "all in, all out" procedure — they keep a meat or egg flock for the most productive part of its life, then remove the whole flock and thoroughly clean the housing before bringing in another flock. Although the natural life of a chicken is about 12 years — with individuals occasionally living

as long as 25 years — rare is the chicken that survives beyond its productive life of about two years.

Well Chickens

The best way to nip a potential disease problem in the bud is to be constantly aware of your flock's state of health. Each time you visit your chickens, take a moment to look around. You'll readily spot problems in the making if you become fully familiar with these characteristics.

Appearance. Healthy chickens are perky and alert. They have full, waxy combs, shiny feathers, and bright eyes.

Activity. Healthy chickens are active. They peck, scratch, dust, preen, or meander almost constantly, except on hot days, when they rest in the shade.

Sound. Well chickens talk and sing throughout the day. To detect atypical sounds, whistle softly whenever you approach your flock. Out of curiosity, your chickens will stop their activities to listen, giving you a chance to hear coughs, sneezes, and other sounds of distress. Listening while they are roosting at night is another good time to hear atypical sounds.

Production. Each flock develops a characteristic laying pattern, based in part on the hens' breed and age and in part on your management style. Keep records to determine your hens' average rate of lay, as well as the seasonal variations you can come to expect. Familiarize yourself with the typical sizes and shapes of your hens' eggs. An inexplicable decline in production or in egg quality (thin or wrinkled shells, thin whites, and so forth) may be a sign of disease.

Eating and drinking. The amount of feed and water a chicken consumes each day depends on its age, size, and production level, as well as on the weather. By paying attention to how much your chickens normally eat and drink, you'll readily notice changes brought on by stress or disease.

Weight. Healthy young chickens gain weight steadily. Healthy mature chickens hold their weight except for possible slight drops during the breeding season or resulting from the stress of exhibition. Any inexplicable loss of weight or failure to gain can be a danger signal.

Odor. A healthy flock has a characteristic odor. Become aware of that odor so you can detect subtle changes that may result from an outbreak of respiratory or intestinal disorders.

Droppings. Chickens expel two different kinds of droppings. Regular intestinal droppings are firm, grayish brown, and capped with white urine salts. Approximately every tenth dropping comes from blind pouches in the intestine, called ceca, where cellulose is digested by fermentation. Cecal droppings tend to be somewhat foamy, smellier than regular droppings, and light brown or sometimes greenish in color. Any change in the normal odor or appearance of either kind of dropping is a pretty good indication of disease.

Chicken Diseases

Infectious organisms of one sort or another are always in the environment. Many of them don't cause disease unless a flock is stressed by such

things as crowding, unsanitary conditions, or changes in feed. Microscopic organisms are spread through the air, soil, and water, as well as through contact with diseased chickens. They may be carried from flock to flock on the feet, fur, or feathers of other animals, especially rodents and wild birds, and on equipment, human clothing (particularly shoes), and vehicle tires.

The disease most likely to strike a specific group of chickens is influenced by the flock's purpose. Exhibition birds are most likely to get a respiratory disease such as laryngotracheitis that spreads through the air. Broilers are most likely to experience diseases related to nutrition and rapid growth — *ascites* (fluid accumulation in the abdominal cavity) or leg weaknesses. A breeder flock is more likely to experience a disease requiring long-term development, such as tuberculosis.

Each disease has unique signs by which it may be identified. Each disease also shares some signs with other diseases. Reduced egg production is often the first general sign of any disease, soon accompanied by depression, listlessness, hunching, hanging of the head, dull or ruffled feathers, loss of appetite, and weight loss. Other signs may be grouped according to the body system affected, as detailed below.

Enteric diseases affect the digestive system and are characterized by loose or bloody droppings, weakness, loss of appetite,

CAUSES OF DISEASE	
Infectious	**Noninfectious**
Bacteria	Chemical poisoning
Mold and fungi	Hereditary defects
Parasites	Nutritional deficiencies
Viruses	Unknown causes

increased thirst, dehydration, and weight loss in mature birds or slow growth in young birds. Enteric diseases include campylobacteriosis, canker, coccidiosis, colibacillosis (*E. coli* infection), necrotic enteritis, salmonellosis (typhoid, paratyphoid, pullorum), thrush, ulcerative enteritis, and internal parasites (parasitic worms).

Respiratory diseases invade a bird's breathing apparatus and are characterized by labored breathing, coughing, sneezing, sniffling, gasping, and runny eyes and nose. Respiratory diseases include coryza, the pulmonary form of fowl cholera, chronic respiratory disease, bronchitis, gapeworm, influenza, laryngotracheitis, Newcastle disease, and wet pox.

Nervous disorders affect the nervous system and are characterized by loss of coordination, trembling, twitching, staggering, circling, neck twisting, convulsions, and paralysis. Typical nervous disorders are botulism and Marek's disease.

Septicemia occurs when any infection reaches the bloodstream. Signs include weakness, listlessness, lack of appetite, chills, fever, dark or purplish head, prostration, and death. Many diseases have the ability to become septicemic — most notably cholera, colibacillosis, and streptococcosis (strep infection).

Acute septicemia hits a bird so fast it literally drops in its tracks. Since most septicemic diseases cause reduced appetite and loss of weight before death, the classic indication of acute septicemia is the sudden death of an apparently healthy bird that has a full crop and is in good flesh.

Not all causes of death are septicemic, however. Death may result from, among other things, degeneration of the intestine due to an enteric disease, blocking off of the airways due to a respiratory disease, inability to eat or breathe due to paralysis caused by a nervous disease, lack of adequate feed or water, or poisoning.

Occasionally finding a dead chicken does not necessarily mean some terrible disease is sweeping through your chickens. Normal mortality among chickens is 5 percent per year. If you find several chickens dead within a short time, however, you have good reason for concern.

Bird Flu

Avian influenza, or bird flu, is included here not to cause panic (too much of that has gone on already) but to provide information so you can reassure family, friends, and yourself the next time the news media try to stir up a panic. Bird flu has been around for many centuries, during which time numerous strains of the virus have evolved. They are divided into two groups: low pathogenicity and high pathogenicity. The *low-path bird flu* strain is somewhat common in the United States, and in most cases causes minor or no signs in chickens and poses little health threat to humans.

High-path bird flu is the one reported in the news as affecting chickens in Asia and Eastern Europe, where chickens roam freely and people have extensive, direct contact with their birds. High-path strains spread more rapidly than low-path strains and are more likely to be fatal to chickens. High-path bird flu has been detected in chickens — and eradicated — in the United States three times: in 1924, in 1983, and in 2004. No humans are known to have become ill in connection with these outbreaks.

A virus is defined as "high path" if it kills 6 out of 10 chicks inoculated in the

PROBIOTICS

The small intestine of a healthy chicken (and also of a healthy human) is populated with a number of beneficial bacteria and yeasts, called *intestinal flora* or *microflora*, that aid digestion and also produce antibacterial compounds and enzymes that stimulate the immune system. If for some reason these good guys get out of balance, disease-causing microbes take over and cause an enteric disease.

A chick acquires some microflora through the egg and gains more from the environment, particularly from properly composting litter.

Microflora are naturally present in certain foods, including grains, meats, and fermented milk products such as yogurt and kefir. A varied and well-balanced diet therefore keeps a chicken's population of intestinal flora strong and healthy.

Chicks raised on improper rations or in a poor environment may not develop microflora fast enough to ward off disease-causing microbes. The use of antibiotics and other antimicrobials in any chicken kills both disease-causing microbes and beneficial microflora.

The once common practice in the commercial poultry industry was to feed chickens antimicrobials to stimulate growth and improve health. But their overuse has contributed to the development of strains of disease-causing microbes that resist antibiotic treatment. So industry has turned to feeding chickens beneficial bacteria, called *probiotics,* to replace the antibiotics that have been misused for too many years, and now poultry suppliers offer probiotic formulations for chickens.

Backyard chickens raised entirely on wire and lacking opportunities to peck out some of their sustenance from the environment may benefit from a probiotic. Any chicken that has been treated with an antibiotic or subjected to extreme stress may also benefit from a probiotic. Chickens that eat a varied diet or are free to peck out some of their sustenance from the environment typically do not need a supplemental probiotic.

Microflora prefer a pH range of 5.5 to 7; disease-causing microbes prefer a pH range of 7.5 to 9. During times of stress vinegar added to the drinking water at a rate of 1 tablespoon per gallon (15 mL per 4 L) — double the dose if the water is alkaline — reduces the pH in the crop to encourage microflora to flourish there, ensuring they make it to the gut to keep the chicken healthy.

laboratory. The high-path strain most often discussed in the news, H5N1, killed 10 out of 10 inoculated chicks.

If you're interested in knowing what the numbers and letters mean, this is it: Bird flu viruses are classified by a combination of two groups of proteins, the hemagglutinin, or H proteins, and neuraminidase, or N proteins. The 16 H proteins are identified as H1 through H16, and the nine N proteins are N1 through N9. These two groups combine to form 144 different strains of the bird flu virus, each of which is identified by the two proteins they display. The H5N1 strain has the fifth hemagglutinin and the first neuraminidase.

H5N1 does not normally affect humans, although since 1997, when it first appeared, sporadic human cases of a serious respiratory infection have occurred during bird flu outbreaks, in which people who became infected had extensive, direct contact with sick birds. Although about half of these people died, the total of human deaths worldwide from bird flu since 1997 remains less than three hundred, compared to annual deaths of about five hundred thousand worldwide for your average run-of-the-mill human flu. Many more people who did not get sick after contact with infected chickens developed antibodies to the virus.

A vaccine has been created to protect chickens against the known high-path strains, for use only in the event of an outbreak to create a buffer zone around diseased flocks and prevent spreading of the flu while infected chickens are being destroyed. But bird flu viruses frequently mutate, just like the human flu viruses. And just as your flu shot doesn't always give you immunity against the flu strain that's going around, existing bird flu vaccines

may not protect chickens against future bird flu strains.

Are your chickens likely to get bird flu? Probably not. The most common way avian flu spreads from chicken to chicken is through direct contact with droppings from infected birds and droppings on contaminated equipment and shoes. Avian influenza is more likely to occur in industrialized factory farms where thousands of chickens are irresponsibly crowded into unsanitary environments than in a small flock enjoying the benefits of sunshine and fresh air.

Are you likely to get bird flu? Probably not. Even if the virus appears again among factory-farmed chickens in this country, it does not readily mutate into a form that spreads from human to human.

Treating Diseases

Some diseases may be effectively treated if you catch them in the early stages and make a positive identification so you'll know precisely what treatment to use. A few illnesses have characteristic signs. With a little knowledge, you can easily identify these diseases yourself. Many diseases, however, have such similar signs that the only way you can get a positive identification, or diagnosis, is by taking a few dead or sick birds to a pathology laboratory for analysis. The path lab won't tell you how to treat the disease, though. For information on treatment, you'll have to take the path report to your regular vet or to your state Extension poultry veterinarian or Extension poultry specialist.

Develop a policy in advance for dealing with disease outbreak. That way, if you ever have an illness in your flock, you'll be ready to act quickly. The moment you suspect a chicken is sick, isolate it from the rest of the flock; then make a choice: either humanely kill the bird and dispose of the body according to local health regulations (burn or deeply bury are two options), or try to find out what's ailing it and begin active treatment. Before any problems occur and while you are unemotional, determine in advance exactly how much time and money you are willing to spend nursing sick chickens.

When you are developing your disease-response plan, keep in mind that avoiding diseases is usually more effective and less frustrating than treating diseases, for these reasons:

- By the time you notice a chicken is sick, chances are it's too far gone to be treated effectively.
- Keeping an unhealthy bird carries the risk that the afflicting disease will spread to others in your flock.
- Chickens that survive a disease rarely reach their highest potential as layers, breeders, or show birds.
- A chicken that gets a disease, even if it fully recovers, will pass its lack of resistance to its offspring.
- A chicken that fully recovers may become a carrier, continuing to spread the disease without showing signs.
- Some diseases are so serious the only way to stop them is to dispose of the entire flock and start over.

Vaccination

Vaccination is a measure you may or may not need. Newly hatched chicks acquire a certain amount of natural immunity from the hen via the egg and continue to acquire new immunities as they grow to maturity. Your chickens may need additional help to develop immunities against diseases commonly encountered in their environment.

Ask your veterinarian or state Extension poultry specialist to help you work out a vaccination program based on diseases occurring in your area. Vaccinate your flock only against diseases your birds have a reasonable risk of getting, including past diseases chickens have experienced on your place or new diseases that pose a serious threat in your area. Do not vaccinate against diseases that do not endanger your flock.

The most common diseases for which vaccines are given to small flocks are listed below. They are all viral diseases.

Fowl pox, unrelated to chickenpox in humans, causes scabby skin, fever, and loss of appetite. It is spread by blood-sucking insects and through injuries resulting from peck-order fighting. This vaccine should be used only if your flock, or your area, has a problem and then all your chickens must be vaccinated and yearly booster shots given.

Infectious bronchitis is one of several viruses that cause coldlike signs, including coughing, runny nose, and swollen eyes. It is so contagious it can spread through the air from one chicken to another 1,000 yards (915 m) away. The signs are similar to Newcastle disease, and the two vaccines are often combined.

Laryngotracheitis is often spread at poultry shows and causes chickens to cough and gasp. It is similar to infectious bronchitis but spreads less readily and is more severe. The vaccine should be used only if your flock, or your area, has a problem. It may be used to prevent the spread of infection, but once you use it you must vaccinate all your chickens now and in the future and give yearly booster shots.

Marek's disease causes leg paralysis and droopy wings and sometimes death. Chickens with this disease shed the virus whether or not they show signs and thereby contaminate the yard for all future chickens. Vaccination does not prevent chickens from becoming infected and shedding the virus, but it does prevent paralysis. Chicks must be vaccinated before being exposed to the virus, and therefore most hatcheries offer to vaccinate chicks before shipping them.

Newcastle disease is brought into the United States by illegally imported birds and spread through droppings on used equipment and on people's shoes. It is easily confused with other respiratory infections. A combination vaccine is available to prevent both Newcastle and infectious bronchitis.

If you wish to administer two or more vaccines that don't normally come in combination, check with the supplier or a veterinarian to determine if the two are compatible. Some vaccines interfere with the effectiveness of others when used at the same time.

Acquiring chickens from different sources and at different times, as well as taking chickens to shows and bringing them back home, increases their risk of getting a disease and spreading it to your other chickens. Because some diseases are easily transmitted from bird to bird, and because recovered birds may become carriers that transmit the disease without showing signs of illness, some states require exhibitors to vaccinate against certain diseases. Before you decide to show your chickens, find out what the local regulations are and determine if you feel comfortable complying.

A vaccine confers immunity only if it is fresh, has been stored and handled properly, and is given exactly as directed on the label. A vaccine transported in a cooler and administered properly should work. A vaccine left in a hot car all day likely won't work no matter how well it's administered. The chickens also must be receptive to being immunized, which they will be if the weather is temperate and the chickens are properly fed and housed, have no internal parasites, and are otherwise in top health.

Vaccines usually come in enough doses for five hundred or a thousand chickens — much too many for a backyard flock. But even if you don't use it all, the cost is low compared to the cost of losing precious chickens in the event of a disease outbreak. Save money by coordinating with neighboring chicken owners to share the vaccine and expense.

SAMPLE VACCINATION SCHEDULE*

Vaccine	Age to Administer
Marek's disease	1 day
Newcastle disease/infectious bronchitis	2 weeks
Repeat at	6 weeks
And at	16 weeks
Laryngotracheitis	6 weeks
Fowl pox	10 weeks

*This sample schedule is offered as an illustration only; develop your own vaccination schedule based on specific conditions prevalent in your area.

Poisons and Other Hazards

Poisoning is relatively unusual in backyard poultry, especially if you use common sense by keeping your chickens away from pesticides, herbicides, rodenticides, fungicide-treated seed (intended for planting), wood preservatives, rock salt, antifreeze, and other known toxins.

Poisoning may be the result of misguided management. Common sense tells you not to put mothballs in your hens' nests in an effort to repel lice and mites, since naphthalene is toxic. And common sense tells you not to spray for cockroaches or other pests where your chickens might eat the poisoned insects. But keeping man-made toxins away from chickens may not be enough to keep them safe. The environment contains plenty of potential poisons.

Natural Toxins

Poisons aren't found only in jugs and boxes in your garage or barn supply closet. They may also be organisms that live and sometimes thrive in and around your flock's home.

Poisonous plants. Some weeds found in a pasture can be toxic, but should not be a problem if your flock has plenty of other good stuff to eat. Most toxic plants don't taste good and therefore are not tempting to eat, except to a starving bird. Some mushrooms are toxic, too, but delicate mushrooms have a hard time getting a foothold where poultry are active and scratching about.

Since birds peck here and there to get a variety in their diet, if they do get a bite or two of a toxic leaf or seed, it's unlikely to create a problem. Even if a bird does get a potentially toxic dose, the effect depends on the bird's age and state of health. And whether or not a specific plant is toxic at any given time

POTENTIAL POULTRY POISONS	
Poison	**Source**
Carbon monoxide	Carrying chickens in trunk of car
Copper sulfate	Antifungal treatment
Ethylene glycol	Spilled antifreeze
Lead	Paint or orchard spray
Mercury	Disinfectant or fungicide
Rock salt	Deicing sidewalks

often varies with its stage of maturity, growing conditions (such as drought), and other environmental factors.

Botulism. Another toxin in the environment is botulism. The organism that causes botulism naturally lives in soil and commonly occurs in the intestines of chickens without causing disease. But when the *Clostridium botulinum* bacteria multiply in the carcass of a dead bird or other animal or in a rotting cabbage or other solid vegetable, they generate some of the world's most potent toxins. Chickens become poisoned after pecking at the rotting organic matter or the maggots that are feeding on it or after drinking water into which the rotting matter has fallen.

A botulism-poisoned bird gradually becomes paralyzed from the feet up. Initially the bird sits around or limps if you force it to move. As the paralysis progresses through its body, the wings droop and the neck goes limp, giving the condition its common name, limberneck. By the time the eyelids are paralyzed, the bird looks dead but continues to live until either its heart or respiratory system becomes paralyzed.

If the chicken isn't too far gone, you might bring it around with botulinum antitoxin available from a veterinarian or by using a laxative *flush* to absorb the toxins and speed up their journey through the intestines (see box). Prevent botulism by promptly removing any dead bird or other animal you may find in the yard and by sorting out rotting fruits or vegetables before feeding kitchen scraps to your chickens.

Toxins in Feed

Fungal poisoning can be the result of by-products generated in moldy feed or forage. A number of poisons, or *mycotoxins*, are produced by molds that grow naturally in grains, and some molds generate more than one kind of poison. Mycotoxin exposure increases a bird's need for vitamins, trace elements (especially selenium), and protein.

Poisoning is difficult to identify and diagnose, in part because the feed may contain more than one kind of mycotoxin. A positive diagnosis usually requires analysis of the feed to identify any fungi present. Backyard poultry keepers generally buy feed in small quantities and would most likely use up a given batch before thinking of having it analyzed. Once the contaminated feed is removed, birds usually recover.

To prevent mold from forming in stored feed, keep it away from humid conditions and use plastic containers rather than metal ones, which generate moisture by sweating. Never give your flock any feed that has gone moldy. If you discover you have bought a bag of moldy feed, take it back and insist on a refund.

Found Objects

Small objects carelessly tossed into a poultry yard can cause distress or death. Cigarette filters, for instance, can cause intestinal

---

FLUSHING

When a chicken suffers from food poisoning or an intestinal disease, you can hasten its recovery by flushing its system with a laxative that absorbs the toxins and removes them from the body. Although a solution of Epsom salts (magnesium sulfate) makes the best flush, birds don't like the taste and won't readily drink it, so they must be treated individually. If a number of birds are involved, or handling them would cause undue stress, use molasses in a complete flock flush. Flush only adult birds, however — never chicks.

Epsom salt flush: 1 teaspoon (5 mL) Epsom salts in ½ cup (118 mL) water, poured or squirted down the bird's throat twice daily for two to three days, or until the bird recovers.

Molasses flush: 1 pint (0.5 L) molasses per 5 gallons (19 L) water, left in the drinker for no longer than eight hours.

---

obstruction. Small shiny objects like nails, pop-tops, and bits of glass or wire attract pecking. When eaten, they may simply irritate the bird and cause depression or they may result in a blockage that interferes with digestion or cause an internal tear that becomes infected.

Prevent such possibilities by meticulously picking up foreign objects you might find in your chicken yard. Remind visitors not to toss cigarette butts and other debris on the ground.

First Aid

Even the healthiest chicken can be injured — by a predator, by another chicken, or by something it does. The two most common types of injuries are skin wounds and broken legs or toes. Assembling a first-aid kit in advance of any incident or injury will let you act swiftly in such situations.

Wounds

The most common causes of bleeding wounds are a rooster's treading on a hen while mating; mauling by a dog or other predator; and getting snagged on a protruding nail, stiff wire, or other barnyard hazard. Peck-order fighting may also result in wounds, but they are rarely deep or serious.

The first thing to do is clean out the wound to assess the damage, prevent infection, and promote healing. If the chicken is tame, you might handle it without restraint. If it is not tame, wrap it in a towel to keep it from struggling and causing further injury to itself or you.

To clean out the wound, use a sterile wash so you don't introduce additional germs. A saline-solution wound wash is available where first-aid products are sold. You can make a saline solution by boiling a quart of water (1 L) and adding 2 teaspoons (10 mL) of noniodized or kosher salt. Be sure to let it cool before using it. In a pinch, you can use bottled or distilled water that hasn't been opened.

Pour the saline solution or clean water over the wound, and dab it dry with gauze pads. Never rub; just dab.

For a deep or really dirty wound, squirt the saline or clean water into the wound, using the pressure to wash out imbedded debris. Saline first-aid wash may have a pressure applicator. Otherwise you'll need a syringe without the needle to squirt the solution. The foaming action of hydrogen peroxide may also be used to lift out debris, but use it only if you absolutely have to, as it is harsh on delicate tissue. Pick out any stubborn debris with a pair of tweezers.

Once the wound is clean, disinfect it by pouring or gently squirting Betadine into it. Then let the wound air dry before dressing it. While it dries, check for feathers that might cling to the wound and inhibit healing. Clip back or pluck out feathers growing at the edge of the wound or tending to hang into it.

Unless the wound is in a place where the chicken picks at it, leave it open, and keep an eye on it to make sure it stays clean. For the first five days, apply a lanolin cream or a thick ointment such as Ichthammol to keep the healing tissue soft and elastic. After five days the wound should be well on the way to healing. If the chicken picks at the wound, cover it with a gauze pad and first-aid tape, and fashion a body wrap with tape to hold the dressing in place and keep the chicken from picking it off. Exactly how to body-wrap the tape depends entirely on the location of the wound. Regularly change the dressing, and examine the wound to make sure it is healing properly and doesn't get infected.

If the wound becomes infected, you'll need to remove the scab so you can treat the underlying infection. An infected wound might drain pus or a murky fluid, develop a yellow scab or a scab that gets larger as time goes by, become increasingly tender to the touch, become surrounded by increasing redness, or fail to heal in two weeks or less. Repeatedly coat such a wound with a thick ointment such as Desitin or Ichthammol for a few days until the scab softens and comes loose. Clean out the infection and re-treat the wound as if it were fresh.

If the wound is really deep or is the result of an animal bite, you'll need an antibiotic to

FIRST-AID KIT

In my barn, I keep a jug of clean water and some Ivory liquid soap, along with several towels folded in a recycled Styrofoam cooler that keeps out dust, bugs, and rodents. I also have a hospital cage handy in case a recovering chicken needs a place to rest, eat, and drink without being pecked or otherwise bothered by other chickens. I used to keep a fully stocked first-aid kit, but I rarely needed to use it, and more often than not the contents expired or otherwise got old and useless by the time I did need them. Since we live in a remote area, our family maintains a well-stocked first-aid kit for ourselves, and it includes many of the same items that would be needed to treat an injured chicken. On the rare occasion when a chicken needs help, I raid our own first-aid kit.

If it makes you feel better to stock a first-aid kit for your flock, or if your chickens happen to be accident prone (which, believe it or not, does happen), your first-aid kit might include the following items:

- ☐ Saline-solution wound wash

- ☐ Hydrogen peroxide for flushing out really dirty wounds

- ☐ Gauze pads to mop out a cleaned wound

- ☐ Tweezers to pick dirt and debris out of a wound

- ☐ Povidone-iodine antiseptic, such as Betadine, for disinfecting wounds

- ☐ A syringe for squirting saline solution or Betadine into a wound

- ☐ Wound powder, such as Wonder Dust, to stop bleeding

- ☐ A thick ointment such as Desitin (zinc oxide) or Ichthammol to keep healing tissue soft and for removing infected scabs

- ☐ Tongue depressors, Popsicle sticks, lollipop sticks, stiff paper or cardboard, or short lengths of water hose for splinting broken legs

- ☐ Pipe cleaners for splinting broken toes

- ☐ Vet wrap or rolled gauze to cushion a splint

- ☐ First-aid tape, vet wrap, or shipping tape to hold splint in place

- ☐ Electrolyte powder to replenish electrolytes in stressed birds (see chapter 5, *Reducing Heat Stress*)

- ☐ Petroleum jelly, such as Vaseline, to protect combs from frostbite (see chapter 5, *Preventing Frostbite*)

- ☐ Water-based lubricant such as K-Y Jelly for treating an egg-bound pullet (see chapter 7, *Pullet Problems*)

- ☐ Anti-inflammatory cream, such as hydrocortisone, for treating prolapse (see chapter 7, *Pullet Problems*)

- ☐ Sandwich or snack-size plastic zip bags for collecting suspicious droppings to take to a vet for examination

- ☐ Paper towels to wipe up whatever needs wiping

- ☐ Old towels to wrap and restrain a chicken that requires treatment

- ☐ A clean container to hold everything

prevent infection, which requires a visit with a veterinarian. A wound that's deep enough or wide enough to require stitching is best attended to by a veterinarian or another person experienced with sterile needles, sutures, and numbing medication and who can determine whether the wound needs a drain tube. However, lots of chicken keepers have successfully stitched up wounds with a sharp sewing needle and thread. A wound that needs stitching is generally one that goes beneath the skin, is too wide to be easily held closed, and occurs in a part of the chicken's body that stretches whenever the bird moves.

Depending on how seriously the chicken is injured, it may go into shock, and the more you handle it, the deeper into shock the bird sinks. The best treatments against shock are working gently, then letting the chicken recover in a warm, quiet, stress-free environment. Clamping a low-watt bulb at one side of the hospital cage allows the chicken to maintain its own comfort level by moving closer to or away from the heat.

Since flies spread bacteria on their feet, keep your recovering patient where flies can't get to the wound. During the chicken's recovery, isolate it from other chickens and provide adequate feed and fresh water. A recovering chicken that is disinclined to eat may be encouraged by frequent feeding or stirring of the feed and might be tempted with a variety of the treats it likes best. Vinegar added to the water at the rate of 1 tablespoon per gallon (15 mL per 4 L) will make the water more appealing to the chicken and encourage it to drink.

Broken Bones

A chicken can break a leg or toe bone by getting its leg trapped or wedged somewhere and struggling to get out, by jumping onto a hard surface from too great a height, or by being stepped on by a horse. The first thing to do is make sure the leg is really broken. If you did not see the action that caused it, consider that a broken bone may not be the problem. Some diseases cause paralysis and lameness. Among fast-growing broilers, lameness is typically a nutritional issue.

Broken toe. A suddenly crooked toe is easily recognized and probably broken, but treating a broken toe is not so easy. Without treatment, it may mend on its own but will remain crooked. If you decide to apply a splint, the chicken will pick persistently at the splint and eventually tear it loose.

Fashioning a comfortable splint is also problematic; one option is to use a pipe cleaner, available from a craft or hobby store, bent into the shape of the chicken's foot. Cushion the

BROKEN TOENAIL OR SPUR CASE

A broken or ripped-off toenail or spur case can bleed profusely. Several good puffs of a wound powder, such as Wonder Dust, will usually reduce or stop the bleeding. In a pinch, possible substitutes include flour or cornstarch.

When the bleeding lessens, hold a gauze pad to the wound with gentle pressure until the bleeding stops, which shouldn't take more than about 10 minutes. Treat the wound with an antiseptic, and isolate the bird until it heals.

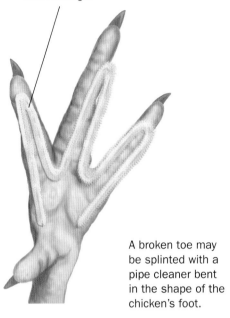

Shaped pipe cleaner goes under foot to hold toes out straight.

A broken toe may be splinted with a pipe cleaner bent in the shape of the chicken's foot.

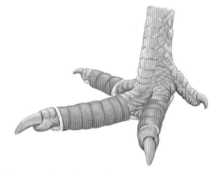

Hold the pipe cleaner in place with Vetrap.

sure sign of a broken bone. Don't try to set the bone, however, as setting it could worsen the injury. Sometimes the best thing to do is let the leg or foot mend on its own. It may grow crooked, but in most cases the bird will get along just fine, although it won't win any prizes at a show.

If the chicken can stand and walk and a splint seems necessary, sometimes just encasing the broken leg with a soft cast made of Vetrap will sufficiently do the trick. Start at the top and work downward, wrapping the leg until you get to the foot. If the break is near the bottom of the leg and no foot bone is broken, cut the ends of the Vetrap and wrap them between the toes to anchor the cast in place.

If a sturdier cast is necessary, add an outer layer of stiff paper or cardboard to create a tube around the leg, and anchor it with first-aid tape. A short piece of irrigation hose also makes a dandy cast. Slit one side, carefully slip it over the wrapped leg, and tape it in place.

A serious leg break may benefit from splinting. A stiff splint must be applied with care, and checked often, to make sure it doesn't slip and press against the chicken's thigh or foot. You also have to be careful not to wrap a leg so tightly that the wrap cuts off circulation to the foot.

Begin by wrapping the leg with gauze or Vetrap to create a cushion. Then apply splints cut to a length suitable for the bird's size — not so long they poke into flesh. Splints may be made of Popsicle sticks, lollipop sticks, tongue depressors, or anything similar. Apply two, to opposite sides of the leg, and anchor them in place with first-aid tape or Vetrap. For a chicken that persists in picking at and removing the wrap, shipping tape, such as Scotch 3M, usually will work. When checking

toes with strips of Vetrap or gauze, place the splint at the bottom of the foot, and wrap a little more Vetrap around the toes to keep them straight. Take care not to make the cast so tight or thick the bird can't walk properly; apply just enough wrap to cushion and protect the foot.

Broken leg or foot. A leg or foot that is twisted out of its normal position is a pretty

the dressing, don't try to pull the shipping tape loose; instead, cut it with scissors to avoid reinjuring the bird's leg.

If the leg persists in twisting despite the cast, you may be able to get it to grow straight by keeping weight off the leg while it mends. Fashion a sling or hammock by nailing together a wooden frame and securely stapling soft netting to the top with two holes for the chicken's legs to go through. When the chicken rests in the sling, the leg holes should not bind and the feet should not touch the ground. Provide feed and water within easy reach. After the bone mends, the chicken's legs will be stiff until it exercises its muscles enough to walk properly. Depending on the severity of the break, the chicken should be able to get around without a splint after about two weeks. While the leg is in the process of mending, change the cast or splint often to make sure nothing has gone wrong. Watch especially for swelling, infection, and broken skin that may signal the need for professional veterinary care.

While the leg is mending, keep the bird isolated and make sure it has proper feed and fresh water. If the hospital cage has a perch, remove the perch so the chicken isn't tempted to reinjure its leg jumping down from the perch. If the bird appears in shock — the skin turns pale, the pulse is rapid but weak, breathing is rapid, and the bird becomes weak or prostrate — keep it in a quiet place and provide the warmth of a low-watt lightbulb at one side of the hospital cage.

Treating a broken leg requires lots of time and patience, and you can't always be sure of the outcome. One of my neighbors splinted a rooster's leg after the bird had been mauled by her dog. Despite the finest nursing a broken-legged chicken could hope for, the cock eventually lost his leg. Even so, he got around amazingly well hopping on one leg, but he was no longer able to breed.

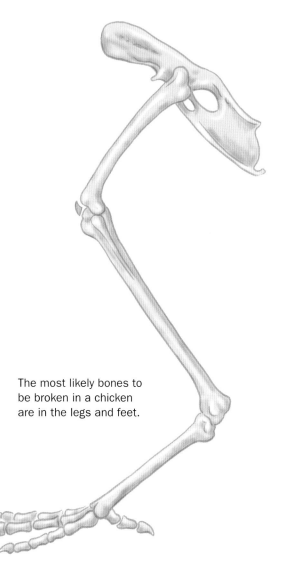

The most likely bones to be broken in a chicken are in the legs and feet.

A sling keeps weight off the broken
leg to prevent twisting while
it mends.

7

LAYING-HEN MANAGEMENT

Each female chick starts life carrying the beginnings of thousands of undeveloped yolks inside her body, not all of which develop into eggs. Ever since chickens were first domesticated, poultry keepers have worked to coax the greatest number of eggs from their hens through selective breeding, improved nutrition, and optimal layer management. Even so, a hen rarely lays more than a small percentage of the total number of eggs she started out with. By keeping your hens healthy and happy, you will be rewarded with the maximum number of eggs they are capable of laying.

Egg Formation

A pullet starts life with two ovaries, but as she grows, the right ovary remains undeveloped and only the left one becomes fully functional. The functioning ovary contains all the undeveloped yolks the pullet had when she was born. If you ever have occasion to examine a hen's innards, you'll find them in a cluster along her backbone, approximately halfway between her neck and tail. Depending on the age of the hen and her stage of lay, the yolks range from head-of-a-pin size to nearly full size. In a pullet or nonlayer,

they are all small because none are developing in preparation for laying the next egg.

When a pullet reaches laying age, one by one the yolks mature, so at any given time her body contains yolks at various stages of development. Approximately every 25 hours, one yolk is mature enough to be released into the funnel of the oviduct, a process called *ovulation*. Ovulation usually occurs within an hour after the previous egg was laid. During the yolk's journey through the oviduct, it is fertilized (if sperm are present), encased in various layers of egg white, wrapped in protective membranes, sealed within a shell, and finally enveloped in a fast-drying fluid coating called the bloom, or cuticle.

Throughout its passage through the oviduct, the egg leads with its pointed end. Just before it is laid, the egg rotates so the blunt end comes out first. The whole process takes about 25 hours, causing a hen to lay her egg about an hour later each day. Since a hen's reproductive system slows down during the night, eventually she'll skip a day altogether and start a new multiple-day laying cycle the following morning.

Providing suitable nest boxes ensures that your hens will deposit their eggs where you can find them.

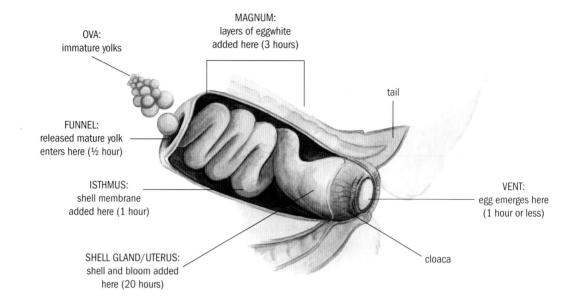

OVA:
immature yolks

MAGNUM:
layers of eggwhite
added here (3 hours)

tail

FUNNEL:
released mature yolk
enters here (½ hour)

ISTHMUS:
shell membrane
added here (1 hour)

VENT:
egg emerges here
(1 hour or less)

SHELL GLAND/UTERUS:
shell and bloom added
here (20 hours)

cloaca

An egg takes about 25 hours to develop in the oviduct, the average
length of which is 25 inches (63.5 cm).

Rate of Lay

The group of eggs laid within one *laying cycle* (which may vary in length from 12 days to nearly a year) is called a *clutch*. Some hens take more time than normal (say, 26 hours) between eggs and therefore lay fewer eggs per clutch than a hen that lays every 25 hours. Conversely, some hens lay closer to every 24 hours and so lay more eggs per clutch. Many production hens are bred to have the shortest possible interval between eggs and therefore lay as many eggs as possible per clutch.

The best heavy-breed hens in peak production lay about 40 eggs in a clutch; a Leghorn lays closer to 80. In 1979, a strain of superior Leghorns developed at the University of Missouri averaged more than one egg per day per hen. One of the hens laid 371 eggs in 364 days and continued laying an egg a day for a total of 448 days without a break.

The Leghorn normally begins laying between the ages of 18 and 22 weeks and averages between 250 and 280 white-shelled eggs during the first year. A commercial hybrid brown-egg layer lays somewhat fewer eggs. The brown-egg hybrids called Red Sex Links average 250 to 260 eggs per year; trade names for this hybrid include Cinnamon Queen, Golden Comet, Gold Star, and Red Star. The brown-egg hybrid called a Black Sex Link, or Black Star, averages about 240 eggs per year, but her eggs are larger than those of the Red Sex Link. The Black Sex Link is gentler than the Red and quicker to mature, beginning to lay at about 16 weeks of age. Although she eats a little more than the Red, at the end of her productive life she weighs 5 pounds (2.3 kg) — a good ¾ pound (0.3 kg) more than the Red — making the Black something of a meat bird as well as a layer.

A laying breed that puts more energy into growing muscle than the commercial strains is considered *dual purpose*, meaning it is suitable for someone who keeps chickens for meat as well as for eggs (although, because these chickens neither lay as well as layer breeds nor grow as well as broiler breeds, some people consider them to be unsuitable for either purpose). These hens generally start laying at the age of 24 to 26 weeks. You won't get as many eggs from the best dual-purpose breed as from a commercial strain, and you'll get fewer eggs still from a dual-purpose strain bred for show than from a strain of the same breed bred for egg production.

All pullets lay small eggs when they first start out, and they lay only one egg every three or four days. But by the time they are 30 weeks old, their eggs will reach normal size, and they will average two eggs every three days; during the spring peak good layers should average five or six eggs a week per hen. At about 18 months, they'll take a break to molt. After the molt they'll lay bigger eggs than before, but not quite as many. As hens age the pattern continues, and each year they lay fewer eggs than the prior year.

Aside from the hen's breed and age, her rate of lay is affected by external factors, including temperature and light. Hens lay best when the temperature is between 45 and 80°F (7 and 27°C). When the weather gets much colder or much warmer than that, production slows down.

Most hens stop laying in winter, not because the weather turns cold, but because daylight hours are shorter in winter than in summer. When the number of daylight hours falls below 14, hens may stop laying until spring.

A healthy hen should lay for a good 10 to 12 years. Occasionally, you'll hear of a biddy laying to the ripe old age of 20, by which time she'd be doing well to pump out an egg a week. Most layers don't manage to live that long. Instead, by the tender age of 3 they have succumbed to the firing squad or falling ax, to be replaced by more efficient fresh, young pullets.

Egg Size

The size of a hen's eggs depends on her breed and age. As a hen gets older, her eggs typically get bigger. The United States Department of Agriculture publishes standards for eggs in six sizes, according to minimum weight per dozen

It is not abnormal for a hen occasionally to produce an unusually small egg.

eggs, to account for slight variations from one egg to the next.

On a per-egg basis, the smallest size is peewee, which weighs about 1.25 ounces (35 g), and the largest is jumbo, which weighs about 2.5 ounces (70 g). Periodically, some poultry keeper will report finding a chicken egg weighing 6 ounces or more (170 g). The heaviest chicken egg on record weighed one pound (0.5 kg).

Egg size is influenced by a hen's weight. A pullet's eggs are usually quite small, but the size continues to increase until the bird is about 12 months old. Pullets that start laying during summer usually lay smaller eggs than pullets that start laying during the cooler months of fall or winter. Pullets that are underweight when they start laying will continue to lay smaller eggs than birds that mature properly.

Hens of any age and weight may temporarily lay small eggs if they're suffering from stress induced by heat, crowding, or poor nutrition, including inadequate protein or salt. Among same-age hens of a given breed or strain, the majority of eggs should be of one size, with only an occasional egg ranging just above or below the majority.

Shell Color

Out of the 25 hours most hens need to lay an egg from start to finish, encasing each egg within a shell takes about 20 hours. Colored

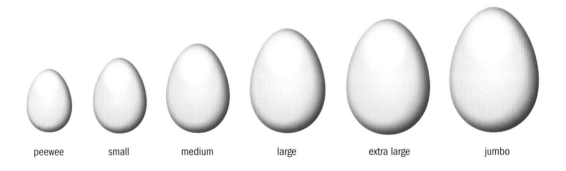

peewee small medium large extra large jumbo

EGG SIZES					
	Weight per Dozen			**Weight per Egg**	
Size	**Ounces**	**Grams**		**Ounces**	**Grams**
Peewee	15	425		1.25	35
Small	18	510		1.50	43
Medium	21	595		1.75	50
Large	24	680		2.00	57
Extra Large	27	765		2.25	64
Jumbo	30	850		2.50	70

Each breed lays eggs of a specific shell color, and each hen's eggs are typically always the same shade. If you have hens of different breeds, you can easily tell which hen laid each egg.

shells are the result of pigments added during shell formation.

Brown-egg layers produce eggs of varying shades ranging from barely tinted to nearly black, thanks to more than a dozen different genes that influence shell color. Most of the pigment of a brown-shell egg is deposited in the *bloom*, the last layer added to the outside of an egg just before it is laid. When you break open a brown-shell egg, the inside of the shell will be paler than the outside or nearly white. Bloom dissolves when wet, and easily rubs off when dry, which explains why cleaning a brown-shell egg removes some of the color.

By contrast, the pigment of a blue-shell egg is deposited throughout the shell, so it is just as blue on the inside as on the outside. Green eggs result from crossing a blue-egg layer with a brown-egg layer, giving you blue-shell eggs with a brown coating. The many different shades laid by so-called Easter egg chickens result from blue shells coated with different shades of brown bloom.

All Asiatics and most Americans (except Holland) lay brown eggs. All Mediterraneans lay white eggs. Other classes are a mixed bag as to white, brown, tinted (pale brown to the point of pinkish), or blue.

As a general rule, hens with white earlobes lay white eggs, and hens with red earlobes lay brown eggs. Exceptions are Crevecoeur, Dorking, Redcap, and Sumatra, which have red earlobes but lay white-shell eggs; Araucana and Ameraucana, which have red earlobes but lay blue eggs; and Penedesenca, which have white earlobes but lay the darkest brown egg of any breed.

Although variations in hue exist within each breed or strain, an individual hen typically lays eggs of a specific color. The use of certain drugs, especially coccidiostats, causes paler or even white-shell eggs. Viral diseases that infect the reproductive system also cause hens to lay pale eggs.

Stress can make brown-shell hens lay eggs that are a lighter-than-usual color. Overcrowded nests, rough handling, loud noises, and anything that makes a hen nervous or fearful can cause her to either lay her egg prematurely, before the brown-bloom coating is completed, or retain the egg long enough to add an extra layer of shell on top of the bloom. An inexplicable lightening of brown-shell eggs is a good indication something is bothering your hens.

As a brown-egg layer ages, her eggs naturally get paler. No one knows exactly why, although one possibility is that as a hen ages and her eggs get larger, the brown-pigmented bloom is spread over a larger surface area. But this explanation does not account for why the tapered end of the egg gets lighter than the rounded end.

SHELL COLOR

White-Egg Layers		Brown-Egg Layers	
Ancona*	**Leghorn**	Aseel	*Langshan*
Appenzeller	**Minorca**	*Australorp*	Malay
Blue Andalusian	Modern Game*	*Barnevelder*	Maran
Buckeye	**Norwegian Jaerhon**	Black Sex Link	*Naked Neck (Turken)*
Campine*	Old English*	Brahma	*New Hampshire*
*Catalana**	Phoenix*	Buckeye	Orloff
Crevecoeur	Polish	*Chantecler*	Orpington
Cubalaya*	*Redcap*	Cochin	*Penedesenca*
Dorking	*Shamo**	Cornish	*Plymouth Rock*
Fayoumi*	Sicilian Buttercup	Delaware	Red Sex Link
Hamburg*	Silkie	*Dominique*	*Rhode Island Red*
Holland	Sultan	*Faverolle*	*Rhode Island White*
Houdan	Sumatra*	Frizzle#	*Sussex*
Kraienkoppe*	White-Faced Spanish	*Java*	*Welsumer*
La Fleche	Yokohama	Jersey Giant	*Wyandotte*
Lakenvelder*			
Blue-Egg Layers			
Ameraucana			
Araucana			

Breeds in bold are superior layers.   *Shells may be lightly tinted.
Breeds in italics are dual purpose.   #The genetic condition of frizzledness can occur in white-egg-laying breeds.

Layer Nutrition

The nutritional needs of laying hens evolve from the time they hatch as chicks. How you feed your pullets will affect their future egg production. And once they mature into laying hens, their dietary needs continue to change with their level of production and with the season.

Feeding Pullets

Nearly all of the protein required for egg production comes from a flock's rations. Pullets that are not in good flesh when they start laying cannot obtain enough dietary protein to continue growing to maturity and to lay eggs. As a result, they lose weight and produce small eggs, as well as too few of them. On the other hand, pullets that grow too rapidly and are fat when they start to lay also tend to lay fewer, smaller eggs and are more likely to have trouble laying them.

Therefore, start chicks with enough protein to give them a good beginning, then back off a bit to give them time to grow without getting fat before they start laying. Restricted feeding is one method of reducing protein intake, but because the pullets spend less time eating, they have more time to get bored and nervous; they may pick on one another and otherwise make each other miserable.

A better way to reduce protein is to continue feeding their ration free choice, but use the Pearson's square method described in chapter 4 to gradually add oats to the ration. Start changing the protein when the pullets are 8 weeks of age until you achieve a level of between 14 and 16 percent.

As your pullets reach the age when they will start laying, their need for protein goes up. Leghorn-type breeds start laying at about 20 weeks, other breeds at 22 to 24 weeks. So when your Leghorn-type pullets reach about 18 weeks (20 weeks for other breeds), gradually reduce the oats until the protein level is between 16 and 18 percent. Do not feed a layer ration to pullets before they reach the age of lay, the higher calcium content may interfere with bone formation and result in weak legs, kidney damage, and possibly death.

To ensure that your pullets get the nutrients their bodies need as they come into production, switch to a complete layer ration by the time they start to lay. By then, you also should discontinue any medicated feed you may have used to avoid problems such as abnormal eggs and drug residues in eggs.

Each lightweight pullet of a commercial layer strain will eat about 15 pounds (6.75 kg) of feed before she starts laying at about 20 weeks of age. Breeds that take a few weeks longer to reach laying age may eat double that amount. The age at which pullets lay their first eggs — and the amount of feed needed to bring them to that point — will vary with the specific breed and strain you choose.

Feeding Hens

Each lightweight layer eats about 4 pounds (1.8 kg) of ration for every dozen eggs she lays, which works out to between 4 and 4½ ounces (124 and 140 g) of feed per hen per day, or just less than 2 pounds (0.9 kg) per hen per week. Dual-purpose hens eat a bit more (about ⅓ pound or 150 g per day), bantams a bit less. Feed your hens free choice, and they'll eat as much as they need.

Since chickens eat to meet their energy needs, expect your layers to eat less in summer than in winter, when they need extra energy to stay warm. If their summer ration contains the same amount of protein as their winter ration, they'll get less total protein in summer and therefore won't lay as well. In warm climates, in addition to regular 16 percent lay ration, some feed stores offer a ration containing 18

percent or more protein for use when high temperatures cause hens to eat too little.

In winter, when the weather is colder and the days are shorter, hens may not get enough to eat to both maintain body heat and continue laying well. One solution is to light their housing to increase daylight hours and give them more time to eat. Another solution is to supplement rations with milk and/or scratch grains. If your hens are losing weight, the better option is to let them take a rest from laying; they will naturally resume egg production in the spring.

Commercial lay ration is supposed to be a complete feed that provides all of a hen's nutritional needs. It comes in either pelleted or crumbled form. Crumbles contain the same ingredients as pellets, since they are simply pellets that have been crushed into smaller pieces. Hens fed pellets waste less feed than those fed crumbles, but they also eat faster and therefore have more time for mischief.

Supplementing rations with table scraps and surplus milk products will increase nutritional variety and may reduce the cost of egg production, but may also reduce egg production itself, unless you take care to maintain dietary balance. Above all, avoid feeding your hens too much scratch, or their energy-protein balance will be thrown way off, they'll get fat, and they'll lay poorly. A good rule of thumb is to feed hens no more grain each day than they can finish within 20 minutes. Pastured hens aren't quite as touchy as confined hens when it comes to grain, since they burn off extra energy while foraging.

Pastured Layer Diets

Pastured hens eat green plants, seeds, and insects, a diet that makes their eggs more nutritious and gives the yolks a darker color.

OMEGA-3s

Omega-3 fatty acids, affectionately known as omega-3s, are polyunsaturated fatty acids required for human growth, development, and good vision; deficiency has been linked to heart disease, cancer, Alzheimer's disease, and other devastating illnesses. Omega-3s are abundant in the green leaves of some plants, as well as in certain oils, nuts, fish, and flaxseed.

Pastured hens lay eggs with yolks that are high in omega-3s.

Although the total fat content remains the same as if the hens were not on pasture, the percentage of polyunsaturated fat increases. As an alternative to pasture, omega-3 content may be boosted in eggs by adjusting the hens' diet to include 10 percent flaxseed. Besides its high omega-3 content, flaxseed is high in protein and a large number of vitamins and minerals; feeding too much, however, can cause eggs to taste fishy.

The National Academy of Sciences recommended daily requirement for omega-3s (1,600 milligrams for an adult male, 1,100 milligrams for an adult female) may be met by eating two meals of fish or six omega-3-enriched eggs per week (or one fish meal and three omega-3 eggs per week). For people who don't like fish or are allergic to it, omega-3 eggs offer a tasty alternative.

Hens with access to grass put more vitamins and omega-3 fatty acids into their eggs than hens without access to pasture, and hens on a pasture with legumes lay more nutritious eggs than hens on grass alone. At times when pasture growth is poor, flaxseed in the ration also increases the eggs' omega-3 content.

Pasturing your layer flock requires a fair amount of land. Although you can range as many as a hundred chickens on a quarter acre (0.1 ha), you need to move them frequently, so the total amount of land required is much higher. Pastured layers are therefore best worked into a pasture-rotation scheme involving other livestock such as goats, sheep, or cattle.

The range house should provide at least 1 square foot (0.1 sq m) of space per hen and should have nests that are equally accessible to birds from the inside and to you from the outside. Move the shelter as often as necessary to keep the chickens on fresh greens and prevent them from grazing in their own droppings. To optimize feed-cost savings, move the shelter

SIGNS OF OBESITY

Signs that a laying hen is too fat include

- Low rate of lay
- Laying at night
- Poor shell quality
- Frequent multiple yolks
- Prolapse

as soon as you notice the layer ration starting to disappear faster. Avoid returning the flock to the same ground twice within a single year.

Pasture plant growth is generally poor during the heat of summer and the coldest months of winter. To maintain your hens' green forage level during these times, feed them *fines* (the bits of leaves that accumulate in a livestock hay feeder) from alfalfa or lespedeza hay. Sprouted grains are another source of good nutrition when pasture growth is poor.

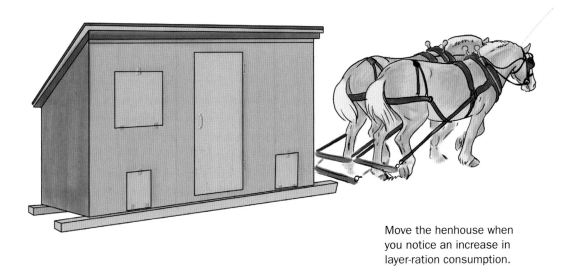

Move the henhouse when you notice an increase in layer-ration consumption.

THE CALCIUM CONNECTION

An egg's shell is made of calcium carbonate. A hen gets some calcium from her diet and draws the rest from her own bones, but the older a hen gets, the thinner the shells of her eggs become. Brittle and broken feathers also indicate the need for more calcium.

Layer rations contain 2.5 to 3.5 percent calcium, enough to meet a pullet's needs if she eats nothing but prepared feed. Older hens, and all hens that eat grain or grass in addition to a layer ration, need supplemental calcium in the form of oyster shell, argonite, limestone, or soluble calcium grit offered free choice.

Hens can get some calcium by eating their own recycled eggshells. Wash the shells, dry them, and crush them before feeding them to your hens. Feed only shells that have been crushed, or you may give your hens the idea of breaking and eating their own eggs. Store and feed only dry shells; shells that are stored while moist will be full of harmful molds and bacteria.

Controlled Lighting

In the natural course of events, chicks hatch in spring, when daylight hours are increasing, and mature during summer and autumn, when daylight hours are decreasing. The following spring, when day length once again begins to increase, they start a new reproductive cycle. Hens continue to lay until either the number of light hours per day or the degree of light intensity signals the end of the reproductive cycle.

To get eggs during winter, you have to trick your hens into thinking the season remains right for reproduction, which you can do by using lights to compensate for decreasing amounts of natural daylight. The farther you live from the equator, where day length is constant, the bigger your seasonal swings in increasing and decreasing day length will be.

Augmenting Daylight

Start augmenting natural light when day length decreases to about 15 hours, which in most parts of the United States occurs in September. Continue the lighting program throughout the winter and into spring, until natural daylight is back up to 15 hours per day.

If you forget to turn the lights on for just one day, your hens may go into a molt and stop laying. To protect yourself from your own forgetfulness, use a timer to turn lights on slightly overlapping natural light. By setting the timer to go on for a few hours at the same time every morning, and again for a few hours in the evening, you can bracket the changing daylight hours to create a constant 15-hour day inside the henhouse.

For convenience, many people leave lights on all the time. Constant lighting has its downside, besides being wasteful — it encourages hens to spend more time indoors during the day stirring up litter dust, scratching in nests, and otherwise engaging in mischief. Round-the-clock lighting also doesn't give hens the 6 to 8 hours per 24 of darkness they need for rest to maintain their immune system.

In an effort to decrease the cost of lighting, you might be tempted to install fluorescent lights. Although fluorescent tubes are cheaper to run than incandescent bulbs, they're more expensive to install, touchier to operate in the dusty henhouse environment, and more difficult to regulate. To adjust the light intensity of fluorescent lights, you have to change the entire fixture; with incandescent lights you just switch to a bulb of different wattage. If in the face of ban-the-bulb realities, however, you wind up using fluorescent fixtures, be sure to use warm-wavelength lights (that produce an orange or reddish light), since cool-wavelength lights (such as those used in offices and households) do not stimulate the hens' reproductive cycle.

If your coop is not outfitted with electricity, you can provide lighting with 12-volt bulbs designed for recreational vehicles, powered by a battery connected to a solar recharger. Since a timer would drain too much power from your battery, 12-volt lights must either be left on all the time or manually turned on and off.

One 60-watt bulb, 7 feet (0.2 m) above the floor, provides enough light for about 200 square feet (18.5 sq m) of living space. Place the bulb in the center of the area to be lighted, preferably over feeders and away from the nesting area. If your shelter is so large you need more than one fixture, the distance between them should be no more than 1.5 times their

height above the birds. Arrange multiple fixtures to minimize shadows, except over the nesting area, which should remain darkened. Installing multiple lights has the advantage that if a bulb burns out, the hens won't be left in the dark.

A reflector behind each light increases its intensity, allowing you to use less wattage than you would otherwise need. In the example above, reflectors would let you substitute 40-watt bulbs for the 60-watt bulbs. Dust and cobwebs accumulating on bulbs decrease their light intensity. To maintain the effectiveness of your controlled-lighting program, dust the bulbs weekly and replace any that burn out.

Lighting for Pullets

Light affects not only the production of hens but also the sexual maturity of pullets — the age at which they begin laying, the number of eggs they lay, and the size of their eggs. Under normal circumstances, pullets mature during the season of decreasing day length. If you raise pullets in the off-season, increasing day length that normally triggers reproduction will speed up their maturity, more so the closer they get to laying age. Pullets that start laying before their bodies are ready will lay smaller eggs and fewer of them, and are more likely

MAXIMUM HEIGHT OF LIGHTS ABOVE BIRDS

| | Age 12–21 Weeks | | Age 21+ Weeks | |
Watts	With Reflector	No Reflector	With Reflector	No Reflector
15	5'	3.5'	3.5'	2.5'
25	6.5'	4.5'	4.5' 3'	
40	9'	6.5'	6.5'	4.5'
60	14'	10'	10' 7'	
75	15.5'	10.5'	10.5'	7.5'

Arrange fixtures to minimize shadows and space them no farther apart than 1.5 times their height above the birds.

to prolapse (described later in this chapter in *Pullet Problems*).

Pullets should be kept either on a constant 8- to 10-hour day or in decreasing light. Pullets hatched from April through July may be raised in natural light. Those hatched from August through March need controlled lighting to delay maturity.

Consult an almanac to determine how long the sun will be up on days occurring 24 weeks from the date of hatch. Add six hours to that day length, and start your pullet chicks under that amount of light (daylight and electric combined). Reduce the total lighting by 15 minutes each week, bringing your pullets to a 14-hour day by the time they start to lay. When they reach 24 weeks of age, add 30 minutes per week for two weeks to increase total day length to 15 hours.

DISCOURAGING BROODINESS

Broodiness, or a hen's instinct to hatch eggs, is triggered by increasing day length. A hen that's thinking of brooding may cluck like a mother hen while she's getting on or off the nest, and while she's on the nest will puff out her feathers, growl, and peck your hand if you reach under her for an egg. When a hen gets broody, her pituitary gland releases *prolactin*, a hormone that causes her to stop laying.

If you keep hens primarily for eggs, or if you raise a rare or valuable breed and prefer to hatch as many of their eggs as possible in an incubator or under other hens, you may wish to discourage broodiness to keep your hens laying. Depending on how serious a hen is about setting, you can try to discourage her, or "break her up," using these techniques:

- Avoid letting eggs accumulate in the nest.

- Repeatedly remove the hen from her nest.

- Move or cover the nesting site so she can't get to it.

- Move the hen to different housing.

- Put the hen in a broody coop.

The function of a *broody coop* is exactly the opposite of what its name might imply. It consists of a hanging cage, with a wire or slat floor, where the hen is housed for as long as necessary to break her up — usually one to three days. The longer the hen has acted broody, the longer she'll take to start laying again. A hen that's broken up after the first day of brooding should begin laying in seven days; a hen that isn't broken up until the fourth day may not start laying for about 18 days.

Hens, like people, don't always react as you expect them to, and a persistent broody may continue no matter what you do. If you are as determined that a hen *not brood* as she is *to brood*, your only remaining option is to cull her. In an extreme case, a stubborn hen that insists on brooding, even with no eggs to hatch, may eventually starve to death and thus cull herself.

Distinguishing Layers from Liars

Within every breed or strain, some hens lay better than others. If your chickens are pets and you consider their eggs a bonus, you're not going to be as concerned about overall efficiency of egg production as the person who keeps hens primarily for their eggs. If eggs are your focus, removing inferior and unhealthy hens that don't lay well has several advantages:

- Reduces feed costs
- Removes potential sources for disease
- Increases available living space for productive hens
- Increases feed and water space for productive hens
- Improves a flock's overall laying average

Culling, or the elimination of so-called unthrifty hens, is an ongoing process involving the removal of injured or sick birds whenever you spot them. It's also a periodic process of determining whether or not each hen is laying, then determining how long she has been in or out of production.

The first time to cull heavily is when your flock reaches peak production, at about 30 weeks of age. A second culling time is toward the end of the first year of production — a good hen lays for at least 12 months; a lazy layer takes an early break. The worst time to cull is during the fall molt, when even your best hens will stop laying for a few weeks.

Culling Criteria

The first step in culling is to look at each hen's overall carriage. A high producer is active and alert. A low producer tends to be lazy and listless. Next look at details and compare:

- The feathers of a good layer are worn, dirty, and broken. Poor layers look sleek and shiny — more like show birds than working gals.
- The skin of a good layer is stretchy and bleached out. The skin of poor layers is tight and in yellow-skin breeds retains full color.
- The comb and wattles of a good layer are large, bright, and waxy. Candidates for culling have small combs and wattles.
- A good layer's legs are wide apart and set back, and its shanks are thin and flat. A poor layer's legs are set forward and close together — with less body capacity for egg-producing organs — and its shanks are round and full.
- The vent of a good layer is large, moist, and oval. Cull candidates have tight, dry, round vents.

The vent of a poor layer is tight and dry.

A good laying hen has a large, moist, oval vent.

- A good layer's abdomen should feel soft, round, and pliable under your hand — never small and hard. But be careful not to mistake an about-to-be-laid egg for a hard abdomen.
- Pelvic bones — the pair of pointy bones located between the keel and the vent — should have enough room between them for three or more fingers for most breeds, at least two fingers for small breeds. The space between the pelvic bones and keel should accommodate at least four fingers; the greater the distance, the better the layer. A lazy layer is tight and nonflexible in these two areas.

Of all these various indicators, the most reliable ones are a moist vent and flexible pelvic bones. A hen with a puckered vent and stiff or thick and inflexible pelvic bones is not laying, period.

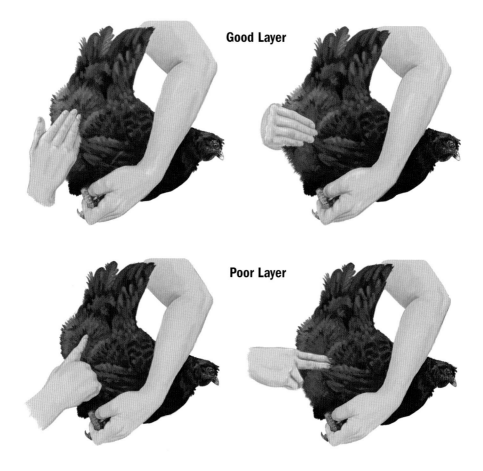

A good laying hen has pelvic bones that are flexible enough for you to put at least three fingers between them and a distance between the pelvic bones and keel of at least four fingers.

LAYER INDICATORS

Body Part	Good Layer	Poor Layer
Carriage	Active and alert	Lazy and listless
Eyes	Bright and sparkling	Dull and sunken
Comb and wattles	Large and bright	Shriveled and pale
Shanks	Thin and flat	Round and full
Back	Wide	Narrow or tapered
Abdomen	Deep and soft	Shallow and hard
Pelvic bones*	Wide apart, flexible	Tight and stiff
Vent*	Large and moist	Puckered and dry
Plumage	Worn, dry, dirty	Smooth, shiny, clean
Molt	Late and fast	Early or slow
Skin	Bleached and stretchy	Yellow and tight

*Most reliable indicators

Bleaching Sequence

Although some laying breeds have white skin, yellow skin is more common. The skin of a yellow-skin pullet contains a considerable amount of pigment, obtained from green feeds and yellow corn. Over time, she uses this pigment to color the yolks of her eggs. As time goes by and the hen's old yellow skin is replaced with new pale tissue, her body parts appear to bleach out (turn pale). After six months of intense laying, a high-producing yellow-skin hen will be completely bleached.

Pigment leaves the body parts in a certain order, based on how rapidly the skin of each part is renewed — the more rapidly the tissue is replaced, the more quickly it appears to bleach out. When a hen stops laying, color returns to her skin in the same order, approximately twice as fast as it disappeared. You can therefore estimate how long a yellow-skin hen has been laying, or how long ago she stopped laying, by the color of her various body parts.

The first color loss in the sequence is in the skin around the vent, which is renewed quite rapidly. Just a few days after a pullet starts to lay, her vent changes from yellow to pinkish, whitish, or bluish. Next to bleach is the eye ring. Within three weeks, the earlobes of Mediterranean breeds bleach out; hens with red lobes do not lose ear color.

The beak's color fades from the corner outward toward the tip, with the lower beak fading faster than the upper beak. In breeds that typically have dark upper beaks, such as Rhode Island Red and New Hampshire, only the lower beak is a good indicator.

The best indicators of long-term production are the shanks, since they bleach last. Color loss starts at the bottoms of the feet, moves to the front of the shanks, and gradually works upward to the hocks.

Molting

Short day lengths serve as a signal to birds that it's time to renew plumage in preparation for migration and the coming cold weather. Like most birds, chickens lose and replace their feathers at approximately one-year intervals.

BLEACHING SEQUENCE

Body Part Sequence	Eggs Laid	Average # of Weeks*
1. Vent	0–10	1–2
2. Eye ring	8–12	2–2½
3. Earlobe	10–15	2½–3
4. Beak	25–35	5–8
5. Bottoms of feet	50–60	8
6. Front of shank	90–100	10
7. Hock	120–140	16–24

*Individual bleaching time depends on a hen's size, her state of health, her rate of production, and the amount of pigment in her rations.

BLEACHING SEQUENCE

Pigment bleaches from the skin in this sequence (and restores itself in reverse order).

The process, called *molting*, occurs over a period of weeks, so a chicken never looks completely naked—although occasionally one comes close. Under natural circumstances, a chicken molts for 14 to 16 weeks during the late summer or early fall.

The best layers molt late and fast. They lay eggs for a year or more before molting — which is why your good layers look so ragged — and take only two to three months to finish the molt.

The poorest layers start early and molt slowly. They may lay for only a few months before going into a molt, and the molt may take as long as six months — which is why the plumage of lazy hens remains shiny and sleek. Culling slow molters is a good way to improve your flock's laying average. These hens are easy to identify because they start molting before September and drop their wings' primary flight feathers one at a time.

Although no one has found a direct connection between molting and laying, common sense tells you that during a molt, nutrients needed to produce eggs are channeled into producing plumage. As a result, most hens stop laying until the molt is complete, so culling during the fall molt on the basis of nonlaying is a bad idea —you could end up getting rid of your best hens.

Molting may occur out of season as a result of disease or stress, such as chilling or going without water or feed. A stress-induced molt is usually partial and does not always cause a drop in laying, while a normal full molt is typically accompanied by at least a slowdown.

Since feathers are 85 percent protein, a chicken's need for dietary protein increases during the annual molt. When your chickens are about to molt, their plumage will take on a dull look. A little supplemental animal protein will help them through it, as well as improve the plumage of show birds. Compared to the protein in grains, animal protein is rich in the amino acids a chicken needs during the molt. Animal protein can come from any of the following:

- High-quality cat food (not dog food, which often derives its protein from grains)
- Raw meat from a reliable source (not chicken; feeding an animal the meat of its own species is a good way to perpetuate diseases)
- Fish (but don't feed fish to a chicken you plan to eat anytime soon or the meat may taste fishy)
- Molting food, sold by pet stores for caged song birds (it's expensive but lets you avoid potentially toxic pet foods and bacteria-laden meats)
- Mashed scrambled or hard-cooked eggs
- Sprouted grains and seeds, particularly alfalfa and sesame seeds (sprouting improves the quantity and quality of the proteins)
- Mealworms
- Earthworms

The molting sequence. Molting occurs in a specific sequence, starting with the head and neck feathers. The newly emerging feathers are called pinfeathers, or sometimes blood feathers because they contain a supply of blood to nourish the growing feather. Blood feathers, especially around the tail and along the back, attract picking at a time when molting chickens crave additional protein, so watch for and deal with picking before it turns into a full-blown case of cannibalism. Once a feather is fully formed, the blood supply is cut off and no further growth occurs, so broken feathers stay broken until the next molt.

Each newly emerging feather is covered in a thin sheath of keratin that comes off the pinfeather when the chicken preens. You may not notice these sheaths, except on the chicken's head and neck, where the bird can't reach with its beak. Eventually, these sheaths fall off or get scratched off by a claw. A tame

MOLTING SEQUENCE

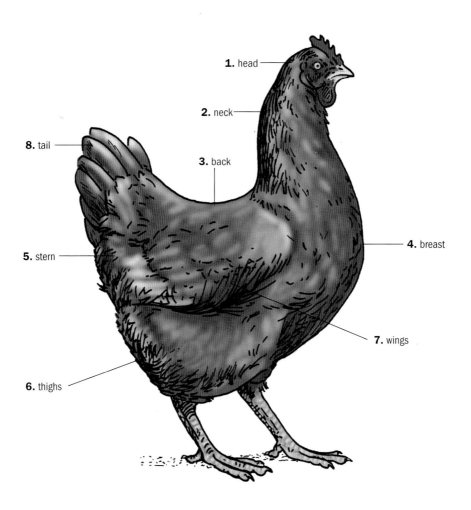

1. head
2. neck
3. back
4. breast
5. stern
6. thighs
7. wings
8. tail

In full molt, feathers renew in sequence, starting with the head and gradually working toward the tail, with some areas molting simultaneously.

CULLING BY THE MOLT

To determine how long a hen has been molting, and how long she will continue to molt, examine her primary, or flight, feathers — the largest feathers, running from the tip of the wing to the short axial feather that separates the primaries from the secondary feathers. The 10 primaries drop out at two-week intervals and take approximately six weeks to regrow.

The primaries of a slow molter drop out one by one, requiring up to 24 weeks for the wing to fully refeather. A fast molter drops more than one feather at a time.

Feathers that fall out as a group grow back as a group, letting the hen more quickly complete her molt and get back to laying eggs.

Fully feathered wing

Slow molt in which primaries drop out one by one. Feather #1 is fully regrown; #2 is 4 weeks old; #3 is 2 weeks old; #4 has just started. This wing has been molting for six weeks and will continue for 18 weeks more.

Fast molt in which primaries have dropped out in three groups. Feathers #1–3 are fully regrown; #4–7 are 4 weeks old; #8–10 are 2 weeks old. This wing has been molting for six weeks and will be finished in only four weeks more.

or pet chicken would appreciate your help removing the hard-to-reach keratin sheaths by gently scraping with your fingers, rubbing the feathers in the direction of their growth with a damp cloth, or lightly misting with a spray of plain warm water followed by gentle wiping with a dry terry cloth.

After your hens molt, their feed efficiency will improve, their eggs will be larger, and egg quality will be better than at the end of the previous laying period. On the other hand, they won't lay quite as well as they once did, and egg quality will decline faster than during the previous laying period.

Egg Issues

The number of eggs a hen lays and their size, shape, and internal quality — as well as shell color, texture, and strength — may be affected by a variety of things, including environmental stress, improper nutrition, medications, vaccinations, parasites, and disease.

Eggs Are Clues

Physical problems, developmental challenges, disease, infections, and everyday upsets your layer may be experiencing sometimes can be detected by closely examining her eggs.

Bloody shells sometimes appear when pullets start laying before their bodies are ready, causing tissue to tear. Other reasons for blood on shells include excess protein in the lay ration and coccidiosis, a disease that causes intestinal bleeding. Coccidiosis does not often infect mature birds, but if it does, you'll likely find bloody droppings as well as bloody shells.

Chalky shells or glassy shells occasionally appear due to a malfunction of the hen's shell-making process. Such an egg is less porous than a normal egg and will not hatch but is perfectly safe to eat.

Odd-shaped eggs or wrinkled eggs may be laid if a hen has been handled roughly or if for some reason her ovary releases two yolks within a few hours of each other, causing them to move through the oviduct close together. The second egg will have a thin, wrinkled shell that's flat toward the pointed end. If it bumps against the first egg, the shell may crack and mend back together before the egg is laid, causing a wrinkle.

Weird-looking eggs may be laid by old hens or maturing pullets that have been vaccinated for a respiratory disease. They may also result from a viral disease. Occasional variations in shape, sometimes seasonal, are normal. Since egg shape is inherited, expect to see family similarities. If you do your own hatching, select hatching eggs only of normal shape and size for your breed.

EGG SHAPE

Most chicken eggs have a rounded or blunt end and a more pointed end, although some eggs are nearly round, while others are more elongated. An egg's shape is established in the part of the oviduct called the *isthmus*, where the yolk and white are wrapped in shell membranes. An egg that for some reason gets laid after being enclosed in membranes, but before the shell is added, has the same shape as if it had a shell. Each hen lays eggs of a characteristic shape, so you can usually identify which hen laid a particular egg by its shape.

Thin shells may cover a pullet's first few eggs or the eggs of a hen that's getting on in age. In a pullet, thin shells occur because the pullet isn't yet fully geared up for egg production. In an old biddy, the same amount of (or less) shell material that once covered a small egg must now cover the larger egg laid by the older hen, stretching the shell into a thinner layer.

Shells are generally thicker and stronger in winter but thinner in warm weather, when hens pant. Panting cools a bird by evaporating body water, which in turn reduces carbon dioxide in the body, upsetting the bird's pH balance and causing a reduction in calcium mobilization. The result is thin-shelled eggs. Thin shells also may be due to a hereditary defect, imbalanced ration (too little calcium or too much phosphorus), or some disease — the most likely culprit being infectious bronchitis.

Soft shells or missing shells occur when a hen's shell-forming mechanism malfunctions or for some reason one of her eggs is rushed through and laid prematurely. Stress induced by fright or excitement can cause a hen to expel an egg before the shell is finished. A nutritional deficiency, especially of vitamin D or calcium, can cause soft shells. A laying hen's calcium needs are increased by age and by warm weather (when hens eat less and therefore get less calcium from their ration). Appropriate nutritional boosters include a calcium supplement offered free choice and vitamin AD&E powder added to drinking water three times a week.

Soft shells laid when production peaks in spring, and the occasional soft or missing shell, are nothing to worry about. If they persist, however, they may be a sign of a serious viral disease, especially when accompanied by a drop in production.

Broken shells often result when a thin or soft shell becomes damaged after the egg is laid. Even sound eggs may get broken in nests that are so low to the ground, the chickens are attracted to scratch or peck in them. Hens and cocks may deliberately break and eat eggs if they are bored or inadequately fed. Boredom may result from crowding or from rations that allow chickens to satisfy their nutritional needs too quickly, leaving them with nothing to do.

A shell may become wrinkled if for some reason it cracks before the hardening process is complete (left). An occasional misshapen egg (right) is no cause for concern, but a hen that typically lays odd-shaped eggs will pass the trait on to her offspring.

A chalky egg (right) occasionally appears as a quality-control glitch in a hen's reproductive system.

If your coop is small and well lit, discourage nonlaying activity in nests by hanging curtains in front to darken them. To allay a hen's suspicions about entering a curtained nest, either cut each curtain into hanging strips or temporarily pin up one corner until the hens get used to the curtains.

Hens may break eggs inadvertently. Such accidents commonly occur if nests contain insufficient litter, eggs are collected infrequently enough to pile up in nests, or nests are so few that two or more hens have to crowd into the same nest at the same time. Sometimes timid birds seek refuge by hiding in nests, and their activities may break previously laid eggs.

Too Few Eggs

Too few eggs being laid can result from so many different causes you practically have to be Sherlock Holmes to determine the reason. For starters, you may have the wrong hens. Although production varies among individuals, strains, and breeds, if you want to collect lots of eggs, you need hens that have been developed for egg production.

Even if you have a breed specifically developed for high egg production, you'll get few eggs if your hens are old. Most hens lay best during their first year, although a really good layer should do well for two years, or even three. As hens age, they generally tend to get fat — especially when fed too much grain by well-meaning keepers — which significantly impairs their ability to lay.

During the molt, all hens slow down in production, and some stop laying all together. Low production as a result of an out-of-season molt is a sure sign of stress. Stress itself, with or without an accompanying molt, can cause hens to slow down or stop laying.

You may get too few eggs if your hens hide their eggs where you can't find them or if a hen lays her eggs and then turns around and eats them. Egg eating, a form of cannibalism, is a management problem that usually starts when an egg gets broken in the nest. An egg eater won't necessarily come from within your flock; it may be a wily predator.

Improper nutrition can cause a drop in laying. Hens may get too little feed or may be fed rations containing too little carbohydrate, protein, or calcium. Imbalanced rations often result from feeding hens too much scratch or from failure to offer a free-choice calcium supplement when the diet includes grain or pasture. Low temperatures increase a chicken's requirement for carbohydrates, and unless rations are adjusted accordingly, low production may result.

Anything that causes hens to eat less than usual also causes them to lay less than usual. To encourage better laying, encourage eating by feeding more frequently or by simply stirring the ration between feedings. Insufficient water can cause a slump in eating, so make sure the drinkers remain filled.

Dehydration due to lack of water for even a few hours can cause hens to stop laying for days or weeks. A hen drinks a little at a time, but often, during the day. Her body contains more than 50 percent water, and an egg is 65 percent water. A hen, therefore, needs access to fresh drinking water at all times for her egg-laying apparatus to function properly. If she is deprived of water for 24 hours, she may take another 24 hours to recover. Deprived of water for 36 hours, she may go into a molt, followed by a long period of poor laying from which she may never bounce back.

Pullets and hens may suffer water deprivation if the water quality is poor or they don't like the taste — reason enough not to medicate water during hot weather. On the other hand, adding vinegar to the water at a rate of 1 tablespoon per gallon (15 mL per 4 L) — double the dose if your water is alkaline — encourages drinking; chickens seem to like the taste of vinegar. In winter you may provide plenty of water, but if it freezes, egg production will drop. In summer, deprivation occurs when water needs increase but the supply doesn't.

A nonlayer will drink 1 to 2 cups (0.25 to 0.5 L) of water each day. A layer drinks twice as much and in warm weather, may drink up to four times more than usual. During the summer put out extra waterers and keep them in the shade, and frequently bring your hens fresh, cool water. They will thank you by continuing to lay those wonderful eggs you prize them for.

Pullet Problems

A pullet that starts laying at too young an age, that is too fat or unhealthy when she starts laying, or that lays unusually large eggs, may experience egg binding or prolapse. Even pullets that start laying without any problems need to learn to put their eggs in the nests.

Egg binding occurs when a too-large egg gets stuck just inside the vent. It can be an extremely serious condition, especially if the pullet goes into shock. If the pullet does not remain bright and alert, take measures to keep her warm.

The first thing to do is to make sure the pullet is truly egg bound. If she's straining to release an egg, and you see the end of egg near the opening, then you know for certain. If you can't see the egg, you can verify egg binding by lubricating a finger with K-Y Jelly or other water-based lubricant and gently inserting it into the vent until you feel the hard shell with the end of your finger. Don't attempt to stretch the vent, as you may tear the pullet's delicate tissue.

Sometimes lubricating the vent area, and as much of the egg as you can reach with a finger, will aid its passage. Gently squirting in warm (not hot) saline-solution wound wash, or warm soapy water, may help get things moving.

Warming up the vent area may relax the muscles enough to release the egg. If the pullet is tame enough not to be frightened by being handled, moisten an old towel, warm it in the microwave (make sure it's not hot), and apply it to her bottom. Reheat the towel as needed to keep it warm, or better yet use two towels and warm them alternately, to maintain moist heat.

An alternative warming method is to put warm, not hot, water in a bucket or basin and stand the pullet in it with the water reaching just above her vent. After warming the pullet's bottom for about 15 minutes, give her a rest, and if she doesn't release the egg soon, try again.

If warmth therapy still doesn't work, maybe you can dislodge the egg. Again lubricate the vent and egg with K-Y Jelly and/or warm soapy water. Gently insert your lubricated finger to help maneuver the egg, while with your other hand push gently against the abdomen and try to work the egg out. Be careful here — you don't want to break the egg, which can cause internal injury.

If all else fails, you may need to collapse the shell to remove the egg. This maneuver is tricky and can injure the pullet unless you work slowly and carefully. First suck out the contents of the egg by piercing the shell with a needle at the end of a syringe. Use a large-bore needle, 18 or lower gauge, or emptying the shell will take forever.

Once the shell is empty of its yolk and white, try to collapse it while keeping the shards together. This part is the trickiest, as you must take great care to avoid injuring the pullet with a sharp shard. For this reason, don't squeeze the pullet's abdomen to crush the egg, but rather work gently with your fingers directly on the shell, or at least one or two fingers on the inside and the other hand gently pressing from the outside.

Using lots of warm saline or soapy water as a lubricant, carefully remove as much of the shell as possible, then rinse away remaining pieces with squirts of saline gentle enough not to wash the shell bits deeper. Don't worry about getting the last little bit; once the egg is out, the pullet is better off left to rest, and any bits left behind should come out on their own.

If tissue protrudes through the vent, treat the pullet as you would for prolapse.

Prolapse is a natural process by which eggs are laid. The oviduct's *shell gland*, or uterus, holds the egg tightly and prolapses, or turns itself inside out, through the cloaca (the chamber just inside the vent) to deposit the egg outside the vent, then withdraws back inside the hen. If an egg is too large, or a pullet is too immature to begin laying, the uterus may not retract back inside. Instead it remains prolapsed, a serious condition in which uterine tissue protrudes outside the vent.

If you catch prolapse right away, you may be able to reverse the situation by applying an anti-inflammatory cream, such as hydrocortisone — if necessary, gently pushing any protruding tissue back inside — and isolating the pullet until she heals. Unless you catch it with the cream in time, the exposed pink tissue will attract other birds to pick, and the pullet will eventually die from hemorrhage and shock. Prolapse that progresses to this stage is called *blowout* or *pickout*.

Prolapse may be largely avoided by ensuring — through seasonal hatching or the use of controlled lighting and proper nutrition — that your pullets don't start laying too young; a pullet that lays before her body is ready is more likely to prolapse. In a mature hen, prolapse is usually a sign of obesity.

Floor eggs, or eggs laid on the floor rather than in nests, are likely to occur in a flock of pullets just starting to lay. Floor eggs get dirty or cracked, making them unsafe to eat. A cracked egg is easily broken, encouraging birds to sample the contents, develop a taste for eggs, and thereafter become egg eaters.

To minimize floor eggs, prepare nests early so your pullets have time to get used to them before they start laying. Place the nests low to, or directly on, the floor until most of the pullets are using them, then raise the nests 18 to 20 inches (45 to 50 cm) above the floor (or have two sets of nests and close off the lower ones) to discourage the pullets from entering nests for reasons other than to lay.

If pullets continue laying on the floor, perhaps you have too few nests; provide at least one nest for every four layers. Or perhaps the nests get too much light, causing pullets to seek out darker corners for laying. Since the primary purpose of laying eggs is to produce chicks, layers have a deep-seated instinct to deposit their eggs in dark, protected places. A nest that is properly designed and located offers just such a place.

A nest egg, or fake egg, left in each nest shows a pullet the proper place to lay. When she sees an egg already in the nest, she says to herself, "Ah-ha, this must be a safe place for my own egg." Old golf balls make good nest eggs. You can find multicolored plastic toy eggs in stores around Easter time; wooden eggs, available year-round at hobby stores, are better than air-filled plastic eggs because they aren't easily bounced out of the nest.

The idea that nest eggs make hens lay more eggs is an old wives' tale. Here's what it does do: by encouraging a hen to lay in a nest, rather than hide her eggs elsewhere, the nest egg lets you find more of the eggs she does lay.

Dirty Eggs

When you wash an egg, you wash off the bloom, the function of which is to seal in moisture and seal out bacteria. A better option is to ensure eggs remain clean until you collect them. Producing clean eggs requires clean housing, clean nests, enough living space for the number of hens you have, and well-placed nests of a sufficient number — one for each four or five hens.

Dirty eggs in the nest are often the result of layers tracking mud or muck on their feet. To keep eggs from getting dirty, clean up the source of mud — most often a muddy entry or damp ground around a leaky waterer — and take measures to ensure the condition doesn't recur.

You may have to redesign your nests to discourage mud tracking. Eggs in nests located on or near the floor are more likely to get dirty than eggs in nests raised above the floor, especially if layers reach the nests by hopping up on a rail or series of rails. Since chickens like to roost on these rails, make sure none is close enough to a nest that a roosting chicken fills the nest with droppings. For most breeds, a rail no closer than 8 inches (20 cm) from the nest's edge should work.

Occasionally, eggs get dirty when birds low in the peck order hide in nests, soiling previously deposited eggs. Avoid crowding your flock, and provide enough environmental variety to allow timid birds to get away from bullies. Darkening the nesting area discourages activities other than egg laying.

Droppings in the nest are the result of activities other than laying, such as roosting on the edge of the nest, hiding in the nest, scratching in nesting material, and sleeping in the nest. Poop in the nest dirties any eggs that get laid there.

At a hen's cloaca, just inside the vent, the reproductive and excretory tracts meet, which means a chicken lays eggs and poops out of the same opening — but not at the same time. As an egg is pushed out into the world, the bottom end of the oviduct turns inside out, wrapping

NEST CLEANUP

Lining nests with shavings, shredded paper, chopped straw, or well-dried chemical-free lawn clippings helps keep eggs clean and protects them from getting cracked. But if an egg does get broken, or a hen poops in the nest, nesting material can get nasty fast, and subsequent eggs will be coated with bacteria.

Cleaning such a nest is a messy job that gets even worse after the mess sticks like glue to the nest bottom. Nest pads are designed to alleviate this situation and are available made of excelsior (wood fiber) or reusable plastic, but buying them adds to the expense of keeping chickens, and cleaning the plastic ones for reuse adds time and frustration to an already messy job.

Corrugated cardboard, cut to fit each nest bottom, works for the short term and is inexpensive to replace. To clean out a nest, fold the cardboard over the nesting material to remove most of the mess easily with the cardboard. Unless the cardboard is replaced often, though, the hens eventually wear or peck a hole through the middle.

Asphalt shingles, the most commonly used roofing material, make ideal nest liners. Cut to fit the nest, and topped with a generous amount of shavings or dried grass, these nest liners are durable and may be easily slid out of the nest at cleanup time, dumped off, and reused.

To make things even easier, nests are available (or may be constructed) with removable bottoms. When messiness dictates, take out the bottom, scrape it down, hose it off, disinfect it, and let it dry in the sun before reinstalling it and furnishing a clean liner and fresh nesting material for your hens.

LAYING SEQUENCE

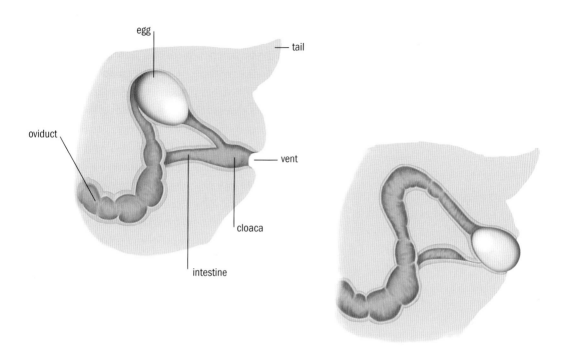

Just before an egg is laid, it rotates 180 degrees so the rounded end emerges first (right); as the egg enters the cloaca, the bottom end of the oviduct turns inside out to wrap around the egg and press shut the intestinal opening.

around the egg and pressing shut the intestinal opening. So the egg emerges clean, and any filth you find on the shell got there after the egg was laid. Bottom line: Clean eggs are the result of good management.

Flock Replacement

If your hens are pets, you may not be concerned about spending more money feeding them than you get back in eggs. But keeping hens primarily for their eggs can be a costly affair if you don't keep an eye on their economic efficiency. Pullets generally reach peak production at 30 to 34 weeks of age. From then on, laying declines approximately 0.5 percent per week until the birds molt or are replaced.

To induce a second round of laying, some producers induce molting at the age of about 60 weeks — a practice involving severe nutritional and environmental stress that, if not done precisely, can result in deaths instead of renewed laying. Another option for maintaining a high rate of lay is to raise a new batch of pullets to replace the older hens when they reach 72 weeks of age.

Keeping layers for a second year has a few disadvantages:

- As hens get older and their production declines, the shells of their eggs get rougher and weaker, the whites get thinner, and the yolk membranes become so weak they break when the eggs are opened into a pan.
- As laying declines, the cost of feeding older hens becomes greater than the value of their eggs.
- If you're selling eggs, production by older hens may fall below the numbers you need to satisfy your customers.
- While you may find a market for 1-year-old layers, by the end of their second year, hens lose nearly all their laying value and are good for little more than stewing.
- The older a chicken gets, the more likely it is to experience disease complications.

Your decision on when to replace your layers will depend, in part, on your chosen breed. Commercially developed hybrids still produce fairly well during their second year, while other hens peter out somewhat faster. Still, breeds that are better known for eggs than for meat may produce at a low but steady rate for years. Dual-purpose hens, on the other hand, tend to run to fat as they age, and fat hens do not lay well.

Your decision may be influenced by whether or not your hens are pastured, or how much of what they eat you are able to raise yourself. Given the savings in feed costs, you may not mind getting fewer eggs.

Your decision will be strongly influenced by the cost of starting new pullets compared to the cost of keeping the old flock. If you must purchase replacement chicks (as you would have to do if you keep hybrids), you may find it more economical to keep hens that are already laying than purchase and raise a new batch. If you keep purebreds, hatching your own chicks will save you the considerable cost of purchasing replacements. And if you realize any salvage value from the older hens (as meat birds or as layers sold to someone with less demanding needs than yours), you can offset the cost of raising new pullets.

In any case, don't be tempted to boost egg production by periodically bringing pullets into your established layer flock. Constantly introducing new chickens disrupts the peck order. The resulting stress can reduce the laying rate of hens and pullets alike. Introducing new birds to an established flock also increases the chance of spreading disease. The new birds may bring in a disease or may catch one to which the established hens have developed an immunity. Finally, as the oldest hens continue to age, their reduced productivity will depress a mixed-age flock's overall laying average.

Egg Sales

If you're thinking of getting into the egg business, first define your *market* — to whom are you going to sell your eggs and how will you reach them? On a small scale, you may earn a nice little income selling eggs to neighbors and coworkers. For serious income, you will have to reach beyond those you know, perhaps working through natural-food stores, farmers' markets, and the like. One schoolboy in my area earns a dandy income peddling eggs to summer campers at a local beach.

If you go beyond, "We've got extra eggs, would you like to buy some?" you'll have to define your product more formally.

You might, for example, market "organic eggs from pastured hens." Check with your county Extension agent or state poultry specialist about local and federal laws regarding claims you wish to make, conditions you have to meet to make those claims, and necessary sales permits.

Before getting into serious egg production, research your market so you'll know whether your customers will want white-shell eggs or brown-shell eggs. In most areas, brown eggs command a higher price than white eggs, even though today's hybrid brown-egg layers are nearly as efficient as white-egg layers. On the other hand, brown eggs are more difficult than white eggs to candle for blood spots (described in the next chapter), yet tend to have a higher rate of such hard-to-detect quality problems.

Some customers are willing to pay a premium for blue-shell eggs or attractively mixed colors.

Pricing may require additional market research. Your customers may accept a fixed price or may expect your price to fluctuate with market prices, which move up and down with feed costs, consumer demand, and seasonal swings in laying. When I lived in California, my customers were happy to pay a fixed price year-round. Here in Tennessee, I've had customers get indignant because I don't adjust my prices to follow the market.

In establishing your price, take into consideration all your expenses, including not only production but any applicable promotion, packaging, and delivery costs. The average market price is the least you should ask for your homegrown eggs. Since yours are fresher

and tastier than store-bought eggs, they are worth at least a few cents more.

Organic Eggs

To label your eggs as organic, you must strictly follow National Organic Program guidelines established by the United States Department of Agriculture (USDA). Information on the latest regulations may be found on the USDA's website.

To qualify as laying organic eggs, your hens must be raised from chicks in a completely organic manner, including parasite control and other health care, housing, and bedding, as well as rations. All feed must have been grown without pesticides, fungicides, herbicides, or chemical fertilizers and may contain no drugs, animal by-products, or genetically modified crops. Any land your chickens have access to must be certified organic, which requires no use of prohibited materials for at least three years. Any products used to clean eggs must be approved.

To qualify as organic, you must document everything you do by keeping careful records regarding such things as the source and age of your pullets, all feed and supplements you use and where you got them, any health products used and their sources, vaccinations, any deaths, what access your layers have to the outdoors, how you clean the housing before bringing in replacements, and your egg sales.

If you wish to be formally certified as organic, you must sign up with a certifying agency selected from a list maintained by the USDA, fill out a farm plan, and be inspected annually. Following all these regulations results in higher costs, making organic eggs more expensive than conventionally produced eggs.

Economic Efficiency

Several different methods may be used to determine the economic efficiency of producing eggs. Since the cost of feed accounts for much of the expense of maintaining a layer flock, many efficiency indicators factor into feed use.

Feed cost is the total amount spent on feed during the production period. Naturally, the lower your feed cost, the lower your cost of egg production. Cost may be kept down by guarding against feed wastage and by seeking economical feed sources, including natural forage.

Feed as a percentage of total cost is the cost of feed consumed during a given production period divided by the total operating cost (including not only feed but also medications, bedding, utilities, and other expenses). The higher this number is, the better you're doing. A typical indicator lies between 46 and 50 percent.

Feed conversion measures the pounds of feed needed to produce a dozen eggs. This indicator is derived by dividing the total amount of feed eaten within a given period by the number of dozen eggs produced during that period. Commercial layer strains always get better feed conversion than other breeds. You can improve this indicator for all chickens by reducing feed use (such as controlling rodents that pilfer feed) and increasing production (culling lazy layers). A good conversion rate is 4 pounds (1.8 kg) of feed per dozen eggs.

Feed efficiency, sometimes called feed cost per dozen eggs, is derived by multiplying the feed-conversion factor by the per-pound cost of feed. The lower your feed-efficiency indicator, the better you're doing. On average, feed cost per dozen eggs represents approximately 65 percent of the total cost of egg production.

Eggs per hen per year measures the flock average in terms of dozens, allowing for losses and culls. A flock average of 20 dozen eggs per hen per year is a pretty good average. The loving keeper may coax out more eggs; the neglectful keeper will get fewer.

Depreciation of layers measures the second greatest cost of producing eggs — the difference between the value of pullets at the beginning of the production period and the value of hens at the end. Culls and deaths during the production period increase depreciation, as does advancing age. Younger hens may have some remaining value as layers; older hens are worth little more than cheap meat. The lower your flock's depreciation, the better you're doing.

Depreciation per dozen eggs is the depreciation of layers divided by the number of eggs laid during the production period. The lower your depreciation per dozen eggs, the better. Keep this indicator low both by keeping down depreciation of layers and by keeping up their production level. Unfortunately, sometimes the same management decision works both ways: culling poor layers improves production but increases depreciation.

Price of eggs to meet costs is the average total cost of maintaining each hen divided by the number of dozen eggs the average hen laid during the production period. This indicator represents your break-even point. If you keep layers strictly for family use, this indicator lets you compare the cost of producing your own eggs to the cost of purchasing eggs at the store. If you sell eggs, anything more you get per dozen must cover both your labor and the profit on your layer-flock investment.

Net return per dozen eggs is the total amount earned from egg sales added to income from other sources (such as the sale of spent hens or composted litter), less total expenses, divided by the number of dozen eggs produced during the production period. Net return is your bottom line.

8

EGGS FOR EATING

Eggs are an extremely versatile food that may be served for breakfast, lunch, or dinner and may be used as the main dish or a side dish or in tasty desserts. Properly collecting and storing those delicious homegrown eggs preserves their freshness for future use in a variety of culinary dishes, and a number of methods are available for prolonging their shelf life for times when your layers take a break.

Egg Collection

Collect eggs often — preferably two or three times a day — so they won't get dirty or cracked and so they won't spoil in warm weather or freeze in cold weather. Carry eggs in a small bucket or a basket in which they can't bang together or roll around. Baskets designed specifically for collecting eggs are available from poultry suppliers, although you can often find nice wire baskets at a hobby or general store. You never know where you might see one — I found a perfect basket for egg collection in the cosmetic department at a grocery store.

Eggs are clean when they are laid, and if your nests are properly designed and managed, the eggs should be clean when you collect them. Occasionally, you might find a really nasty egg, such as one laid on the floor or in a nest where a chicken has roosted on the edge and soiled the litter. Such an egg is covered with bacteria and therefore not safe to eat — discard it.

A slightly dirty egg may be brushed off or rubbed with a sanding sponge or nylon scouring pad. A shell with egg white or yolk smeared on it from a broken egg may be rinsed in water that's slightly warmer than the egg; water that's cooler than the egg can cause bacteria to be drawn through the shell into the egg. Dry a washed egg before placing it in the carton, and use it as soon as possible.

Avoid getting into the habit of routinely washing eggs, since water rinses off the natural bloom that helps preserve an egg's freshness.

Store eggs in clean cartons. An enclosed carton keeps eggs fresh longer than a carton with part of the top cut away to make the eggs more readily visible. Orient eggs with their pointed end downward to keep the yolks nicely centered. Since an egg left at room temperature ages the same amount in a day as a refrigerated egg ages in an entire week, refrigerate eggs as soon as possible if you plan to eat them or sell them for eating.

Egg Quality

Commercial eggs are sorted — according to exterior and interior quality — into three grades established by the United States Department of Agriculture: AA, A, and B. For all grades, the shell must be intact. Nutritionally, all grades are the same.

Grades AA and A are nearly identical, the main difference being that Grade A eggs are slightly older than Grade AA eggs. Grade AA eggs therefore have firmer, thicker whites that hold the yolks up high and round, whereas the white of a Grade A egg is "reasonably firm," meaning it spreads a little farther when you break the egg into a frying pan. Grade A are the eggs you are most likely to see at a grocery store. Both grades are suitable for frying, poaching, and other dishes in which appearance is important.

Grade B eggs have stained or abnormal shells, minor blood or meat spots (described in the section *Abnormalities* later in this chapter) and other trivial defects. They are used in the food industry to make liquid, frozen, and powdered egg products, so you are unlikely to find them at a grocery store. Homegrown Grade B eggs are best used for scrambling, baking, and similar recipes in which the eggs are stirred.

Any egg that does not fit into one of these three categories is unfit for human use and consumption. Although you needn't worry about grading your homegrown eggs, the USDA grading system offers a guideline for assessing the quality of the eggs your hens produce.

Exterior Quality

Exterior quality refers to a shell's appearance, cleanliness, and strength. Appearance is important because the shell is the first thing you notice about an egg. Cleanliness is important because the shell is the egg's first defense against bacterial contamination; the cleaner the shell, the easier it can do its job. Strength influences the egg's ability to remain intact until you're ready to use it.

The shell accounts for about 12 percent of the weight of a large egg. It is made up of three layers:

- The inner, or *mammillary*, layer encloses the inner and outer membranes surrounding the egg. Between these two membranes is the air space that develops at the large end as the egg ages.
- The spongy, or *calcareous*, layer is made up of tiny *calcite crystals* consisting of 94 percent calcium carbonate with small amounts of other minerals. Viewed through a microscope, these crystals look like thousands of thin pencils standing on end. The spaces between them form pores connecting the surfaces of the inner shell and outer shell so moisture and carbon dioxide can get out of the egg and air can get in to create the air space.
- The bloom, or *cuticle*, is a light coating that seals the pores to preserve the egg's freshness by reducing evaporation and preventing bacteria from entering through the shell. Sometimes you'll find a freshly laid egg before the bloom has dried. Most of the pigment that gives the shell its color is in this layer.

When you wash an egg, the bloom dissolves, making the egg feel temporarily slippery. To replace natural bloom, commercial producers spray shells with a thin film of mineral oil, which is why store-bought eggs sometimes look shiny. If you wash an egg, rubbing

EGG STRENGTH TEST

Use a bathroom scale half supported by books to measure egg-shell strength.

the dried egg with clean vegetable oil somewhat replaces the bloom.

An eggshell's strength is naturally influenced by the vitamins and minerals in a hen's diet, especially vitamin D, calcium, phosphorus, and manganese. Shell strength is also influenced by a hen's age — older hens lay larger eggs with thinner, weaker shells.

A shell gets strength from its shape as well as from its composition. The curved surface is designed to distribute pressure evenly, provided the pressure is applied at the ends of the egg, not at the middle. The middle of a shell must be weak enough to allow an emerging chick to peck all around and break out of an incubated egg. Some chefs take advantage of this characteristic to make a big show of breaking an egg with one hand — what you don't see is the thumb they press against the middle of the shell. By contrast the ends of an egg must be quite strong so a newly laid egg won't crack when it plops into a nest, blunt end down.

One way to test the strength of an egg is to press the ends between the palms of your hands. For a more precise measurement, use an ordinary bathroom scale. Stack some boards or books on the floor to equal the height of the egg when it's standing on end. With paper towels, fashion a ring around the bottom of the egg to stand it on end, next to the books. Rest one edge of the scale on top of the books and the other edge on the top of the egg. Press on the part of the scale that's just above the egg. A well-formed shell should support up to 9 pounds (4 kg) before it breaks.

Except for preserving the freshness of eggs, shells have no culinary use (although I was once served a blended health-food drink containing a raw egg, shell and all, and I must admit it tasted pretty good). Shells have plenty of other uses. They may be:

- Dried, crushed, and fed back to hens as a calcium supplement
- Added to compost to sweeten the soil
- Placed in tomato planting holes to prevent blossom-end rot
- Decorated for a variety of arts and crafts

Interior Quality

Interior quality refers to the appearance and consistency of an egg's contents, which may be determined easily by breaking the egg into a dish for examination. In doing so, you will discover the egg has more than one kind of egg white, or *albumen*. Two clearly visible kinds are the firm white around the yolk and the thinner white closer to the shell. A less obvious second, or inner, thin layer lies between the outer thick white and the yolk.

The outer thin egg white contains anti-microbial chemicals that together with its alkaline pH help prevent contamination by reducing the availability of nutrients that support bacterial growth. The firm or thick egg white surrounding the yolk cushions the yolk, and its composition includes defenses against bacteria. The older an egg gets, the more thin white and the less thick white it has.

Another kind of white is the *chalaziferous*, sometimes called the inner thick (in contrast to the other thick egg white, which is called the outer thick), layer made up of dense albumen surrounding the yolk. During its formation, as the egg travels through the oviduct and rotates, the ends of this layer become twisted together to form a cord of sorts, or *chalaza* (pronounced *kah-LAY-za*), on each side of the yolk. These two cords anchor the chalaziferous layer and protect the yolk by centering it within the white.

When you break an egg into a dish, the chalazae snap away from the shell membrane and recoil against the yolk. Misinformed cooks sometimes mistake the resulting two white blobs at opposite sides of the yolk for the beginnings of a developing chick.

A chick develops instead from a round, whitish spot on top of the yolk called the *germinal disc* or *blastodisc*. When an egg is infertile, the blastodisc has an irregular shape. If the egg has been fertilized, the blastodisc becomes the *blastoderm* and organizes into a set of tiny rings, one inside the other.

Egg yolks get their yellow-orange color from xanthophylls and carotenes, plant pigments in corn and other feedstuffs that also color the skin of yellow-skin breeds. The exact color of the yolk depends on the proportions of xanthophylls and carotenes in the feed. For example, the high xanthophyll content in alfalfa produces a yellowish yolk, while the

ANATOMY OF AN EGG

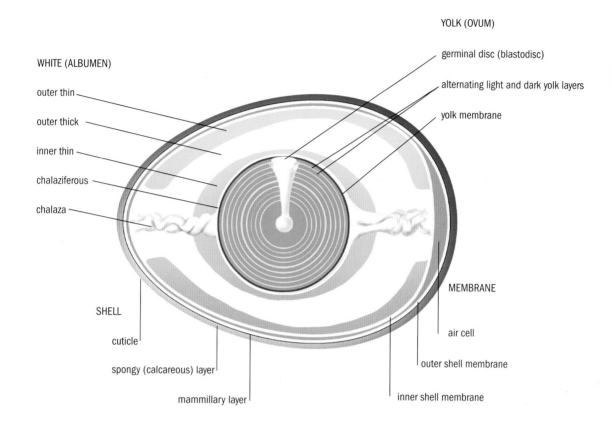

YOLK (OVUM)

germinal disc (blastodisc)

alternating light and dark yolk layers

yolk membrane

WHITE (ALBUMEN)

outer thin

outer thick

inner thin

chalaziferous

chalaza

MEMBRANE

air cell

SHELL

cuticle

outer shell membrane

spongy (calcareous) layer

inner shell membrane

mammillary layer

predominance of carotenes in corn gives yolks a reddish orange color.

Excessive amounts of certain pigmented feeds can affect yolk color. Alfalfa meal, clover, kale, rape, rye pasture, and certain weeds including mustard, pennycress, and shepherd's purse make yolks darker. Too much cottonseed meal can really throw off yolk color, causing it to be salmon, dark green, or nearly black.

The yolk is not of uniform color throughout. Look closely, and you will see that it consists of concentric rings. At the center is a ball of white yolk, around which are alternating layers of thick dark yolk and thinner white yolk. Although you might never see it — except maybe in a hard-cooked egg — a neck of white yolk extends from the center to the edge of the yolk, flaring out and ending just beneath the blastodisc.

As an egg ages, both its white and yolk deteriorate. Their quality may not have been all that great to start with, depending on the hen's age and health, the use of medications, the weather, and hereditary factors. The better an egg's starting quality, the better it keeps.

YOLK COLOR ORIGINS

Color	Cause
RAW	
Green	acorns
	shepherd's purse
Reddish orange to dark yellow	green feed
	yellow corn
Reddish, olive green, black green	grass
	cottonseed meal
	silage
Yellow, dark	alfalfa meal
	marigold petals
Yellow, medium	yellow corn
Yellow, pale	coccidiosis (rare)
	wheat (fed in place of corn)
	white corn
COOKED	
Gray or green surface	iron in cooking water
	overcooked yolk
Green rings	iron in hen's feed or water
Greenish when scrambled	egg left too long on a steam table
	overcooked egg
Greenish when soft-cooked	fried (or served with blueberry pancakes!)
Yellow rings	normal layers of yolk

Fresh-Egg Tests

Occasionally, you may find an egg, or a cache of eggs, and not know how long ago they were laid. Several different methods allow you to estimate an egg's age.

Candling is the process by which an egg's contents are viewed in front of a light, even though no one actually uses a candle anymore. Poultry-supply outlets offer a handheld light designed specifically for the purpose, but a small, bright flashlight works just as well.

If you have never candled an egg, practice with white-shell eggs before you tackle colored ones, which are more difficult to see through. In a darkened room, grasp an egg at the bottom between your thumb and first two fingers. With the egg at a slant, hold the large end to the light. Turn your wrist to give the egg a quick twist, sending the contents spinning.

The albumen of a fresh egg is fairly dense. If the yolk looks vague and fuzzy, the thick white albumen surrounding it is holding it properly centered within the shell. As the egg ages and the white grows thinner, the yolk moves more freely. When you twirl the egg, the deteriorating albumen lets the yolk move closer to the shell. A yolk that's clearly visible indicates albumen that has thinned.

CANDLING

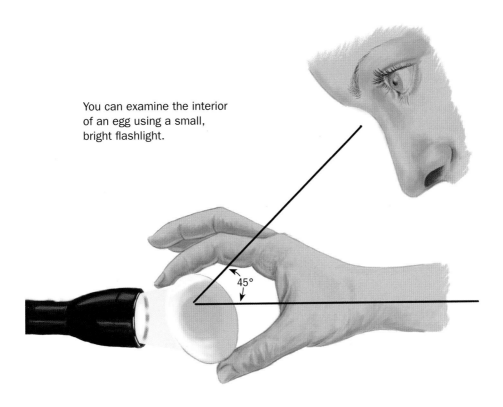

You can examine the interior of an egg using a small, bright flashlight.

45°

Air-cell size increases as an egg ages. Unlike a slightly older egg, a freshly laid egg has no air cell under the shell at the large, round end. As the egg cools, its contents shrink and an air space develops. The inner shell membrane pulls away from the outer shell membrane, forming a cell, or pocket. As time goes by, moisture evaporates from the egg, its contents continue to shrink, and the air cell grows.

Candling to measure the air cell will give you an idea of the egg's age. The cell of a freshly laid cool egg is no more than ⅛ inch (3 mm) deep. From then on, the larger the cell, the older the egg. Just how fast the air space grows depends on the porosity of the shell and on the egg's storage temperature and humidity.

Floating an egg in plain water lets you gauge its air-cell size without candling. A fresh egg will settle to the bottom of the container and rest horizontally. The larger air cell of a 1-week-old egg will cause the big end of the egg to rise up slightly from the container bottom. An egg that's 2 to 3 weeks old will stand vertically at the bottom of the container, big end upward. When the air cell grows large enough to make the egg buoyant, the egg will float. Although a floating egg is quite old, it's not necessarily unsafe to eat.

Smell is the quickest way to detect the age of an egg that's too old to eat. A rotting egg emits foul-smelling hydrogen sulfide, otherwise known as rotten-egg gas. Any egg with an off-odor, whether raw or cooked, should be discarded.

Breaking an egg and examining its contents is another way to estimate the egg's age.

The albumen of a fresh egg contains carbon dioxide that makes the white look cloudy. As an egg ages, the gas escapes and the albumen looks clear or transparent.

A fresh egg's albumen is also firm and holds the yolk up high. A stale egg has watery albumen that spreads out thinly around the yolk.

AIR-CELL GAUGE

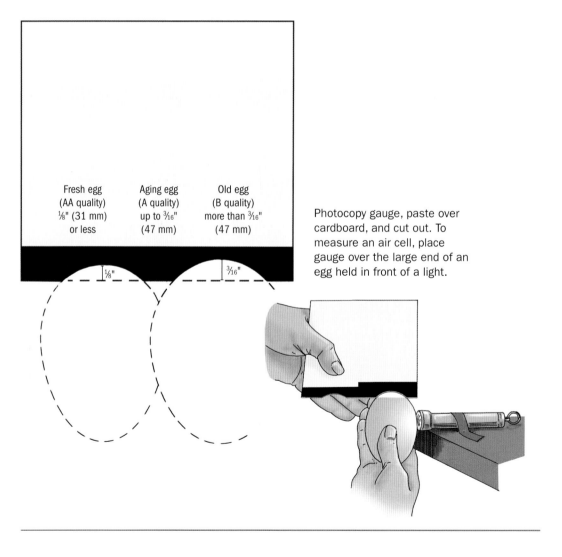

Fresh egg
(AA quality)
⅛" (31 mm)
or less

Aging egg
(A quality)
up to ³⁄₁₆"
(47 mm)

Old egg
(B quality)
more than ³⁄₁₆"
(47 mm)

⅛"

³⁄₁₆"

Photocopy gauge, paste over cardboard, and cut out. To measure an air cell, place gauge over the large end of an egg held in front of a light.

FLOAT TEST

An easy way to determine an egg's age is by placing it in water. Older eggs have larger air cells and will therefore float.

very old

stale (2–3 weeks old)

1 week old

fresh

As an egg ages, water migrates from the albumen to the yolk, stretching and weakening the yolk membrane. The older the egg, the greater the likelihood that its yolk will break.

Even the freshest egg occasionally has a watery white or an easily broken yolk. But if an egg has a mottled yolk or otherwise doesn't look right to you, discard it.

Abnormalities

You may occasionally find an egg that is abnormal due to an accidental occurrence, a hen's hereditary tendencies, or environmental or management factors. Some egg abnormalities are little more than nonrecurring curiosities, and others may require corrective action on your part. You can detect these abnormalities by candling and by inspecting broken eggs. Trapnesting (as described in chapter 9) lets you identify hens that habitually lay problem eggs.

No-yolkers are called *dwarf eggs* or *wind eggs*. They most often occur as a pullet's first effort, produced before her laying mechanism is fully geared up.

In a mature hen, a wind egg is unlikely but can occur if a bit of reproductive tissue breaks away, stimulating the egg-producing glands to treat it like a yolk and wrap it in albumen, membranes, and a shell as it travels through the oviduct. In place of a yolk, this egg will contain a small particle of grayish tissue.

In the old days, no-yolkers were called cock eggs. Since they contain no yolk and therefore can't hatch, people believed they were laid by roosters.

Double yolkers appear when ovulation occurs too rapidly or when one yolk for some reason moves too slowly and is joined by the next yolk. Double yolkers may be laid by a pullet whose production cycle is not yet well synchronized. They're occasionally laid by heavy-breed hens, often as an inherited trait.

Sometimes an egg contains more than two yolks. One time I found an egg with three. The greatest number of yolks ever found in one egg is nine. Record-breaking eggs are likely to be multiple yolkers: the *Guinness Book of Records* lists the world's largest chicken egg (with a diameter of 9 inches [23 cm]) as having an amazing total of five yolks and the heaviest egg (1 pound/0.5 kg) as having a double yolk and a double shell.

An egg within an egg, or a *double-shell egg*, appears when an egg that is nearly ready to be laid reverses direction and gets a new layer of albumen covered by a second shell. Sometimes the reversed egg joins up with the next egg, and the two are encased together within a new shell. Double-shell eggs are so rare no one knows exactly why or how they happen.

Blood spots occur when a small blood vessel ruptures and a bit of blood is released along with a yolk. They appear in less than 1 percent of all eggs laid and do not mean the eggs are unsafe to eat.

Each developing yolk in a hen's ovary is enclosed inside a sac containing blood vessels that supply yolk-building substances. A mature yolk is normally released from the only area of the yolk sac free of blood vessels, called the *stigma* or *suture line*. Occasionally a yolk sac ruptures at some other point, causing vessels to break and blood to appear on the yolk or in the white. As an egg ages, the blood spot becomes paler, so a bright blood spot is a sign the egg is fresh.

Spots may appear in a pullet's first few eggs but are more likely to occur as hens get older. They may be triggered by too little vitamin A in the diet or may be hereditary — if you hatch replacement pullets from a hen that typically lays spotty eggs, your new layers will likely do the same.

Meat spots are even less common than blood spots, but like blood spots, they may be hereditary. They appear as brown, reddish brown, tan, gray, or white spots in an egg, usually on or near the yolk. Such a spot may have started out as a blood spot that changed color due to a chemical reaction or may be a bit of reproductive tissue. Meat spots look unappetizing but do not make the egg unsafe to eat.

Wormy eggs are extremely rare, occurring only in hens with a high parasite load. Finding a worm in an egg is not only unappetizing but is a clear indication that you are not doing a good job of keeping your hens healthy and parasite free.

Off-flavor eggs may result from something a hen ate or from environmental odors. Hens that eat onions, garlic, fruit peelings, fish meal, fish oil, or excessive amounts of flaxseed will lay eggs with an undesirable flavor. Eggs can also absorb odors that translate into unpleasant flavors if they're stored near kerosene, carbolic acid, mold, must, fruits, and vegetables.

Developing embryos are a sign that eggs have been partially incubated, which occurs when hens hide their nests and the eggs accumulate. When laying subsequent eggs, the hen may warm the earlier eggs enough for them to start developing. Or a hen may have started setting and later gave up.

If you find a bunch of eggs and you can't be sure how long they've been there, inspect them either by candling or by breaking them into a separate dish before using them. You will not get developing embryos if you have no rooster to fertilize eggs or your housing is set up so the hens can't readily hide their eggs.

Egg Safety

A freshly laid egg is warm and moist and therefore attractive to a variety of bacteria and molds in the environment. As an egg cools and its contents shrink, these microbes may be drawn through the shell's six thousand-plus pores. Your first line of defense in keeping eggs safe to eat is to keep nests clean and lined with fresh litter. Eggs produced in a clean environment, collected often, and promptly refrigerated contain too few microbes to cause human illness.

An egg that's cracked — called a *check egg* — is safe to eat, provided the membrane is intact, the egg is refrigerated promptly, and it's used right away. If a cracked egg leaks, indicating the membrane has also been broken, discard it.

Discard any eggs that are seriously soiled. Although moderately soiled eggs may be washed, in doing so you'll rinse away the bloom that seals the pores and keeps out bacteria. If you do wash an egg, sanitize the cleaned shell by dipping the egg for 30 seconds in a solution of 1 teaspoon of chlorine bleach in 1 quart of water warmed to 101°F. Wipe the egg dry with a clean paper towel or soft cloth before placing it in a storage carton. Rubbing the shell with clean vegetable oil before refrigerating the egg will prolong its shelf life.

Clean eggs stored at 45°F (7°C) and 70 percent humidity will keep well for at least three months. In a standard household refrigerator, where foods tend to dry out thanks to the refrigerator's defrosting mechanism, eggs will remain edible for up to five weeks.

Certain species of bacteria (such as *Streptococcus* and *Staphylococcus*) tolerate dry conditions and are therefore able to survive on an egg's shell. These bacteria dwindle during storage but are replaced by other species (including *Pseudomonas*) that cause eggs to rot. The presence of *Pseudomonas* may be detected by a pink, iridescent, or deep green color in an egg's white and perhaps by a sour smell emitted when the egg is broken open. A variety of molds (*Penicillium, Alternaria, Rhizopus*) that get on the shells of eggs washed in dirty water or stored in humid conditions may also penetrate the shell and cause spoilage.

Salmonella bacteria may be either on the shell or inside the egg when it is laid. These bacteria can cause serious illness in humans but only if allowed to multiply, which happens when a contaminated egg is held for too long at room temperature. The problem gets worse if the egg is combined with other eggs in a mixture left on the counter and then undercooked or served as is.

Although you are unlikely to get *Salmonella* poisoning from your own carefully homegrown eggs, cooking them puts you on the safe side by destroying any *Salmonella* that might be present. A thoroughly cooked egg has its white cooked through and its yolk at least beginning to thicken — the yolk need not be hard, but it should not be runny. Cook eggs slowly to make sure they're heated all the way through.

Keep hot foods containing eggs at 140°F (60°C) or warmer and cold foods at 40°F (4°C) or cooler. Including acid ingredients, such as lemon juice or vinegar, in foods and recipes containing raw eggs retards bacterial growth.

Even so, eat such foods immediately or keep them refrigerated.

To avoid food poisoning follow these tips:

- Collect eggs often, and refrigerate them promptly.
- Discard seriously dirty or cracked and leaking eggs.
- Wash hands and utensils after handling raw eggs.
- Immediately cook or refrigerate foods prepared with raw or undercooked eggs, and cook them within 24 hours.
- Cook eggs and egg-rich foods to 160°F (71°C) and serve them immediately, or cool quickly and refrigerate.
- Promptly refrigerate leftovers, and use them within four days.

Nutritional Value

Eggs have often been called the perfect food. One egg contains almost all the nutrients necessary for life, lacking only vitamin C. It also contains the antioxidants *lutein* and *zeaxanthin*, which help prevent the eye disorder *macular degeneration*, and *choline*, which aids the functioning of your brain and helps reduce your risk of heart disease.

A large egg is approximately 31 percent yolk, 58 percent white, and 11 percent shell. Aside from the shell, an egg has about 75 percent water, and the remainder is about 12 percent protein and 12 percent fat.

The yolk is basically an oil-water emulsion containing proteins, fats, pigment, and a number of minor nutrients. Yolks contain *lecithin*, which acts as an emulsifier for making mayonnaise and hollandaise sauce. Some evidence shows that lecithin may affect the human brain function by improving memory, as well as help

reduce cholesterol. Unfortunately, most of an egg's cholesterol and calories are also in the yolk. The exact caloric value of an egg depends on its size.

The white is made up of water combined with several different kinds of protein. Egg protein is complete, since it contains all the essential amino acids. It is among the highest-quality protein found in food, second only to mother's milk. One large egg contains approximately 6.25 grams of protein, roughly equivalent to 1 ounce (28 g) of lean meat, poultry, fish, or legumes.

One of the proteins in raw eggs, *avidin*, ties up the vitamin *biotin* as part of an egg's defenses against bacteria, since most bacteria can't grow without biotin. Pets are sensitive to the effects of avidin and should not routinely be fed raw eggs. A human would have to eat two dozen raw eggs a day to be affected. Cooking an egg inactivates the avidin.

Many people have the mistaken belief that fertile eggs are more nutritious than infertile eggs. The idea is encouraged by unscrupulous sellers who cater to health-conscious consumers and find they can charge more by claiming that their fertile eggs are more nutritious than the infertile eggs commonly sold in supermarkets. In truth, a sperm contributes an insignificant amount of nutrients to a fertilized egg.

Another idea promoted in health-food circles is that eggs with colored shells are more nutritious than white-shell eggs. Although eggs from backyard or pastured hens are likely to have brown or blue-green shells, in contrast to the white shells of commercial eggs, the shell color itself has nothing to do with an egg's nutritional value.

The nutritional difference is not in the shell color but in how the eggs are produced.

CALORIE CONTENT OF EGGS

Size	Calories	Compares to
Peewee	47	1 cup (0.25 L) strawberries
Small	54	½ large grapefruit
Medium	63	1 large peach
Large	72	1 medium apple
White	16	1 small tomato
Yolk	56	1 small sweet potato
Extra large	81	1 cup (0.25 L) blueberries
Jumbo	90	1 large orange

White-shell eggs produced by hens on pasture will be more nutritious than eggs with colored shells laid by caged hens. The eggs laid by hens with access to pasture or other green feeds, no matter what color the shell, contain less cholesterol and saturated fat; more vitamins A, D, and E; and more beta carotene, folic acid, and omega-3 polyunsaturated fatty acids than factory-farm-produced eggs laid by caged hens. Eggs from hens fed flaxseed also have more omega-3s.

Cholesterol

Cholesterol is a fatty substance found in the bodies of chickens and humans alike. It is required for both the synthesis of vitamin D from sunshine and the production of sex hormones. But it can also collect in the bloodstream and clog the arteries. Thanks to media scare tactics, many people have the impression that all the cholesterol you eat goes straight into your bloodstream. Not true.

Most of the cholesterol in your body is manufactured within your body. The amount of manufactured cholesterol is regulated by your liver, based on your dietary intake. Some people have better controlling mechanisms than others.

Cholesterol in the foods you eat does increase your blood cholesterol somewhat, but a healthy body compensates by producing less or excreting more cholesterol to maintain the proper blood level. Medical practitioners who don't buy into the cholesterol panic point out that even if all the cholesterol in an egg went into your bloodstream — which isn't likely — you'd have to eat five jumbo eggs a day to raise your cholesterol level from 150 to 152.

Saturated fat is the major dietary culprit in increasing your blood cholesterol. (Other factors such as age, gender, and genetics can play a bigger role.) Many foods that contain cholesterol are high in saturated fat. An egg is the rare exception — it is high in cholesterol but contains little saturated fat. The yolk of one large egg contains about 210 milligrams of cholesterol and 1.5 grams of saturated fat — compared to no more than 22 grams of saturated fat recommended for a person eating two thousand calories per day.

A study reported in *Medical Science Monitor* showed that eating one or more eggs a day does not increase the risk of heart disease or stroke among healthy adults but may decrease blood pressure. And a review of 30 years of cholesterol research published in the *Journal of the American College of Nutrition* concluded that eating eggs has little relationship to high blood cholesterol or the incidence of heart disease.

Whether or not eggs are harmful seems to depend on who's eating them. An estimated two-thirds to three-quarters of the population can handle a moderate number of eggs. Perhaps one reason is that eggs contain lecithin, and as nutritional researchers at Kansas State University have learned, lecithin interferes with the body's absorption of cholesterol.

Cholesterol Reduction

Some egg sellers charge outrageous prices for blue-green eggs, claiming they're lower in cholesterol than the more familiar white-shell eggs. The fact is that eggs laid by hens of heavier breeds (including Araucana) are likely to contain slightly more cholesterol than eggs laid by commercial Leghorn strains raised under similar circumstances.

In their efforts to explain these findings, researchers reason that the more often a hen lays, the less time she has to put cholesterol into each egg. Unfortunately, the difference among breeds is so slight it's insignificant. Otherwise, commercial producers wouldn't have such a hard time trying to reduce the cholesterol level of eggs through genetic selection.

A greater influence on the cholesterol and saturated-fat content of eggs is what the hens eat. Pasturing layers, regularly feeding them greens, or altering their rations to include 10 percent flax seed, can reduce both the cholesterol and the saturated-fat content of their eggs by some 25 percent. Because the eggs of hens that eat green plants have darker yolks than those of hens without access to greens, yolk color is a good reflection of an egg's lower levels of cholesterol and saturated fat.

The only other property known to influence the total percentage of fat is an egg's size — the larger the egg, the more white it has in proportion to yolk, and therefore the lower its percentage of saturated fat and cholesterol. Most of the fat and all of the cholesterol are in the yolk. And for the record, cooking an egg affects neither fat nor cholesterol.

According to some nutritionists, you can eliminate more saturated fat from your diet by eating chicken without the skin than you can by not eating eggs. Still, if you wish to reduce fat in an egg dish (such as a quiche or an omelet), substitute two egg whites per whole egg for half the eggs called for in the recipe. To reduce saturated fat in baked goods, substitute two egg whites and 1 teaspoon of vegetable oil for each whole egg. If the recipe already has oil in it, leave out the extra oil.

So-called egg substitutes are made from egg white. The yolk is replaced by such ingredients as nonfat milk or tofu, along with a variety of emulsifiers, stabilizers, antioxidants, gums, and artificial coloring. You have to wonder if the more healthful choice wouldn't be to eat the real thing.

Preserving Eggs

At some times of the year, you'll have more eggs than you can use, while at other times you'll have too few. It's logical to preserve surplus eggs in times of plenty to use in times of need. Preserve only eggs with clean, uncracked shells; keep in mind that dirty eggs that have been cleaned by washing or dry buffing do not keep well under prolonged storage.

THE HEALING EGG

In folk medicine, egg whites are used to heal wounds and inhibit infection. Egg white contains *conalbumin*, a substance that binds iron and thereby inhibits bacterial growth, protecting a developing embryo from infections. To treat a wound and speed healing, the protein-rich membrane is peeled from inside the shell and bandaged over the wound. Raw eggs are also used as beauty aids — whites in facials, yolks in shampoos and hair conditioners.

Throughout the ages, different methods have been devised to extend the storage life of eggs. The ancient Chinese stored them for years in various materials ranging from clay to wood ashes to cooked rice. You probably wouldn't want to eat one of those eggs, with its greenish yolk and gelatinous brown albumen.

In modern times many processes have been developed for preserving eggs. Although none is an outright substitute for cold storage, these methods can let you extend the shelf life of your homegrown eggs. They also offer short-term ways to prolong storage without electricity, which is handy if you live in or are planning a trip to the outback. Even in the backwoods you can often take advantage of a cold running stream or an ice bank to keep eggs fresh. Many cellars offer suitable egg-storage conditions. The cooler the temperature, the longer eggs will keep without spoiling.

Clean eggs may be safely stored for the short term (two to three months) at temperatures up to 55°F (13°C), where the relative humidity is close to 75 percent. The moisture level is important, since at low humidities eggs dry out and at high humidities they get moldy. A little air circulation helps retard mold growth.

At 30°F (–1°C) eggs will keep for as long as nine months. The temperature must not get below 28°F (–2°C), though, or the eggs will freeze and burst their shells. Relative humidity of 85 percent is best for such long-term storage. To prevent mold growth and condensation, seal egg cartons in plastic bags. To minimize drying out at lower humidities, use the thermostabilization, oil, or water-glass methods described below.

Given the infinite number of combinations of possible temperature and humidity conditions, it's not possible to list definitive storage times under all conditions. The accompanying guidelines (see box, Egg Storage Shelf Life), in combination with the fresh-egg tests offered earlier in this chapter, give you a good starting

EGG STORAGE SHELF LIFE

Method	Maximum Storage Time*
Refrigerated	
Yolks, covered with water	2 days
Whites, in tightly sealed jar	4 days
Hard-cooked, peeled, in water	1 week
Hard-cooked in shell	2 weeks
Whole in carton	5 weeks
Hard-cooked, peeled and pickled	6 months
Water glass	6 months at 34°F (1°C)
Oiled	7 months at 31°F (–0.5°C)
Thermostabilized	8 months at 34°F (1°C)
Oiled and thermostabilized	8 months at 34°F (1°C)
Cold storage	9 months at 30°F (–1°C)
Frozen	12 months at 0°F (–18°C)

*These guidelines are for fresh, home-produced eggs only.

This sturdy, two-piece, clear-plastic egg carton may be washed between uses, making it ideal for storing homegrown eggs.

place. The longer you store eggs, the more likely they are to develop a stale or off-flavor that makes them less suited for breakfast than for mixing with other ingredients in recipes.

Refrigeration

Refrigeration is usually the quickest, most convenient way to store eggs. But the worst place to store them is on an egg rack built into the refrigerator door, where they'll get jostled every time the door is opened. And if the rack lacks a cover, the eggs will be exposed to lost moisture and blasts of warm air whenever the door is opened.

Storing them in closed cartons on a lower shelf keeps them cooler, thus fresher longer, and reduces evaporation through the shell. On the lowest shelf, where the temperature is coolest, eggs in a closed carton will keep for up to 5 weeks.

The biggest problem with a household refrigerator is its low humidity, especially in a self-defrosting (frost-free) model. If you wrap egg cartons in plastic bags to prevent moisture loss (as well as absorption of flavors from other foods), you can safely refrigerate eggs for two months.

Egg cartons may be reused only so many times before they become unsanitary, so much so that in some places it's illegal to sell eggs in recycled cartons. A washable, reusable egg carton is a better option, but good ones are hard to find. Plastic egg carriers of the sort used by backpackers and craft hobbyists make an inexpensive option that may be readily washed before each refill, but unfortunately the flimsy plastic hinges don't hold up well under constant use.

If you need only the whites or only the yolks for a recipe, you may refrigerate leftover whites in a tightly covered container for up to four days. Leftover yolks, covered with cold water, may be stored in a sealed container for two days.

Once whole eggs have been refrigerated, they need to stay that way. A cold egg left at room temperature sweats, encouraging the growth of bacteria. Never leave a refrigerated egg on the counter for more than two hours.

EGG STORAGE SAFETY

Temperature	°F	°C	Humidity
Quality is rapidly lost at these temperatures; the higher the temperature the faster the loss.	100	38	
	95	35	
	90	32	
	85	29	
	80	26.5	Molds appear and grow at relative humidities above wet bulb 52°F (11°C) or 85 percent relative humidity.
	75	24	
	70	21	
	65	18	
Short Storage Periods	60	15.5	
	55	13	
Long-Term Storage Quality is retained best at these temperatures.	50	10	**Safe:** Wet bulb 42°F to 52°F (5.5°C to 11°C) or 75 to 85 percent relative humidity.
	45	7	
	40	4.5	
	35	1.5	
	30	−1	
Eggs break at temperatures below 28°F (−2°C).	25	−4	
	20	−6.5	Eggs lose quality as a result of moisture loss when relative humidity is below wet bulb 42°F (5.5°C) or 75 percent.
	15	−9.5	
	10	−12	
	5	−15	
Freezing	0	−18	

Adapted from: "Table Egg Room Safety Chart," University of Georgia

Freezing

Freezing lets you keep eggs longer than any other method—up to one year at 0°F (−18°C). Freeze only raw eggs; hard-cooked eggs will turn rubbery. Since eggs that are frozen intact will expand and burst their shells, the shells must be removed before freezing.

Eggs that are frozen in ice-cube trays should be removed as soon as they are frozen and wrapped in freezer bags with as much air removed as possible—a handheld vacuum pump comes in handy here. To compare the size of a frozen cube with the normal size of an egg (or its yolk or white), use a measuring spoon to determine how much each cube in the tray holds, then consult the "Frozen Egg Equivalents" table (see below). If you use metal trays with removable grids, measure the total amount the tray holds and divide by the number of cubes per tray to determine the amount per cube.

For use in recipes calling for several eggs, freeze eggs (or yolks or whites) in airtight freezer containers, each holding just enough for one recipe. Leave a little head space to allow for expansion; otherwise the lid may pop or the container may split. Place a square of freezer paper on top of the eggs to minimize the formation of ice crystals. Label each container with the date, contents, and recipe for which it is intended.

To store whole eggs, break the contents into a bowl, stir just enough to blend the yolks with the whites (taking care not to whip in air), and press the eggs through a sieve to break up the thick albumen. You can get by without adding sugar or salt if you take great care to mix the whites and yolks thoroughly. Otherwise, to each cup of eggs add ½ teaspoon salt (for use in a main dish) or ½ tablespoon honey, corn syrup, or sugar (for use in a dessert). Pour the mixture into trays or containers for freezing.

To store whites separated from the yolks, break the egg carefully to avoid getting any yolk into the whites; otherwise you'll have trouble whipping them later. Press the whites through a sieve to break up the thick albumen, and freeze them in trays or containers. Thawed whites may be whipped just like fresh ones if you warm them to room temperature for 30 minutes before beating them.

To store yolks separately, include sugar or salt to prevent gumminess. Add either ⅛ teaspoon salt or ½ tablespoon honey, corn syrup, or sugar per four yolks (approximately ¼ cup or 60 mL). Freeze the egg yolks in ice-cube trays or in airtight containers.

Thaw frozen eggs overnight in the refrigerator or in airtight containers placed in cool water, 50 to 60°F (10 to 16°C). Thaw only as many as you will use within three days.

Use thawed eggs only in foods that will be thoroughly cooked.

Pickling

Pickling is a good way to preserve hard-cooked eggs. Pickled eggs may be used in place of plain hard-cooked eggs in salads or in place of cucumber pickles in sandwiches. They also make a nutritious snack.

Over the years, I have prepared many dozens of pickled bantam eggs, packed hot in boiling spiced vinegar and processed in pint jars in boiling water for 10 minutes. Various eggsperts I have consulted can't agree on whether or not such eggs are safe for long-term storage out of the refrigerator. A problem would arise if the pickling solution did not penetrate all the way through the eggs, which is why I pickle only bantam eggs and the early small eggs laid by pullets.

For the pickling solution, you could mix your own vinegar and spices or use the leftover flavored vinegar from your favorite prepared pickled cucumbers or beets. The eggs will be more tender if you pour the boiling solution over them while it's hot, rather than letting it cool first.

The fresher the eggs, the better. Select your smallest eggs so the pickling solution can penetrate easily. Half a dozen bantam eggs will fit into a wide-mouthed pint jar. One dozen pullet eggs will fit into a wide-mouthed quart jar.

FROZEN EGG EQUIVALENTS		
This	**Equals**	**This**
2 tablespoons thawed white	=	1 large fresh white
1 tablespoon thawed yolk	=	1 large fresh yolk
3 tablespoons thawed whole egg	=	1 large fresh egg

Pickling is a time-honored way to preserve hard-cooked eggs.

Let the eggs season for two to four weeks before serving them. The acidity in the pickling solution keeps bacteria from growing but also causes the eggs eventually to deteriorate. Stored in the refrigerator, pickled eggs keep well for six months.

Oiling

Coating eggs with oil seals the shells to prevent evaporation during storage. Eggs should not be oiled until 24 hours after being laid so some of their carbon dioxide can escape and the whites won't have a muddy appearance.

Into a small bowl pour white mineral oil, available at any drugstore. The oil must be free of bacteria and mold, which you can ensure by heating it to 180°F (82°C) for 20 minutes. Cool the oil to 70°F (21°C) before dipping the eggs.

The eggs must be at room temperature (50 to 70°F/10 to 21°C) and completely dry. With tongs or a slotted spoon, immerse the eggs in the oil one by one. To remove excess oil, place each dipped egg on a rack (such as a rack used for cake cooling or candy making), and let the oil drain for at least 30 minutes. Catch the dripping oil for reuse, being sure to reheat it before using it. Discard oil that contains debris or water or that changes color.

Oiled eggs may be used like fresh eggs except when it comes to recipes requiring whipped whites — oiling interferes with the foaming properties of the whites, so they won't whip up as well as fresh ones. Experiments in Australia prove that oiled eggs will keep for as long as 35 days at tropical temperatures. Stored at 50°F (10°C) for eight weeks or 70°F (21°C) for five weeks, they retain their flavor better than untreated eggs.

In clean, closed cartons in a cool place, eggs dipped in oil will keep for several months.

Like all eggs stored for the long term, however, they'll eventually develop an off-flavor. The longer the eggs are stored, the greater the flavor intensity compared to untreated eggs. This flavor change is pronounced in eggs stored at 34°F (1°C) for more than four months, and by six months most people find the flavor quite unacceptable.

Thermostabilization

Thermostabilization of eggs was regularly practiced by housewives during the late 19th century. In this process, heating the eggs destroys most spoilage-causing bacteria on the shell and seals the shell by coagulating a thin layer of albumen just beneath it. When the egg cools, the coagulated albumen sticks to the egg membrane and cannot be seen when the egg is cracked open. Unlike oiling, this method does not affect the egg's foaming properties.

Process eggs the day they are laid. Heat tap water to exactly 130°F (54°C). Use a thermometer, since the temperature is critical — the water must be just warm enough to destroy spoilage organisms but not hot enough to cook the eggs. Place eggs in a wire basket (such as a vegetable steamer or pasta cooker). Submerge the eggs in the water for 15 minutes if they are at room temperature or 18 minutes if they have been refrigerated. Lift the basket, and thoroughly drain and dry the eggs. Thermostabilized eggs will keep for two weeks at 68°F (20°C), and eight months at 34°F (1°C).

Thermostabilization with Oiling

Thermostabilization destroys bacteria and protects the quality of the egg white. Oiling minimizes weight loss from evaporation and preserves the quality of the yolk. Combining the two improves an egg's keeping qualities compared to either method alone.

You can thermostabilize eggs and then oil them or combine the two procedures into one. For a combination operation, heat the oil to 140°F (60°C) and hold it at that temperature. Using a pair of tongs, rotate each egg in the hot oil for 10 minutes, then set the egg on a rack to drain.

As with simple oiling, albumen-foaming properties are reduced by this process. These eggs are therefore unsuitable for making any recipe calling for whipped whites.

Water Glass

Submerging eggs in water glass was the preferred method of storage during the early part of the 20th century. *Water glass* is a syrupy, concentrated solution of sodium silicate, available from some drugstores and on the Internet. Its purpose as an egg preservative is to minimize evaporation and inhibit bacteria. The water glass imparts no taste or odor, and although it causes a silica crust to develop on the outside of the shell, it does not penetrate the shell.

Put eggs in water glass the same day they are laid. Candle them and eliminate any with blood spots or meat spots. As with the other processes, use only clean (not cleaned) eggs that are free of cracks. Place the eggs in a scalded glass jar with a tight-fitting lid. A 1-gallon (3.8 L) jar will hold about three dozen eggs.

Combine 1 part water glass to 10 parts boiled water. If the solution is not diluted enough, it will become a gel that makes egg handling difficult. Mix the solution thoroughly, and let it cool. Slowly pour the cooled liquid over the eggs until the solution covers the eggs by at least 2 inches (5 cm). Do not save leftover solution. Screw the lid onto the jar to prevent evaporation. If you don't have many eggs at one time, continue adding eggs and fresh solution until the jar is full, always making sure the solution is at least 2 inches (5 cm) above the eggs at the top.

Store the jar in a refrigerator, basement, or other cool place where the temperature is preferably not higher than 40°F (4.5°C). At 35°F (1.7°C), eggs in water glass will keep for six months or more. If you wish to hard-cook an egg, use a pin or tack to poke a tiny hole in the big end to keep the shell from cracking as a result of the silica crust.

Even under the best storage conditions, water glass causes eggs to lose their fresh flavor and become bland tasting. The whites will eventually get thin, and the yolks will flatten when cracked into a pan, making them less suitable for frying or poaching than for scrambling or incorporating into recipes. But even at a temperature as high as 55°F (13°C), eggs in water glass will remain satisfactory for cooking for several months.

9
MANAGING A BREEDER FLOCK

You needn't concern yourself with maintaining a breeder flock if you are content purchasing pullets every two or three years to raise as replacement layers, or you like the idea of spending only a few weeks per year growing your annual broiler supply. But if you want to hatch chicks from your own eggs, you'll need to manage your chickens in a way that optimizes the fertility and hatchability of their eggs.

Most people who maintain a breeder flock do so for one of four main reasons:

- For the self-sufficiency of producing the household's poultry meat and replacement layers
- To enjoy the challenge of developing prize-winning chickens for show competitions
- To help preserve the genetic uniqueness of one or more of the endangered breeds
- To enjoy the experience of seeing baby chicks hatch from eggs laid by your own hens

Rise and Fall of Breeds

The chicken, as we know it today, is a man-made creature. All the various breeds were developed by human design from the wild jungle fowl of Southeast Asia. Although genetic differences distinguish one breed from another, exactly when a breed becomes a breed is purely a human invention. A *chicken breed* is commonly accepted as a family of genetically related individuals having the following:

- Shared physical characteristics, including the size and the shape of body, head, and comb
- Shared tendencies, such as broodiness or the laying of large or small eggs or brown or blue eggs
- The ability to reproduce offspring with the same distinguishable characteristics as the parent stock

In other forms of livestock, an animal's breed is identified by its papers. Chickens have no registry and therefore no papers, so some people argue that chickens have no breeds, only types. On the other hand, until DNA testing and blood typing came along, the papers of a registered animal were based on little more than the honesty of the person who registered the animal, which is no different from accepting the honesty of someone who sells you a chicken.

Breeding for show involves paying close attention to the details of both conformation, and plumage color.

Some chicken keepers are excited by the challenge of developing new breeds or varieties, but most are content with preserving or improving an existing breed or variety. Some people enjoy working with an established strain; others prefer to leave their mark by developing a unique strain.

Genetic Stocks

A *strain* is a family of chickens having recognizable characteristics that readily distinguish them from others of their breed and variety and the ability to transmit those characteristics to their offspring. A strain is the result of one person's (or organization's) vision and is developed by working for many generations with a single family of birds. If you work with a group of chickens so long that other poultry people begin recognizing a bird as having come from your flock, you have developed your own strain.

Every strain belongs to one of five main groups of genetic stock: exhibition breeds, commercial layers, commercial broilers, dual-purpose farmstead chickens, or sport breeds. Each of these groups has little contact with the others, and each carries emphasis on a different set of traits.

In any planned breeding program, a trait that's important to the breeder's goal is emphasized. An irrelevant trait is ignored and thus may eventually disappear. Any birds with a detrimental trait would be removed from the breeding flock.

In exhibition strains, emphasis is on traits involving form. In commercial laying and meat strains, emphasis is on function. Dual-purpose and sport strains combine form and function. While their goals differ, owners of dual-purpose and sport strains share two important characteristics:

- Both groups maintain low profiles — dual-purpose flock owners because they engage in the quiet business of home meat and egg production, sport-bird owners because they engage in the illegal business of cockfighting.
- Both groups take on responsibility for genetic stocks with the greatest degree of sustainability.

Form versus Function

Keepers of exhibition and industrial strains have done irreparable damage to the sustainability of the genetic stocks in their care — the former by breeding for extremes in appearance (form), the latter by breeding for extremes in production (function). To understand the degree of damage done, let's look at nine characteristics that may be important, irrelevant, or detrimental within each of the five basic genetic groups.

Broodiness — a hen's desire to hatch eggs — is detrimental to egg production because once a hen starts setting she stops laying. It is irrelevant to meat production because broilers don't live long enough to reach laying age, although it can be an issue for breeders who produce commercial broilers. It can be a nuisance in an exhibition strain: the more eggs you get from a valuable hen, the more offspring you can hatch and therefore the greater your chance of producing a perfect specimen.

Among sport strains, however, a hen's purpose is to perpetuate her strain by laying and hatching eggs. And part of the charm — not to mention labor reduction — in keeping a dual-purpose breed is having a broody hen hatch her own chicks, although the tendency toward broodiness varies even among dual-purpose farmstead breeds. As with industrial strains, breeds that are valued for their superior laying ability tend less toward broodiness than the heavier, broodier breeds.

Fecundity, or the ability to lay copious quantities of eggs, is all-important in commercial laying strains but of lesser importance among meat strains. In exhibition birds, laying ability would seem important for the production of hatching eggs, but in reality laying ability rates well behind characteristics of appearance, so much so that some exhibition strains have an abysmal track record in the laying department. In most dual-purpose flocks and among sport strains, poor layers are considered freeloaders and are culled so they won't reproduce more of the same.

Fertility is important in industrial breeding flocks but irrelevant in layer flocks (which don't include roosters). It is ostensibly important in exhibition stock, yet inbreeding small populations to focus on conformation leads to fertility loss, as does breeding for extremes of size. In a commercially developed hybrid dual-purpose flock, fertility is irrelevant but is essential for the perpetuation of a dual-purpose pure breed. Among sport birds, fertility is emphasized because a good brood cock makes a good battle cock and vice versa.

Foraging ability is irrelevant to industrial and exhibition stock, since both are kept in confinement — the former for reasons of labor efficiency and isolation from disease, the latter to protect flesh and plumage. In dual-purpose farmstead flocks, foraging has traditionally been important as a means of maintaining good health and reducing feed costs. Foraging is a significant source of exercise and nutrition

for sport birds, whose owners were among the first chicken keepers to plant special grasses for the grazing pleasure of their stock.

Plumage color is irrelevant in commercial layers, although most strains happen to be white. In meat birds, white feathering creates a cleaner finished appearance. The greatest variety in feather color occurs among exhibition birds, yet little variation is tolerated within each color variety. Furthermore, standard colors are sometimes contrary to natural tendencies, being derived by perpetually crossing different strains or even different breeds.

In traditional farmstead flocks and among sport strains, plumage color retains its original survival purpose — any color other than white offers camouflage for foragers and setters. Among sport enthusiasts, plumage color also identifies established bloodlines, called breeds in sport circles, but in reality they are different varieties of Old English Game.

Size is important among all five genetic stocks. For commercial layers, small size promotes efficiency. Among commercial meat strains, emphasis is on large size and unnaturally rapid growth. Exhibition birds must conform to the sometimes-arbitrary standard sizes and weights designated for each breed, with extremes in either direction tending to mitigate against fertility and fecundity. Dual-purpose flocks are, by definition, midsize, as a compromise between laying efficiency and meaty flesh — although the various farmstead breeds cover a broad range from lighter-weight layer types to heavier meaty types. Among sport strains, size relates to agility, 5 pounds (2.25 kg) being considered ideal.

Temperament takes a backseat to other traits in the selection of industrial strains. As a result, meat birds are prone to panic and piling (see page 305), and layer strains are notoriously nervous and flighty. Backyard dual-purpose flocks are usually bred for good temperament to enhance the keepers' enjoyment and ensure the safety of less nimble family members. Among exhibition strains, good temperament is essential, since calm birds show better than flighty ones. Game birds, too, are bred to be good-natured and gentle around people. In show and sport circles, a nervous, flighty bird is referred to as "wild," in contrast to a calm, gentle bird, which is "tame."

Conformation, or type, is important to all five genetic stocks, although it holds the least importance for owners of farmstead flocks. These breeders typically don't select against variations but rather tend to embrace them, often having a soft spot for the odd or unusual bird. As a result, the greatest diversity in type exists within farmstead strains. Industry, as you would expect, emphasizes strict conformity to traits specific to production, resulting in cookie-cutter layers and peas-in-a-pod broilers. In exhibition and sport strains, strict adherence to type is also an essential trait.

Although exhibition promotes the greatest number of overall types, it is intolerant of variations within each type, resulting in little diversity within each strain. Developing a strain with good show conformation involves inbreeding and selection, which often leads to loss of fecundity, fertility, and the self-sufficient ability to forage. In addition, the quest for uniqueness in type has allowed certain traits to flourish that would otherwise inhibit survivability. Frizzledness and silkiness, for example, offer less protection from the elements than smooth, webbed feathers, and vision-restricting crests hinder a bird's ability to get away from predators and to catch mates.

Vigor is a complex trait that embodies not only resistance to disease but also adaptability to the environment, freedom from lethal genes, and the ability to produce fertile, hatchable eggs. While everyone agrees that vigor is important, the trait is not high on everyone's selection list.

Broilers need only short-term vigor, since most are dispatched by 8 weeks of age. Texts have been written listing ways to keep unhealthy broilers alive until slaughter. Their health, like that of commercial layers, typically relies more on pharmaceuticals and other means of protection from disease-causing organisms than on inherent constitutional vigor and natural resistance to disease.

Exhibition birds must be hardy to stand up under the rigors of show, yet vigor is too often not a breeding priority. Farmstead flocks tend to have inherent vigor as a result of their longevity — only the strongest individuals survive to the second year and beyond and pass on their vigor to their offspring.

The most vigorous genetic stock is among chickens bred for sport. I saw a particularly awe-inspiring example at a major poultry show, entered by an unwitting sport breeder. The cock — which everyone who had gathered to take a look agreed was the best at the show — was the epitome of alertness and good health, was exceptionally well groomed and well tempered, yet it didn't stand a chance of winning because its size and type did not conform to the rigid criteria set forth in the *Standard of Perfection*. Its superiority as a representative of its species did not translate well into exhibition.

Vanishing Gene Pool

The incredible diversity represented by the five different genetic groups is a phenomenon of bygone days, when chickens were raised primarily in backyards and people had more time for experimentation. The trend toward diversity was reversed with the advent of industrial production, which concentrates genetic resources into a few strains that lay, or grow and convert feed, uniformly well.

The old lower-yielding breeds were left in the hands of backyard keepers. Unfortunately, as the interests of small-flock owners shift in other directions, or we lack heirs willing to carry on with poultry, the old varieties have been slipping away. In a process called *genetic erosion*, the gene pool is becoming more uniform and less diverse.

Genetic erosion is occurring not just among chickens but also among all livestock and plant crops, as well as in wild populations that cannot withstand destruction of their natural habitat. Although genetic erosion in general is accelerating at an alarming rate, by some accounts the loss of classic poultry breeds is far worse than losses among other livestock.

At the same time, no other form of livestock is so completely concentrated in so few genetic lines; a mere handful of companies, most of which lack long-term goals or a backup plan, maintain the industrial gene pool in a limited number of highly selective strains whose genealogy is a closely guarded secret. As a result, industrial chickens—which make up the greater percentage of the world's total chicken population — lack more than half of the total genetic diversity native to the species, leaving them vulnerable to newly emerging diseases and putting into serious doubt their long-term sustainability.

Yet it's no secret that the greater the genetic diversity, the better the odds are of finding individuals with the potential to improve characteristics or resist disease and other stresses that change with our changing environment. A growing number of enlightened souls, some of them within the industry, see the older breeds as a sort of insurance policy, since their traits may prove genetically useful in the future. Because poultry sperm and embryos, unlike those for cattle and other valuable stock, cannot be readily preserved through freezing, the only way to perpetuate poultry genetics is through the living flocks kept in backyards around the world.

If at some future date commercial producers turn to backyard breeders for help, it won't be the first time. In the 1940s, the broiler industry sought a broad-breasted breed to incorporate into their meat strains. They found what they needed in backyard flocks of exhibition Cornish. Ironically, at that time the Cornish had become nearly extinct — being among the many breeds that had been ignored in the scramble toward industrial egg production.

Pessimists say it's too late to save some of our endangered varieties, believing the numbers have already dwindled well below viable breeding populations. Optimists feel that any variety may be preserved, so long as you can find one pair to breed.

Breeding Plan

In small-scale poultry circles, people who collect and hatch eggs from their flocks are divided into two camps: the so-called *propagators* or *multipliers*, who emphasize quantity, and the *breeders*, who emphasize quality. Both groups hatch lots of chicks. To the propagator, the end goal is the large numbers of chicks. To the breeder, a large number of chicks is merely the means to an end — the more chicks you have, the more heavily you can cull; the more heavily you cull, the better the genetic quality of your stock. Breeding is therefore a long-term investment. The best breeders have been at it for decades.

Breeders look down on propagators because they know that leaving the matings to chance gets you nowhere, genetically speaking. Indeed, if you mix chickens of several different breeds and let them mate freely, eventually their descendants will begin to look like the wild jungle fowl from which they came.

All the domestic chicken breeds we know today were developed from jungle fowl of Southeast Asia.

To strengthen your strain's desired traits and maintain its quality, you need a well-thought-out breeding plan. Your plan should include these steps:

- Begin with the best birds available.
- Establish a long-range goal.
- Make deliberate matings to meet that goal.
- Keep meticulous breeding records.
- Mark chicks to track their parentage.
- Ruthlessly cull any bird that does not bring you closer to your goal.

Your breeding goal will depend on the quality of the stock you start with, compared with what you want. When selecting breeding stock, look for two things: individual superiority and good lineage. Avoid breeding a bird with poor ancestry, no matter how great it may appear.

Once you decide on a long-range goal, break it down into a series of short-term goals that will help you periodically gauge your success. Set a quality line, and don't mate any bird that falls below the line. Each year, raise your quality line a little higher. Concentrate your efforts on improving one trait at a time but not to the exclusion of others. It wouldn't do, for example, to concentrate on improving type to the exclusion of fecundity.

The most successful breeders specialize in one breed and one or only a few varieties within that breed. Each breed offers so much to learn genetically that concentrating your efforts increases your chance of success. After spending several years mastering one breed or variety, you may feel ready for the challenge of taking on a new breed or an additional variety within your chosen breed.

Breeding for Show

Breeding exhibition chickens involves mating for type, since type defines breed. It also involves mating for plumage color, since in most cases color defines the variety. Pay special attention to the color of your cocks; they have greater influence on the plumage color of offspring because cocks have two genes for color compared to only one in hens.

Select cocks and hens that closely resemble the ideal for their breed and variety as described in the relevant standard. Since every bird has both strong and weak points, avoid mating birds with the same fault. Instead, look for mates with opposite strong and weak points, but never breed a bird with great strong points if it also has serious faults for its breed.

In a show bird, temperament is nearly as important as type. A less typey bird that's tame will almost always win over a typier bird that's wild. Tame doesn't mean lacking in spirit, though. A good show hen is perky and likes to sing; a good exhibition cock is a show-off. Unfortunately, the more he likes to show off, the more aggressive he is likely to be. The typiest Silver Sebright I ever raised was so aggressive, he attacked me every time I fed him. Despite his fine looks, I removed him from my breeding program to avoid creating more of the same.

Not all chickens used to produce show winners are of show quality themselves. I visited a breeder who was known in his time as the man to beat at major poultry exhibits. He showed me a number of outstanding chickens he was offering for sale for prices I couldn't begin to think about paying. I asked if he was

Chickens bred for show should closely resemble the ideal for their breed and variety as described in the relevant standard.

bothered by the idea of selling a bird that later beat one of his at a show, and he said no — because both would be from his line. He kept steering me away from some birds at one side of the building that weren't nearly of the same fine quality as the others. When I asked how much he wanted for one of those, he smiled and said they weren't for sale. Why not? Because those were the breeders he used to produce the chickens that were winning at shows.

Breeding Layers

Small populations of inbred birds invariably decline in egg production, which is why show birds generally don't lay well. On the other hand, production responds well to crossbreeding, which is why commercial layers are bred by crossing birds from separate genetic lines. Breeding for commercial production is quite complex, requiring the skills of highly trained specialists. Taking a tip from them, you can markedly improve a strain's laying ability by crossing two different inbred strains.

Good laying ability is not a highly *heritable* trait, meaning it is less influenced by genes passed directly from a hen to her daughters than by environmental factors. A better indication of a hen's worth as a breeder is the average rate of lay of all the hens in her family line. If the family, in general, consists of good layers, the hen is likely to pass the ability to her offspring, even if she herself is not a particularly outstanding layer. At the same time, an outstanding layer is not likely to pass her ability to offspring if her rate of production is not typical of her family's average.

Hatching eggs from hens that are at least 2 years old gives you plenty of time to evaluate their family track record, and hens that are still laying well at 2 years are likely to pass along to the next generation not only their laying ability but also their vigor and longevity. As they age and their egg production declines, the hens that laid well during their first and second year will continue to pass their superior qualities along to their chicks.

If the color of your eggshells is important, breed only hens that lay eggs of the desired hue or shade. Araucanas that lay eggs with shells of any color other than blue can, in several generations, be brought back to laying eggs with blue shells. Persistently hatch only eggs with the bluest shells, use as breeder cocks only those that hatched from a blue-shell egg, and take great care not to breed the offspring back to birds that lay eggs with shells that are not blue.

For all breeds, select cocks from your most productive hens. To maintain the depth of body needed in a good laying hen, look for breeder cocks with wide backs and an ample space between the keel and pelvic bones.

Breeding Broilers

Breeding for commercial meat production shares much in common with breeding commercial layers: both are complex and highly specialized, concentrate genetics in a limited number of strains, and involve crossing strains. Among broiler breeders, some specialize in raising sire lines, others in dam lines; many work with both. Commercial broilers may carry two-part hyphenated names, such as Peterson-Cobb. The first name represents the source of the sires; the second, the source of the dams.

The resulting broiler chicks grow faster than any you could hatch yourself. But if one of your purposes in keeping chickens is to produce meat as well as eggs, it makes sense to select your breeders for the good qualities of meat producers as well as layers.

The characteristics of good meat birds — rapid growth and efficient feed conversion — are unlike laying ability; these traits are quite heritable. Since a fast-growing bird passes the trait directly to its offspring, breeders are selected on the basis of having the greatest weight among flock mates at 8 weeks of age. At the same time, avoid using as breeders any bird that grows significantly faster than is typical for the breed — those extremely fast growers tend to be the least healthy.

At maturity, each bird selected as a breeder should weigh within ½ pound (0.25 kg) of the standard weight for the breed. Since meat is muscle, along with size goes a body type that accommodates good muscling — a broad breast; a spacious heart girth; a deep body; and a wide, flat back.

Breeding for Recovery

Trying to recover a rare or endangered breed differs from other breeding plans primarily in the number of chickens available to select. You may have to include inferior birds in your initial breeder flock, if they are all you can find of the breed you choose to work with.

To develop a sensible breeding plan, you must first identify the characteristics typical of your chosen breed. In selecting future breeders, The Livestock Conservancy suggests you focus on six basic qualities:

MEAT-BIRD CHARACTERISTICS

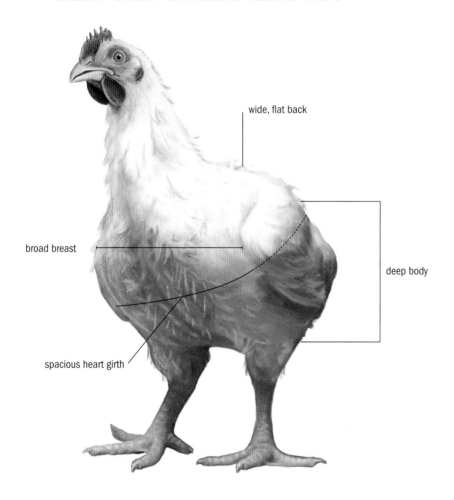

A chicken used to produce future generations for meat should have a broad breast, wide back, deep body, and spacious heart girth.

Rate of growth can influence the immune system. Extremely fast- or slow-growing chickens have less-robust immune systems than chickens that mature at a normal rate for the breed. Maturity ranges between 12 and 18 weeks of age for most breeds; few mature earlier, but some breeds mature at a much later age. Jersey Giants, for instance, take 26 weeks to mature, and some longcrowers stretch it out to 18 months.

Mature weight should fall within ½ pound (0.25 kg) above or below the breed's ideal weight. In most, but not all, breeds the ideal weight differs for cocks and hens.

Egg-laying ability is of primary importance for laying breeds, of significance for dual-purpose breeds, and of some importance for meat breeds. Regardless of their expected rate of egg production, most breeds begin laying by 6 months of age.

Breed type affects the size and shape of the internal organs and the distribution of flesh, thus affecting the breed's suitability for the purpose for which it was originally developed. Type also influences adaptability to the environment, including climate.

Plumage color is an indication of breed purity. Slight variations among family lines sometimes help identify the strain.

Fertility and vigor give chickens the ability to withstand inbreeding, disease, and other challenges. Good indications of vigor are dominance in the peck order for young birds, longevity and fertility in mature chickens, and liveliness at all ages.

One of the major problems in working with a limited number of breeders is the loss of fecundity and fertility that typically occurs with inbreeding. The Livestock Conservancy suggests you can retain the greatest degree of genetic diversity by keeping as future breeders the best representatives of each mating, rather than taking the traditional approach of keeping the best individuals overall. A five-year breeding plan for the recovery of endangered dual-purpose breeds is included in *Chicken Assessment for Improving Productivity*, available from the The Livestock Conservancy both in print and as a free download from their website.

Methods of Breeding

Not all improvements occurring from one generation of birds to the next are due to inheritance. Some may result from good management or from what is sometimes called favorable accident. The amount of improvement attributable to inheritance is called heritability, and not all inherited characteristics are equally heritable.

In general, conformation characteristics are highly heritable, while production characteristics are not. To perpetuate traits with low heritability, you must select your breeders on the basis of family averages. To perpetuate traits with high heritability, you must select your breeders on the basis of individual superiority of those traits. Hence the evolution of two different breeding methods: *flock breeding*, which emphasizes the overall performance of a production flock; and *pedigree breeding*, which emphasizes individual characteristics.

Flock breeding is often practiced commercially and is also suitable for small-scale dual-purpose and layer flocks. Pedigree breeding is most often used for exhibition and sport strains and to implement the recovery of endangered breeds. Pedigree breeding itself has two common methods, individual mating and pen breeding.

Individual mating may be accomplished either by mating a single cock with a single hen or by rotating one cock every two days among three or four individually penned hens. Individual matings may be tracked, or pedigreed, with certainty, since you know exactly which cock fertilized which hen's eggs.

Pen breeding involves mating a cock with a small number of hens together in one pen. It is similar to flock breeding except that smaller numbers of hens are mated to only one cock at a time. Unless you can identify which egg comes from which hen, pen-bred birds may be pedigreed only on the cock's side. You may be able to positively identify parentage if each hen lays an egg of unique size, color, or shape. The only way to be absolutely certain of each hen's identity is to trap the hen when she goes into a nest to lay.

Linebreeding

The most common form of pedigree breeding is *linebreeding*, in which the influence of a superior sire or dam is concentrated by mating the bird to his or her best descendants. Pullets are mated to their sires or grandsires, cockerels are mated to their dams or granddams.

A good linebreeding plan includes four or more related families, starting with the best cock and four best hens available. Each family line consists of all the female offspring from one hen. The advantage to maintaining several lines is that if one of them fails to live up to your ideals, you can easily scrap it and start over with a new foundation female from within the same strain. If you are working with an endangered breed, maintaining several lines — including those that might initially produce inferior offspring — preserves genetic diversity.

In most cases, you won't know whether the matings you have chosen will prove successful until your third year of linebreeding, when both desirable and undesirable genes become concentrated. The third year is therefore the time when novice breeders are most likely to get discouraged and quit. The more separate lines you keep, the more you decrease your chance of getting discouraged by increasing your chance of producing one or more successful lines.

Double mating is used for breeds or varieties of show stock in which the male and female differ in type or color. It involves maintaining two different breeding lines, which may or may not be related to one another. To get good show cocks, match your best male to females that tend toward cock color or type, even though they themselves aren't suitable for showing. To get good show females, mate your best females to a cock with hen color or type, even though he would not be suitable for showing. Exhibit only the top cocks from your male line and top hens from your female line.

Breeding Bird ID

Pedigreeing requires keeping track of each chick's ancestry by giving every breeder an identification code and marking every hatching egg with the identity of the mating that produced it, so each hatched chick may be traced back to a specific mating. Breeding birds are identified by means of numbered wing badges, wing bands, or leg bands available through poultry-supply catalogs.

A **wing badge** is similar to the plastic ear tag worn by four-legged livestock. It attaches to the wing and has numbers big enough to be read without handling the chicken. Because wing badges are unsightly, they are not often (if ever) used for backyard flocks.

TRAPNESTING

A *trapnest* is a nesting box fitted with a trapdoor. When a hen enters to lay, the door shuts and locks. The hen remains trapped in the nest until you let her out, at which time you collect her egg and mark it with the hen's identification code and the date, using a grease pencil, china marker, or felt pen. Some people use a soft pencil, but they run the risk of piercing a shell, and if the eggs are hatched under a hen, the penciled codes rub off.

To accustom hens to entering the trapnests, install the doors in advance and secure them in an open position. When you start trapnesting, check nests often — every 20 or 30 minutes during prime laying time. A hen confined for too long in the nest may soil or break her egg while trying to get out, and in warm weather she may suffer from being enclosed too long.

You can make wooden trapnest doors or buy nest fronts made of heavy-gauge wire.

Although ready-made fronts are designed to fit industrial metal nests, you can easily fit one to a wooden nest by fastening it to a wooden frame designed to fit the front of your nest and attaching it to the nest with screws.

Trapnest fronts mounted on industrial metal nests.

Traps may be folded up out of the way when not in use.

Trap is set and awaiting entry by a hen.

Trap has been tripped by hen.

The open construction of a store-bought all-wire trapnest front lets you easily see the hen in the nest and provides her with good ventilation in warm weather; this wire trap front is fitted to a homemade wooden nest.

Homemade Trapnests

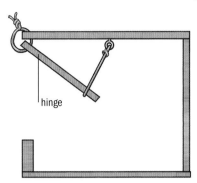

HOOK STYLE: A lightweight wooden door is hinged with #9 wire to swing freely. A hook made of #9 wire is set to hold the door open 6⅜ inches (16 cm) above the nest floor. A hen entering the nest brushes the hook loose with her back. The nest must be about 6 inches (15 cm) deeper than normal, so the trapdoor can close behind the hen.

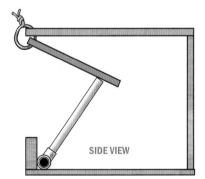

SIDE VIEW

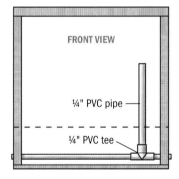

FRONT VIEW

¼" PVC pipe

¼" PVC tee

PROP STYLE: A lightweight wooden door is propped open by a PVC pipe with a T at the bottom end. A bolt or heavy-gauge wire through the T holds the prop in place. The prop falls when a hen enters the nest, causing the trapdoor to close behind her.

prop

hook

With either style trapnest, a sill prevents the hen from pushing the door open to get out. After laying her egg, she remains confined until she is set free. Holes drilled in the trapdoor are needed for ventilation.

A **wing band** is an aluminum strip that is attached through a chicken's wing web, usually when the bird hatches, as described on page 292. The embossed numbers of a wing band offer a permanent and reliable means of identification for the life of the bird. One size fits all, so unlike a leg band, the wing band needn't be changed as the chicken grows.

A **leg band** is plastic or aluminum ring that wraps around a chicken's leg, and must be fitted to the chicken's size. Leg bands are the most common means of identifying grown chickens, but when used on a young bird, must be constantly changed as the bird's leg grows. Leg bands come in different sizes and colors, with or without numbers.

Plastic spiral bands, similar to key rings, are the least expensive option, but they tend to break easily and come in only a limited number of colors, although many breeders use them in color-code combinations. They generally come in lots of 50, all of one size and color. We use them to identify certain characteristics, such as hens that make good broodies, a group of pullets resulting from a specific hatch, or a chicken we suspect (but aren't yet certain) might be eating eggs. Some exhibitors band their birds according to show wins, applying a blue spiral to any bird that takes first place, a red one for a second place winner, and so forth (in Canada it's the other way around, red denoting a winner, blue for the runner-up). Others assign a different color spiral to each year, so they can tell the age of each chicken at a glance.

Zip ties, or cable ties, are used in place of spirals by breeders looking to save money. Like spirals they come in various colors, but unlike spirals they are not reusable. They are also not as flexible, so you have to be extremely

careful not to get one too tight. On growing birds, that means checking often and replacing the zip ties as the birds grow; cutting one off after it becomes too tight is difficult at best. To ensure the chicken doesn't pick on the end and tighten the zip tie, after fitting the tie to the leg, cut off the excess portion. Spirals and zip ties may be used alone or in combination with bands that identify individual birds by number.

Numbered plastic bandettes, like spirals, break after a while, so a good plan is to have two sets and band both legs with bandettes of the same color and number. They generally come in lots of 25, all of one size and color but sequentially numbered. Different colors might be used to denote different years or family lines. Breeders who double-mate typically use blue bandettes for the male line and red (or pink) ones for the female line.

Adjustable aluminum bands come in sequentially numbered colored metal strips, as well as plain aluminum, and have the advantage over single-size plastic bandettes of being adjustable to fit legs of more than one size. One style has a narrower end that fits into a slot at the wider end, and then is folded back on itself. No applicator is needed, although with a pair of ordinary needle-nose pliers you can tighten down the fold. Another style has a rivet at one end and a series of holes at the other end; the rivet is inserted into one of the holes and an applicator tool is used to clinch the rivet. This style is meant to be permanent, so is not used on birds that are still growing.

Butt-end bands, so called because the two ends butt together rather than overlap, are the most expensive leg bands and also the most elegant looking. They come in aluminum, nickel alloy, or stainless steel and may be customized with such information as a bird's pedigree data or your name and address. Sealing them requires a different applicator tool for each size. Like riveted aluminum bands, these bands are designed as a permanent means of identification and may be applied only to birds reaching full size.

To ensure that a leg band is neither so small it binds nor so big it falls off, you'll need to get the right size for your birds. Different breeds may require different sizes, and even within a breed the cocks usually require a larger size than the hens. Sizes range from #2 to #16, not all of which are suitable for use with chickens. Most bantams fall between #5 and #9, and most large breeds fall between #9 and #12. The size denotes the band's inside diameter in sixteenths of an inch; the diameter of a #7, for instance, is $\frac{7}{16}$ inch (11 mm) and a #9 is $\frac{9}{16}$ inch (14 mm).

Whenever you apply any kind of leg band, make sure it's loose enough that you can move it up and down on the shank but not so loose that it falls off when the chicken folds its toes together. If you band a growing bird, check frequently to make sure it doesn't get too tight as the bird matures. As you might imagine, a band that becomes imbedded in a leg causes serious problems.

Radio frequency identification (RFID) devices are available on both leg and wing bands. They are rarely used by backyard breeders, because accessing RFID information requires a reader, or scanner, similar to those used to read price tags at the checkout counter, and not all scanners work with all RFID bands. Similarly, wing and leg bands come barcoded, but require a barcode scanner.

Progeny Testing

The primary goal of pedigreeing is *progeny testing*, or measuring the value of breeders by the quality of their offspring. Pedigreeing makes no sense unless you keep meticulous mating records. Prepare a separate sheet for each mating, identifying the sire and dam (or pen number, for pen-bred birds). Then list the number of eggs you set, the number that were fertile, the number hatched, and the date they hatched.

List each chick individually so you can track it to the next generation. As the chicks grow, tabulate information that's important to your breeding program, such as gender, variations in type and color, size, temperament, age at first lay, laying ability (number of eggs per year), and the date the bird was sold or culled. If the bird was culled, note why.

When you find an individual or mating that produces superior offspring, repeat the mating as often as possible. Avoid repeating a mating that produces unhealthy or otherwise substandard chicks — unless you are engaged in a recovery program that requires preserving the greatest possible degree of genetic diversity, in which case retain as future breeders the best offspring resulting from each mating.

Progeny testing is a long-term process that involves keeping track of all birds (not just the occasional superior bird) from each mating. You may find that certain matings produce outstanding traits in one area, while others produce outstanding traits in another area. To produce superior birds combining both traits, you might set up separate breeding lines for each.

AVERAGE LEG-BAND SIZE BY BREED

Breed	Bantam Hen	Bantam Cock	Standard Hen	Standard Cock
Ameraucana	6	7	9	11
American Game Bantam	6	7	—	—
Ancona	5	6	9	11
Andalusian	7	7	11	12
Appenzeller	5	6	9	9
Araucana	6	7	11	12
Aseel	—	—	9	11
Australorp	6	7	11	12
Barnevelder	—	—	11	12
Bearded/Booted Bantam	6	7	—	—
Belgian	7	7	—	—
Booted	9	9	—	—
Brahma	11	11	12	12
Buckeye	9	9	11	12
Campine	5	6	9	9
Catalana	7	7	11	12
Chantecler	7	7	11	12
Cochin	9	9	12	12
Cornish	7	9	12	14
Crevecoeur	6	7	11	12
Cubalaya	6	7	9	11
Delaware	7	7	11	12
Dominique	6	7	11	12
Dorking	7	7	11	12
Dutch Bantam	5	5	—	—
Faverolle	6	7	11	12
Fayoumi	—	—	9	9
Frizzle, clean leg	7	7	9	11
Frizzle, feather leg	9	9	12	12
Hamburg	6	7	9	9
Holland	7	7	11	12
Houdan	7	7	11	12
Japanese	6	7	—	—
Java	7	7	11	12
Jersey Giant	9	11	12	14
Kraienkoppe	—	—	9	11
La Fleche	6	7	11	12

AVERAGE LEG-BAND SIZE BY BREED

Breed	Bantam Hen	Bantam Cock	Standard Hen	Standard Cock
Lakenvelder	5	6	9	11
Langshan	9	9	12	12
Leghorn	5	6	9	11
Malay	7	7	11	12
Maran	7	9	11	12
Mille Fleur	9	9	—	—
Minorca	6	7	9	11
Modern Game	5	6	9	11
Naked Neck	7	7	11	12
Nankin	5	6	—	—
New Hampshire	7	7	11	12
Norwegian Jaerhon	—	—	9	9
Old English Game	6	7	9	11
Orloff	7	7	11	12
Orpington	7	7	11	12
Penedesenca	—	—	9	11
Phoenix	6	6	9	9
Plymouth Rock	7	7	11	12
Polish	6	7	9	11
Redcap	6	7	11	12
Rhode Island Red	6	7	11	12
Rhode Island White	6	7	11	12
Rose Comb	6	7	—	—
Sebright	5	7	—	—
Shamo	7	7	12	12
Sicilian Buttercup	5	6	9	11
Silkie	9	9	—	—
Spanish	7	7	11	12
Sultan	6	7	9	11
Sumatra	6	7	9	11
Sussex	7	7	11	12
Vorwerk Bantam	6	7	—	—
Welsumer	7	7	11	12
Wyandotte	7	7	11	12
Yokohama	5	6	9	11

The Gene Connection

To interpret the results of your breeding efforts and make necessary adjustments, you have to know a little about *genes* — the hereditary units that transmit characteristics from parents to offspring. Each chick acquires two sets of genes, one from its sire and one from its dam. The genes match up into pairs of like function; for example, controlling comb style or feather color.

Paired genes that are identical to each other are called *homozygous* (homo- from the Greek word *homo* meaning "same"; -zygous from *zygos* meaning "pair"). A bird with a large number of pairs of identical genes yoked together is also described as homozygous. The more closely birds are related, or inbred, the more homozygous they become and the more predictable their offspring.

When the genes in a pair differ from each other, the pair is called *heterozygous* (hetero- from the Greek word *heteros* meaning "different"). The same word describes a bird with a large number of paired dissimilar genes. When heterozygous chickens are mated, or an outcross has been made, the genes in their offspring can pair in many different combinations, making the results highly unpredictable.

Dominant versus Recessive

Each chicken has a combination of dominant and recessive genes. If a dominant gene pairs up with a recessive gene, the dominant gene overshadows or modifies the recessive gene, and the dominant trait prevails.

As long as you work with heterozygous birds, recessive genes can remain hidden to pop up at any time. Old-timers use the word "throwback" to describe a bird displaying traits that have been hidden for several generations. A recessive trait shows up when two genes in a pair control the same recessive trait. Since a homozygous chicken is more likely than a heterozygous chicken to have a large number of matched pairs, the more inbred a bird is, the more likely it is to display recessive traits.

Revealing recessive traits can be a good thing or a bad thing. If the recessive is desirable, you want to encourage it. If it is undesirable, you want to weed it out, which is possible only if you maintain sufficient genetic diversity to prevent the concentration of undesirable recessives in all your breeders.

Not all traits are controlled by dominant or recessive genes but rather by combinations of genes. An example is *rumplessness*, a genetically complex feature of Araucanas determined by an interaction among many different genes. One of the best guides to the genetic complexities of heredity in chickens is *Genetics of the Fowl* by F. B. Hutt.

EXAMPLES OF DOMINANT AND RECESSIVE TRAITS

DOMINANT		RECESSIVE	
· 5 toes	· side sprigs	· 4 toes	· wry tail
· feathered legs	· frizzledness	· stubs	· silkiness
· crest		· single comb	

Rumplessness is a feature of Araucanas determined by the interaction of many different genes.

Lethal Genes

Among the recessive traits concentrated by inbreeding are *lethal genes*. A chick that acquires the same lethal gene from both parents dies early, often in the embryo stage. More than 50 different lethals have been identified in chickens. They are readily recognized because they are usually accompanied by such quirks as stickiness at hatch, shaking, winglessness, twisted legs, missing beaks, and extra toes.

When you mate two birds carrying the same lethal recessive, 25 percent of their offspring will display the lethal trait. One well-known lethal is the so-called creeper gene carried by short-legged Japanese chickens, once prized as broodies because their short legs keep their bodies close to the ground. Other well-known lethals are carried by Araucana, dark Cornish, New Hampshire, and white Wyandotte. Fortunately, lethal genes are relatively rare.

A good way to find out about lethals and other problem genes in your chosen breed is to chat with experienced breeders at poultry shows. Even if you don't intend to show your chickens, you can glean a wealth of genetic information about your breed from people who do. Other ways to obtain information are by joining a breed club and by participating in discussion groups on the Web. Some breeds have entire books written about them. You can find out if a book has been written about your chosen breed through the breed club, poultry suppliers, experienced breeders, or an online search.

Inbreeding Depression

By concentrating genes, inbreeding not only creates uniformity of size, color, and type, but it also brings out weaknesses such as reduced rate of lay, low fertility, poor hatchability, and slow growth — a phenomenon called *inbreeding depression*. Inbreeding doesn't cause these problems but does accentuate any tendency toward them.

To minimize inbreeding depression, avoid brother-sister and offspring-parent matings and instead mate birds to their grandsires or granddams. Retain as future breeders those with the best fertility, hatchability, chick viability, disease resistance, and body size. Never breed birds with any tendency toward infertility. By inbreeding gradually and choosing your breeder cocks and hens carefully, you can improve such traits as laying ability and disease resistance.

Some strains are less susceptible than others to the effects of inbreeding depression. And popular breeds with lots of varieties offer more opportunities for avoiding inbreeding depression than less popular breeds with fewer numbers, and breeds with few varieties.

Outcrossing

Outcrossing means increasing heterozygosity by introducing a bird that is not directly related to your bloodlines. In an inbreeding program, red flags indicating that an outcross is needed include the following:

- Unexpected appearance of an undesirable trait
- Rapid or drastic reduction in fertility, hatchability, chick viability, or general health
- Continuing lack of improvement, indicating that your birds simply do not carry the right genes

When you bring in new blood, you may not see the changes you desire until the second generation. Meantime, you run the risk of introducing new weaknesses, a hazard you can minimize by crossing your strain with distantly related birds, called a *semi-outcross*. A semi-outcross is essential if you're trying to improve type, since conformation cannot be improved by crossbreeding.

To further reduce the risks of outcrossing, select new blood from a strain that is not deficient in any of the properties you have been working to establish and that has been properly inbred. A properly inbred sire or dam is likely to be *prepotent*, or able to pass on its attributes to the majority of its offspring. Prepotency can result only from homozygosity.

If you work with a breed or variety that has few distinct bloodlines, the time may come when you have no choice but to outcross to a different breed or variety. If you're breeding show birds, you'll then have to work to bring them back to type. To avoid, or at least delay, the need for such an outcross, retain as future breeders the best offspring from each mating rather than the best offspring overall.

Hybrid Vigor

Outcrossing results in *hybrid vigor*, the opposite of inbreeding depression. Hybrid vigor is a phenomenon whereby a chick is better than either of its parents. Traits with low heritability that show the greatest degree of inbreeding depression — such as reproductive performance and chick viability — react the most favorably to hybrid vigor.

To realize the greatest benefits of hybrid vigor, you must maintain a high degree of heterozygosity by continually outcrossing or semi-outcrossing, which entails a constant search for new blood. If your goal is to preserve genetic diversity, not only shouldn't you cross different breeds but you shouldn't mix established strains within a breed.

In deciding whether to create homozygosity through inbreeding or heterozygosity through crossbreeding, consider these two points:

- How important is predictability to your breeding program?
- Do you prefer to hide your birds' genetic weaknesses and hope they never surface or force them to the surface so you can eliminate them?

Culling for a Healthy, Hardy Flock

A breeding flock can degenerate rapidly if you make no effort to select in favor of health, vigor, hardiness, and good reproduction. Birds that don't measure up should not be used to produce future generations and should be culled. Some backyard poultry keepers use culling as an opportunity to fill the freezer. Others sell their culls to people with less demanding needs, such as those who keep chickens as pets or wish to produce eggs for eating rather than for hatching.

Problem birds, especially those with health issues or a mean streak, should be humanely killed rather than being passed along — together with their problems — to someone else. In pursuit of your breeding goals, keep only your best offspring or, if you're working with limited genetics, the best from each mating. Get rid of the rest, even though they may be a large percentage of each hatch. A good rule of thumb for developing a critical eye is to retain for future breeding only about 10 percent of each year's progeny.

To ensure that your initial breeders, or *foundation stock*, are worth devoting years of work to, acquire birds from someone who specializes in the strain you're interested in, has worked with it for a long time, freely offers details about its background, and has a good reputation among fellow poultry breeders. Once you acquire your foundation stock, improving the stock to meet your goals is a matter of mating and culling.

Culling is an ongoing process. It starts when chicks hatch, and any deformed chick or runt is removed. As the birds grow, cull in favor of good health and resistance to disease. If you choose to nurse a sick chicken back to health, do not include that bird in your breeder flock, as it will pass its weak immunity to future generations. Since evidence shows cannibalism is heritable, all other things being equal, cull birds that feather-pick or otherwise exhibit cannibalistic behavior.

Cull birds that develop slower than is normal for your breed, are not energetic, or might otherwise be described as unthrifty. The keepers are the ones that show some improvement over their parents. Don't just visually inspect the birds, but handle each individual to check for skeletal irregularities or outright deformities. Any bird that does not measure up should not be retained as a breeder.

Aside from removing obviously ill, deformed, or other undesirable individuals, double-check your decisions by noting which ones you plan to keep and which ones you plan to cull. Then revisit your decisions later to make sure you still feel the same way. I make my first pass when my chicks are about 8 weeks of age and continue reviewing my decisions as the birds mature.

At 8 to 12 weeks, depending on how rapidly your breed matures, cull slow-growing birds and any with poor conformation. If you're breeding for egg production or for show, the first heavy culling should occur when your flock reaches about 30 weeks of age. The pullets will be reaching peak production and the cockerels will look their best at that age.

A second culling time for layers is toward the end of the first year of production, since a good hen lays for at least 12 months, while a poor layer takes an early break. If you're raising exhibition birds, select your future breeders according to the relevant standard, culling more heavily against disqualifications than defects.

Regardless of your purpose, cull in favor of good temperament — it's no fun raising chickens that are wild, aggressive, or downright mean. Although you'll hear all manner of advice on how to cure meanness, the only sure cure is to breed for good disposition.

After the first generation, you can start identifying and culling problem breeders so your flock will include not only good birds but also birds that transmit their good qualities to their offspring. The older your breeder flock, the better the chances they will pass along to future generations such essential qualities as vigor and disease resistance.

Feeding Breeders

Assuming your breeder flock is healthy and free of both internal and external parasites, good nutrition is the greatest factor in promoting fertility and hatchability. Poor breeder-flock nutrition can arise because rations are

- Poorly balanced
- Insufficient
- Nutritionally deficient

Ration Balancing

The main cause of poorly balanced rations is feeding breeders too much scratch grain or other treats. To ensure the proper protein/carbohydrate balance, reduce your flock's grain ration about a month before the hatching season begins.

See that your breeders get enough to eat for their size, their level of activity, and the time of year. Feed either free choice or often so those lowest on the totem pole will get a turn at the hopper. Examine each bag of feed to make sure it isn't dusty, moldy, or otherwise unpalatable, which will cause your chickens to eat less.

Periodically weigh a sampling of both cocks and hens — a weight loss of more than 10 percent can affect reproduction. Underfed cocks produce less semen; underfed hens don't lay well.

A hen's diet affects the number and vitality of her chicks and the quality of carryover nutrients the chicks continue to absorb for several weeks after they hatch. Nutritional deficiencies that may not produce symptoms in a hen can still be passed on to her chicks. Feeding the breeder hen therefore includes feeding her not-yet-hatched chicks.

The same ration that promotes good egg production won't necessarily provide embryos and newly hatched chicks with all the elements they need to thrive. *Layer ration* contains little animal protein and too few vitamins and minerals for proper hatching-egg composition and high hatchability. Feeding lay ration to a breeder hen may result in a poor hatch or nutritional deficiencies in her offspring. The older the hen, the worse the problem becomes.

To improve your hatching success, feed your flock a bona fide *breeder ration*, rich in necessary nutrients, starting two to four weeks before you intend to begin hatching. If you're lucky, you'll find breeder ration at your feed store. Be sure it's fresh, and use it within two weeks after it was mixed. Even if you feed your flock the best breeder ration in the world, nutritional deficiencies will result if the feed is stored so long that the fat-soluble vitamins are destroyed by oxidation.

If you can't find a ready source of breeder ration at your feed store, the closest alternative may be a game-bird ration. If neither is available, six weeks before you begin collecting hatching eggs, supplement rations with animal protein as you would for molting (see page 195) and add a vitamin/mineral supplement to the drinking water.

Supplements

Excessive embryo deaths during mid- and late hatch may be a sign of low vitamin levels in the breeder ration. If you have reason to question the freshness of the ration you use, feed your flock natural supplements or add vitamins to their drinking water prior to and during the breeding season.

Vitamin A is essential for good hatchability and chick viability. It comes from green feeds, yellow corn, and cod liver oil. If you use cod liver oil, keep it fresh by mixing it into rations at each feeding.

Vitamin D is related to the assimilation of calcium and phosphorus needed for egg production. Deficiency causes shells to become thin. Since an embryo takes calcium from the shell, thin-shelled eggs may produce stunted chicks. Two signs of deficiency are a peak in embryo deaths during the nineteenth day of hatch and chicks with rickets. Vitamin D can be supplied by cod liver oil and sunlight.

Vitamin E affects both fertility and hatchability. It comes from wheat germ oil, whole grains, and many fresh greens.

Riboflavin, one of the B vitamins, is often deficient in poultry rations, resulting in embryo deaths in early or mid-incubation, depending on the degree of deficiency. Chicks that do hatch may have curled toes and may grow slowly. Riboflavin comes from leafy greens, milk products, liver, and yeast.

Calcium deficiency can cause thin-shelled eggs and reduced laying. Calcium excess can reduce hatchability. Supplying a calcium supplement free choice rather than mixing it into rations allows for differences in the needs of individual hens.

For fast results, use aragonite (calcium carbonate) or agricultural limestone (calcitic limestone, which is mostly calcium carbonate), not dolomitic limestone (which contains a higher percentage of magnesium carbonate), as the latter is detrimental to egg production and hatchability. Crushed oyster shell is a good long-term supplement.

The touchy problem of providing breeders with adequate nutrition will be compounded if you hatch early in the season, before your hens have access to fresh greens. If you live in a northern area, you can measurably improve your hatching success by supplementing with sprouted grains, or alfalfa meal or pellets, or by delaying incubation until your birds have access to fresh forage and plenty of sunlight.

Mating Logistics

When a cock mates with a hen, sperm travel quickly up the oviduct to fertilize a developing yolk. If the hen laid an egg shortly before, the mating will likely fertilize her next egg. The number of additional eggs that will be fertilized by the mating varies with the hen's productivity and breed. Highly productive hens generally remain fertile longer than hens that lay at a slower rate, and single-comb breeds remain fertile longer than rose-comb breeds — possibly as long as a month, but that's pushing your luck. The average duration of fertility is about 10 days.

However, if you switch cocks, the eggs laid after making the switch are more likely (but not guaranteed) to be fertilized by the new cock than by the old one.

To be reasonably certain eggs are fertilized by the new mating, wait at least two weeks before collecting them for hatching.

Although an egg must be fertilized to hatch, not all fertilized eggs do hatch. Eggs fail to hatch for a variety of reasons, not all of them easy to determine. One possible reason is that the embryo died before incubation began. This phenomenon, commonly known as *weak fertility*, may have nothing to do with fertility at all but may be due to other deficiencies within the egg.

Like all reproductive qualities, fertility has low heritability. Aside from problems related to inbreeding depression, management factors

MALE/FEMALE MATING RATIOS		
For 1 Cock of the Breed	**Optimum Number of Hens**	**Maximum Number of Hens**
Bantam	18	25
Light breeds	12	20
Heavy breeds	8	12

(rather than inheritance) are more likely to play a significant role. The many possible reasons for low fertility include the following:

- The flock is too closely confined.
- The weather is too warm.
- Breeders (both cocks and hens) get fewer than 14 daylight hours.
- The cock has an injured foot or leg.
- Breeders are infested with internal or external parasites.
- Breeders are diseased; especially troublesome for hatching eggs are chronic respiratory disease, infectious coryza, infectious bronchitis, Marek's disease, and endemic (mild) Newcastle disease.
- Breeders are too young or too old.
- Breeders are stressed due to excessive showing.
- Breeders are undernourished.
- The cock-to-hen ratio is too high or too low.

Battle of the Sexes

The ideal cock-to-hen ratio for pen breeding is influenced by the cock's physical condition, health, age, and breed. On average the optimum ratio for heavier breeds is 1 cock per 8 hens, although a cock in peak form can handle up to 12 hens. The optimum ratio for lightweight laying breeds is 1 cock for up to 12 hens, yet an agile cock may accommodate 15 to 20. The mating ratio for bantams is 1 cock for 18 hens, although an active cockerel might handle as many as 25. An older cock or an immature cockerel can manage only half the hens of a virile yearling.

If you have too many cocks, the fertility rate will be low because the cocks will spend too much time fighting among themselves. If

you have too few cocks, the fertility rate will be low because the cocks can't get around to all the hens. If you have only one cock and more than half a dozen hens, the cock will favor some hens and ignore the others.

Since cocks tend to play favorites, it pays to switch them periodically. The more cocks you have to switch off, the less likely you are to experience inbreeding problems; it's far better to have chicks sired by several different cocks than hatch the same number of chicks all with one sire. And by keeping extra cocks, you won't be left high and dry if your favorite rooster becomes incapacitated during the breeding season.

If you use only one cock at a time, house the extras in separate pens or cages. My friend learned this lesson the hard way. He knew one rooster was enough for his dozen hens, but he kept a second one as insurance. The two cocks fought so incessantly that feeders and perches were soon painted with blood, and the hens produced no fertile eggs. The exasperated fellow sold one cock to a neighbor, whose coop was within earshot. Whenever the neighbor's cock crowed, the homeboy flew up to the rooftop to answer in defiance. One day when the cock jumped down from the roof he broke his foot, abruptly ending my friend's breeding season. This unfortunate experience would have been avoided if the fellow had alternately penned each proud cock alone and given it a turn with the hens.

Ironically, cocks are less apt to fight if you keep three or more, instead of just two. If you use several cocks for flock breeding, keep them in two or more groups and rotate the groups, rather than rotating individuals — the less you disturb their peck order, the less fighting will occur. If you lose one cock out of a group

TREADING

A cock intent on mating grabs feathers on the back of a hen's head with his beak to help balance himself while he attempts to stand on her back. More often than not, his feet slide on her smooth feathers and he makes a few quick movements of his feet to get a good hold. This movement of the feet is called *treading*, and over time results in the loss of feathers from the hen's back. A hen with missing feathers has little protection from the cock's sharp claws during future matings and as a result may be seriously wounded. Bleeding wounds lead to pecking by other chickens, and deep wounds become infected, possibly resulting in the death of the hen.

Before the situation goes that far, take measures as soon as you notice hens are missing feathers because of treading — or even before feathers go missing. The first step is to keep the cock's toenails properly trimmed, taking care to round off the corners.

If you have several cocks, you might house them in separate coops and let each run with the hens for a few hours a week. Each cock will have a different set of favorite hens, offering the others some relief. If you have only one rooster, you might divide your hens into two flocks and alternate the cock between the two groups.

Dress a hen for action. As a temporary measure, dress each hen — or at least those the cock favors — in a jacket, also known as an apron or saddle. You can buy them ready-made in a variety of sizes, or make a quick and inexpensive version from two pieces of canvas or old-fashioned stiff cotton denim stitched together and fitted with elastic straps. Make a preliminary pattern from the legs of old jeans and adjust the pattern until it properly fits your hens. A jacket that is too tight will chafe, rub off breast feathers, injure the wings, or strangle the hen. A too-loose one will flop to the side, making it useless.

Once you get your pattern sized just right to fit your hens, if you want to spend more time perfecting it, you might add some refinements. You could stitch the pieces together inside out to avoid raw seams around the edges, or slightly curve the bottom inward to make room for the tail. To readily identify each hen, you might add her band number using paint, embroidery, machine stitching, or iron-on patches.

Apply the jacket when the hen's feathers start disappearing, not after she's already wounded. To dress a hen, put her head through the center opening between the two elastic straps, then put one wing through each of the other openings so a strap runs beneath each wing. When first dressed, the hen will try to back away from the jacket (please refrain from wounding her dignity by laughing), but soon enough she'll get used to it.

A jacket is not intended as permanent clothing for your hens but to get them safely through the period during which you are collecting fertile eggs. One also comes in handy in the event a hen does get wounded, to protect her back while the feathers regrow and the gashes heal. When the hen is with other chickens, you must leave the jacket on 24 hours a day until this fertile egg collecting period is over or until she has healed.

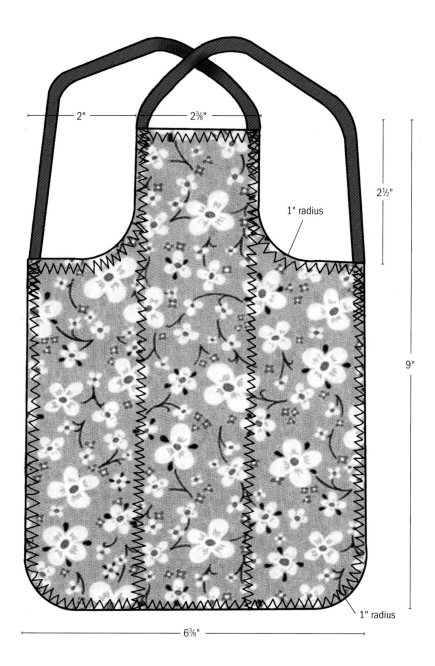

Making a jacket proportioned for a Rhode Island Red hen requires two pieces of denim or canvas 6⅜" by 9" (162 mm by 229 mm), two 8½" (216 mm) lengths of ¼" (6 mm) elastic, and a spool of matching or contrasting thread.

during the season, chance a possible slight drop in fertility rather than replacing him and running the risk that peck-order fighting will cause fertility to plummet.

Pedigree breeders commonly mate chickens in pairs or trios, the latter consisting of one cock and two hens. The fewer hens you keep with a cock, the more you have to watch out for damage due to treading (see page 258).

Housing, Privacy, and Fertility

Breeder-flock housing plays an important role in the fertility and hatchability of eggs. Facilities with lots of environmental variety help give a cock privacy to mate undisturbed by other cocks. Excessive fighting among cocks may be a sign of poor facility design, since cocks that are lower in peck order have no place to hide from dominant birds.

If you have more than one rooster, furnish each with at least one feeding station. Each cock will gather his harem of hens around a feeder, and fighting will be reduced, especially if you space the feeders well apart and locate them where each cock can reach his station without having to pass through another cock's feeder territory.

Floor space for the breeder flock should offer a minimum of 3 square feet (0.3 sq m) per bird for large breeds, 2 square feet (0.2 sq m) for smaller breeds, and 1.5 square feet (0.15 sq m) per bantam. Include at least one nest for every four hens and frequently change nesting material.

Housing should protect the flock from extremes of climate, since sudden changes can cause a decrease in laying or fertility. During the early part of the season (late winter), provide lighting not only for warmth and to stimulate egg production but also to stimulate the flow of semen. You can improve fertility by exposing cocks to 14 hours of light, artificial and daylight combined, for four to six weeks prior to collecting eggs for hatching.

Outside disturbances may upset a flock and interfere with mating. Protect your birds not only from predators but also from unruly dogs and teasing children. You may also have to protect them from ignorant adults who need to be educated that their running, screaming kids or playful, barking dogs may not be doing visible damage to your chickens but nevertheless are causing them stress and harm.

Breeder Flock Age

In general, expect maximum fertility and hatchability from mature cockerels and pullets. Most cockerels reach sexual maturity around 6 months of age, although early-maturing breeds may be ready to mate sooner, while late-maturing breeds may not be ready until they are 7 to 8 months old. Comb development is the best indication of maturity.

You can start collecting hatching eggs when pullets are about 7 months old and have been laying for at least six weeks. Eggs laid earlier tend to be low in fertility. The few that hatch are likely to produce a high percentage of deformed embryos — perhaps because of the relatively small yolks of early eggs. As time goes by, fertility and hatchability improve, leveling out by the sixth week.

After about six months of laying, fertility and hatchability begin to decline — most gradually among bantams and more rapidly among the heavier breeds than among the lighter breeds. In industry, broiler breeders are kept for 10 months or less, compared to 12 months or more for layer breeders.

A much-debated question among small-flock owners is whether it's better to hatch eggs from hens that are older or younger than 2 years of age. Hatching from older hens will improve the health and vigor of future generations. Two-year-old hens that are laying well must be relatively disease resistant and are likely to pass that resistance to their young. Furthermore, older birds tend to be the more valuable breeders, since they have proven their ability to pass desirable traits along to their offspring, the less desirable breeders having long since been culled.

On the other hand, using only older birds as breeders increases the generation gap, thereby decreasing the rate at which you can make genetic progress. Furthermore, hatchability declines slightly but significantly after a hen's first year and continues to decline as the hen gets older. After the second year, you'll see a greater percentage of early embryo deaths and failure of full-term embryos to emerge from the shell. If you breed older birds, take care to keep them stress free; for example, leave the show circuit to the younger generation.

Showing and Fertility

The frequent showing of breeders can result in poor fertility and inferior chicks. Birds become stressed by travel, inconsiderate spectators, peculiar feed and feeding schedules, and perhaps lack of water due to oversight or simply because the birds don't like the strange taste. Lack of water is particularly a problem with layers.

Hens, in general, are more greatly stressed by showing than are cocks, and older birds of either sex are more strongly affected than younger ones. Keep valuable breeders away from the showroom and the consequent exposure to stress and potential disease, not to mention the possibility of theft. Alternatively, minimize your risks by hatching a hefty number of chicks before showing your breeder stock.

Breed-Related Fertility Problems

Low fertility may be breed related, as a result of hereditary problems. The most common hereditary trait that influences fertility is comb style. For some reason single-comb breeds tend to have higher fertility than rose-comb breeds, and the sperm of single-comb or heterozygous rose-comb cocks live longer than the sperm of homozygous rose-comb cocks. So although purists insist that the occasional single-comb chick commonly hatched from rose-comb parents should not be used for breeding, old-timers keep them, albeit in low numbers, to ensure heterozygosity and good fertility.

Breed-related mechanical problems may also result in low fertility. Such mechanical problems include these:

- **Comb size.** Breeds with large single combs have trouble negotiating feeders with narrow openings, and the resulting nutritional deficiency affects fertility. Large combs may also suffer frostbite, resulting in reduced fertility or even sterility.
- **Crests.** Houdans, Polish, and other heavily crested cocks may not see well enough to catch hens. A quick fix is to clip back their crest feathers.
- **Heavy feathering.** Brahmas, Cochins, Wyandottes, and other heavily feathered breeds have trouble mating. Fertility may be improved by clipping their vent feathers.

- **Foot feathering.** Booted bantam cocks and males of other breeds with heavy foot feathering have trouble getting a foothold when treading hens.
- **Rumplessness.** Araucanas (and occasionally birds of other breeds) have no tail; as a result, the feathers around their vents can't separate properly for mating. The quick fix is to clip the vent feathers of both cocks and hens, with more attention to the feathers above the hens' vents and those below the cocks' vents. A better solution is to select breeders with the least vent feathering.
- **Heavy muscling.** Cornish cocks and other heavy-breasted males have trouble mounting hens because of the wide distance between their legs. These breeds are typically bred through artificial insemination.

Artificial Insemination

Artificial insemination is used by breeders of Cornish chickens, by exhibitors who wish to keep hens in show condition, and by owners of valuable cocks that tend to be shy or otherwise low in sex drive. The technique is not difficult, but like everything else it takes practice — on the part of both you and the birds involved. As you might expect, tame birds are easier to work with than wild ones.

House cocks you wish to collect semen from, or milk, should be kept in separate coops, away from other birds but preferably within sight of hens. Hens may be kept together but should be separated from other cocks if you wish to pedigree the chicks.

Semen Collection

For semen collection, or milking, you'll need a small glass or cup about the size of an eyecup and a 1-cc eyedropper or syringe (without a needle). The procedure is easiest if you have a helper to collect the semen while you do the milking. Hold the cock in the palm of one hand, with your fingers toward the tail, and gently massage his abdomen.

With your free hand firmly stroke the cock's back several times, then move your hand beneath the tail so the palm pushes the tail feathers up out of the way while your thumb and index finger apply pressure on both sides of the vent, ready to squeeze (but not yet). Stop massaging the abdomen with your other fingers, and press upward.

As soon as the cock's organ appears, gently squeeze out the creamy-white semen while your helper collects it in the cup. You should get at least 0.2 cc of semen from a bantam and 0.4 cc from a large bird, although you may get as little as 0.1 cc or as much as 0.8 cc from either. If the cock is fertile, the concentration will range between 3 and 4 billion sperm per cc.

Should the semen become contaminated with fecal matter or chalky white urates, discard it and try again another day. If the cock persists in discharging feces at the same time as semen, withhold feed and water for four to six hours before collection. A healthy cock may be milked once every three days.

Among naturally breeding chickens, the greatest frequency of mating is in the late afternoon. Since this time of day coincides with the greatest yields of semen and the greatest numbers of sperm, you'll have the greatest insemination success by scheduling collection for that time of day.

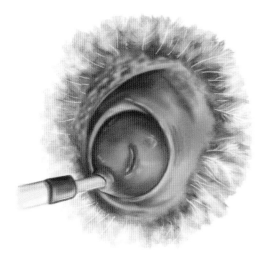

Insert the semen in the left opening, which leads to the oviduct (the opening on the right goes to the intestine).

Insemination

Be ready to inseminate hens right away, since fresh semen loses its fertilizing capacity after about an hour. Have the semen ready in the syringe or eyedropper. If you are inseminating more than one hen, use 0.1 cc for each.

Inseminate only a hen that has been laying; otherwise you run the risk of injuring her (and anyway, what would be the point?). Feel her abdomen to make sure a finished egg isn't coming through to block the passageway. Hold the hen, and massage her the same way you did the cock. When you apply pressure on her abdomen, her vent will open up. You'll see a fold of skin at the opening to the oviduct on the hen's left (the opening on the right comes from the intestine).

Your helper should quickly insert the syringe 1 inch (2.5 cm) into the oviduct and inject the semen. As soon as the semen is injected, release your pressure on the hen's abdomen to let the vent close and draw in the semen.

You should start getting fertile eggs two to three days after the first insemination. Early in the season, inseminate hens at least every seven days, preferably every five days to ensure good fertility. As the season progresses and fertility drops, inseminate more often.

Sex Determination

All the genetic information that's transmitted from a chicken to its offspring is organized on chromosomes. A cock has 39 pairs of chromosomes. One pair, called the sex chromosomes, contain the information that determines gender. The other 38 pairs are called autosomal chromosomes. Like a cock, a hen has 38 pairs of autosomals, but unlike a cock, she essentially has only one sex chromosome.

Every fertilized egg contains a sex chromosome from the cock, but a hen transmits her sex chromosome to only about 50 percent of the eggs she lays. If a fertilized egg contains sex chromosomes from both the cock and the hen, it will hatch into a cockerel; if it contains only the single sex chromosome contributed by the cock, it will hatch into a pullet.

Since each egg has a 50/50 chance of containing two sex chromosomes, eggs hatch in approximately a 50:50 ratio of cockerels to pullets. The two most common reasons for significant deviations from this ratio are

- Random death of embryos and chicks
- Sex-linked lethal genes

Despite all sorts of old wives' tales to the contrary, determining in advance which sex will hatch from a given egg is not a simple matter; if it were, the poultry industry would be using the technique. After a chick hatches, the traditional way to learn its sex is by the Japanese method, also known as cloacal sexing or vent sexing. Accuracy depends on the skill of a trained observer in examining minor differences in the tiny cloaca just inside a chick's vent.

A hatch that yields cockerels with white down and pullets with red down results from a sex-linked cross between a white (Rock or Leghorn) hen and a red (New Hampshire or Rhode Island Red) rooster.

Because a pullet does not acquire her dam's sex chromosome, she cannot acquire any of the genetic information it contains. A cockerel, on the other hand, always acquires genetic information contained on its dam's sex chromosome. All characteristics that are controlled by genes on the hen's sex chromosome are called *sex linked*. Some sex-linked characteristics are the silver color pattern (white plumage with black hackles, wings, and tail feathers), albinism, dwarfism, nakedness, barring, late feathering, and broodiness.

When a hen with a certain sex-linked trait is mated to a cock without it, the trait is acquired by all the resulting cockerels but not the pullets. Since all the pullets are like their sire and all the cockerels are like their dam, this so-called *crisscross inheritance* allows the sex-linked sorting of chicks according to such things as their color or the speed of their feather growth.

Color sexing takes advantage of the sex-linked gene that controls feather color. It is commonly used to produce hybrid brown-egg layers. Numerous variations are possible.

Mating a white Rock or white Leghorn hen with a Rhode Island Red or New Hampshire cock, for example, results in white cockerels and red pullets. Examples are Red Star and Golden Comet. Crossing a barred Rock hen with a Rhode Island Red or New Hampshire cock will give you black chicks, but each cockerel has a white spot on its head. Examples are Black Sex Link and Black Star.

Feather sexing involves crossing a slow-feathering hen (such as a Rhode Island Red) with a rapid-feathering cock (such as a white Leghorn) to get slow-feathering cockerels and rapid-feathering pullets. The chicks may be sexed with fair accuracy by the appearance of well-developed flight feathers on the wings of pullets at the time of hatch. Feather sexing is commonly used in the broiler industry, where only white-feathered birds are preferred, to separate slow-growing pullets from their faster-growing brothers.

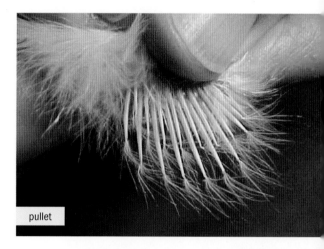

pullet

cockerel

A hatch that yields pullets whose primary feathers grow faster than the cockerels' results from a sex-linked cross between a hen of a slow-feathering breed and a rooster of a rapid-feathering breed.

10

HATCHING EGGS

Between the fun of maintaining a breeder flock and the joy of raising your own chicks comes a 21-day period of anticipation while you wait for the chicks to hatch. Even old pros sometimes have pangs of anxiety during this waiting period, as they know — and have probably experienced — all the things that can go wrong. But you can look forward to the enormously rewarding experience of hatching your own chicks by paying careful attention to detail, starting with when and how you collect eggs destined for hatching.

Egg Collection

A breeder flock is strongest and healthiest in spring, making spring chicks the strongest and healthiest as well. Chicks hatched in cool weather also have time to develop immunities before encountering germs that proliferate in the warm, humid weather of summer; chicks hatched later in the year don't have that luxury.

Pullets hatched in spring begin laying by fall and continue to lay for about a year. Pullets hatched in winter will begin laying by midsummer but may molt and stop laying in the fall and won't start again until the following spring.

Large breeds mature in 8 to 10 months, bantams in 6 to 7 months. If you raise show stock and want to have young birds in prime condition for the fall round of shows, hatch large breeds no later than December and bantams no later than March.

For general health and vigor, in most areas the best months to collect eggs for hatching are February and March. In the far north, where the weather stays cold long into spring, March and April are the best hatching months. Collect eggs at least three times a day to minimize their contact with dirty surfaces and to keep them from getting chilled or overheated, which reduces their hatchability.

Egg Storage

An egg need not be rushed into the incubator the moment it is laid. Saving up eggs is, in fact, a natural part of the incubation process. In nature just-laid eggs go dormant to give a hen time to accumulate a full setting before she starts to brood.

Even under optimum storage conditions, the longer eggs are stored, the longer they will take to hatch. Keep in mind, their ability to hatch also decreases with time. You can store eggs for up to six days without noticing a significant difference. For each day thereafter, hatchability will suffer by approximately 1 percent. I don't feel comfortable storing eggs for hatching for longer than 10 days.

No matter how many times you witness the event, seeing a chick emerge from an incubated egg is always a thrilling experience.

When a hen selects a place to make her nest, she instinctively seeks out conditions that are optimum for storage and incubation. When you gather and store eggs for hatching, try to duplicate those conditions.

Store eggs out of sunlight in a cool, relatively dry place but not in the refrigerator. The best temperature is 55°F (13°C). Humidity should be low enough to prevent moisture from condensing on the shells, which would attract molds and also encourage any bacteria already on the shell to multiply.

Excessive dryness, on the other hand, increases the rate at which moisture evaporates through the shells. The less moisture that evaporates from eggs during storage, the better chance they have of hatching. Small eggs laid by bantams and jungle fowl have a relatively large surface-to-volume ratio and therefore higher evaporation rates than larger eggs. Late-summer eggs of any size have thinner shells because the hen has been calling on her calcium reserves all summer. These thinner shells allow more rapid evaporation than occurs in early-season eggs. To minimize evaporation of eggs stored longer than six days, seal cartons in plastic bags, or better yet, wrap the eggs individually in plastic wrap.

Store eggs in clean cartons. Hard-plastic egg cartons (of the sort backpackers use) can be easily disinfected for reuse; recycled Styrofoam or cardboard cartons from the grocery store accumulate bacteria over time. When I use such cartons for hatching eggs, I discard them after one use.

Place eggs in the carton with their large ends up to keep the yolks centered within the whites. If the eggs will be stored longer than six days, keep yolks from sticking to the inside of the shell by tilting the eggs from one side to the other. Instead of handling eggs individually, elevate one end of the carton one day, and raise the opposite end the next day.

Egg Selection

Select eggs for hatching that are of normal color, shape, and size for your breed. Small eggs laid by pullets or old hens will give you smaller, less vigorous chicks. Excessively large eggs hatch poorly, and those that do hatch may result in chicks that have unabsorbed yolks (soft bellies that don't heal) or inconsistent

SEXING AN EGG

Wouldn't it be nice to save incubator space by hatching eggs that produce mostly pullets? Unfortunately, despite old wives' tales to the contrary — that cockerels hatch from elongated or pointy eggs or eggs over which a key swings longitudinally — there is no sure way to determine the sex of a fertilized egg or a developing embryo by examining the outside of the egg.

If it were that easy, the layer industry wouldn't persist in spending time and money seeking a way to sex eggs prior to hatching them.

To date, the best they've come up with is a tedious and expensive process involving inserting a needle into incubating eggs to remove and assay a sample to determine which ones contain the estrogen compounds found within female eggs.

growth rates. Eliminate eggs that are round, oblong, or otherwise oddly shaped, and those with shells that are wrinkled, glassy, or abnormal in any other way.

Candle the eggs to eliminate any with two yolks, which rarely hatch (see page 216 for candling instructions). If you're raising replacement layers, you should also candle for blood spots, as they can be hereditary. During candling, you might find some shells with hairline cracks, which appear as white veins in the shell. Since cracks open the way for bacteria to enter, eliminate these eggs from your selection for hatching.

Avoid incubating dirty eggs. Incubation temperature and humidity provide an ideal environment for the growth of bacteria, making nest and egg-storage hygiene essential. Bacteria can cause eggs to explode during incubation or can infect developing embryos.

This nest is designed
to ensure clean eggs.

If dirty eggs are extremely valuable, wash them in water that's warmer than the eggs, then sanitize the cleaned shells by dipping the eggs for 30 seconds in a solution made by adding 1 teaspoon of chlorine bleach to 1 quart of water warmed to 101°F.

If eggs consistently get dirty in the nest, reconsider your nest design. One option is to build nests that allow eggs to roll to the outside for collection. Another is to use excelsior pads, which are manufactured from wood shavings and brown paper. If you choose to use them, keep extra pads on hand to replace dirty ones. Introduce pads at the pullet stage, since hens that have grown used to nesting in natural bedding may not take readily to something new.

Natural Incubation

Some people prefer to hatch eggs the natural way — under a broody hen — and in fact may keep hens of a separate breed specifically and solely for the purpose of hatching breeder-flock eggs. Silkies are particularly known for easily going broody and having the tenacity to see it through to successful completion.

Because laying stops when setting starts, throughout the ages people who kept hens primarily for their eggs have culled persistently broody hens. As a result, hens of the breeds known for superior laying ability are less apt to brood than heavier hens, and hybrids developed for commercial production are less apt to brood than purebred hens. At the other end of the scale are backyard breeds like my cute little Cochin bantams that laid a scant few eggs each spring and promptly went broody. Just as broodiness has been bred out of certain strains, this trait may be improved in your flock by culling your breeders in favor of broodiness.

But a hen that gets broody won't necessarily continue for the full 21 days. Some hens lose interest midway and simply walk off. Sometimes a pullet hasn't gotten the hang of it yet, but some hens, like some people, don't have the attention span to stick with it until the job is done and therefore cannot be trusted with valuable hatching eggs.

Feather-legged breeds in general are likely to brood but are not always successful because their leg feathers may flick eggs out of the nest and bowl over newly hatched chicks. The

BREEDS MOST LIKELY TO BROOD	
Ameraucana	Jersey Giant
Araucana	Kraienkoppe
Aseel	Langshan
Australorp	Modern Game
Brahma	Naked Neck
Buckeye	New Hampshire
Chantecler	Old English Game
Cochin	Orpington
Cornish	Penedesenca
Cubalaya	Phoenix
Delaware	Plymouth Rock
Dominique	Shamo
Dorking	Silkie
Fayoumi	Spanish
Holland	Sumatra
Houdan	Sussex
Java	Yokohama

problem may be solved by clipping the broody's feathers close to her leg.

Hatching eggs under a hen offers the advantage of being less time-consuming than using an incubator. The hen handles all the logistics of temperature and humidity control, turning the eggs at exactly the right intervals, and keeping the newly hatched chicks safe and warm. You don't have to worry about a thermostat malfunctioning, a dried-out water pan, the power going out, or some klutz tripping over the electric cord and pulling the plug out of the wall.

On the downside, the hen — not you — decides when she's going to set. No one has yet devised a foolproof way to make a hen brood. People who try to encourage a reluctant hen to set by enclosing her in a small space with a pile of eggs usually end up with scattered eggs, smashed up by a frantic hen trying to get out.

You can, however, smoke out a potential broody by letting eggs accumulate in the nest. Among hens that tend to set at the slightest whim, just the sight of a few eggs accumulating in a nest is enough to set off the maternal instinct. Because you don't know exactly how long that might take, instead of putting your good hatching eggs in the nest, use half a dozen eggs of lesser value or use plastic or wooden eggs from a hobby store. The best setters will hatch several broods a year, especially if you take away the chicks. Be sure to give such a hen plenty of time off to eat and rest between clutches. A setting hen takes in more water than solid food. She'll eat about one-fifth of the amount she normally eats, and on some days she won't eat at all. During the 21 days of incubation, she'll lose as much as 20 percent of her normal weight. At that rate, a persistent broody that's encouraged to hatch out clutch after clutch without a break would soon starve to death.

Broody Signs

Just because a hen is sitting on a nest doesn't necessarily mean she's setting. She may still be thinking about the egg she just laid, or she may be hiding from some bully that's higher in the peck order.

To test a hen for broodiness, gently reach beneath her, and remove any eggs you find there. If she runs off in a hysterical snit, she's not broody. If she pecks your hand, puffs out her feathers, or growls, things are looking good. Within two or three days, she'll likely settle down to serious business.

As the hen sets, she develops defeathered brood patches on the sides of her breast. These bare spots serve two purposes: to bring her body warmth closer to the eggs and to keep the eggs from drying out too fast by lending moisture from her body.

Clucking is one sure sign of broodiness. Many broodies won't cluck until their eggs are ready to hatch, but some start clucking almost as soon as they start setting. The hen's habit of clucking to reassure her chicks has led to the nickname "clucker" for a broody hen. Some broodies practice for motherhood by ruffling their feathers and clucking whenever they hop off the nest for a quick bite to eat.

Brooding Nests

Separating a setting hen from the rest of the flock is a good idea, since another hen may enter her nest while she's off eating, causing her to resettle in the wrong nest. If she does manage to hatch out a brood, other chickens may kill the fuzzy intruders, much as they would destroy a mouse or a frog that wanders into their yard.

Move the hen after dark so she's less likely to try to get back to her old nest. Continue with the fake eggs, or eggs of little value, until you're sure she'll stick with it. Then replace the faux eggs with real hatching eggs. They needn't be the hen's own eggs — she won't know the difference.

A suitable brooding nest is darkened, well ventilated, and protected from wind, rain, and temperature extremes. It should be 14 inches (36 cm) square and at least 16 inches (41 cm) high, with a 4- to 6-inch-high (10 to 15 cm) sill at the front to hold in nesting material. Clean, dry wood shavings make the best nesting material. Straw and hay are less suitable because they readily mold, possibly infecting developing embryos or newly hatched chicks.

Be sure both litter and hen are free of lice and mites that can make a hen restless enough to leave the nest. These parasites can take enough blood to kill a setting hen or her newly hatched chicks. The use of cedar shavings discourages parasites, as does sprinkling a poultry-approved insecticide in the nest before adding litter at the start of incubation. Before chemicals became widely available, old-timers put tobacco leaves in a broody nest as an insecticide.

Eggs are likely to break or roll away from the hen if she's trying to cover more than she can handle. Most hens can cover 12 to 18 eggs of the size they lay. A bantam can hatch only 8 to 10 eggs laid by a large hen, while a larger broody might cover as many as 24 banty eggs. Watch the nest for the first few days, and remove any egg that sticks out around the edges so it doesn't get rotated back in, leaving a different egg out in the cold and spoiling both.

Give the hen a helping hand by hollowing the nesting material at the center, making the bowl small enough to keep eggs from rolling away but big enough so they won't pile up and crush each other beneath the hen. Broken eggs attract bacteria and ants, so check at least occasionally, and if necessary, remove any egg-soaked nesting material and yolk-smeared eggs.

If you have more than one hen brooding at a time, separate them from each other. Otherwise the hens may accidentally switch nests, shortening the incubation period for one hen and prolonging it for the other. Or they may pile on top of each other in one nest and abandon the other.

Both setters may follow the first chicks that hatch, leaving the remaining eggs to chill. Even if each hen successfully hatches her eggs and tends to her own chicks, one may capture the feed and watering stations for her brood and chase the other's chicks away. You will avoid all these problems by separating your broody hens.

Broody Management

On most days, a setting hen will get off the nest for a few minutes to eat. A hen rarely poops in the nest; she knows to avoid contaminating the eggs by holding her droppings until she leaves the nest. When she does come off the nest, the first thing she does is relieve herself by dropping one big blob. In case of an accident, you can keep droppings solid — and therefore easier to clean out and less likely to stick to eggs — by feeding scratch grain instead of lay ration, which also helps the hen maintain her weight.

A setting hen should start out at the peak of health with a good layer of body fat as indicated by a creamy or yellow hue to the skin; a hen with reddish or bluish looking skin has insufficient fat. Because she won't eat much during the 21 days she's on the nest, she needs fat reserves to see her through.

Put food and water near the nest, but don't worry if the hen doesn't eat for the first few days. After that she should get off the nest for 15 or 20 minutes at about the same time each day to eliminate, grab a few kernels of grain, maybe take a quick dust bath, then zip back onto the nest.

If your hen doesn't seem to be getting off the nest to eat, encourage her by either lifting her off the nest and putting her down near the feeding station or sprinkling a little scratch in front of the nest to pique her interest. If she remains indifferent, chances are she's been eating when you're not around.

Whenever you handle a broody hen, first gently raise her wings. She may be holding an egg between a wing and her body. If the egg drops into the nest as you lift the hen, it might break and in the process crack other eggs as well.

After the sixteenth day, do not disturb the broody or her eggs. See that she has plenty to eat and drink, then contain your impatience until the little peepers pop into the world. While the chicks are hatching, keep an eye on things but don't interfere unless your help is absolutely necessary — such as to rescue early chicks wandering away to get chilled while Mama stays on the nest to hatch the rest of the brood.

Normally, all the chicks will hatch within a few hours of each other. The hen will continue covering her brood on the nest for another day, maybe two, before venturing forth. If for some reason the hatch is slow, the hen may hop off the nest to follow the earliest chicks, leaving the rest to chill and die. In that case, gather up and care for the first chicks, returning them to the hen after the hatch is complete.

Occasionally a hen will be so horrified by the appearance of interlopers beneath her that she'll attack the little fuzzballs. Be ready to rescue the chicks and brood them yourself.

If you're hatching particularly valuable chicks, you might wish to brood them yourself in any case. Some breeders of exhibition birds jump the gun by moving term eggs to an incubator for the hatch. That's tricky business, though, since any delay while moving the eggs, or an incorrect setting of the incubator, can ruin the hatch. On the other hand, a properly functioning mechanical hatcher can be the safer option, especially if your hen doesn't have enough of a track record for you to be sure of her competence in mothering chicks.

Incubating Eggs Artificially

The decision to use a mechanical incubator is not necessarily an either/or one. I let my hens set whenever one gets the urge because I enjoy the sight of a hen with chicks, but I use an incubator when I need a large number of chicks — to raise pullets as replacements for old layers while at the same time replenishing our family's broiler supply with the excess cockerels. Consider artificial incubation if

- You keep a breed that doesn't tend toward successful brooding
- You want chicks out of the normal brooding season
- You are trying to produce the perfect show bird
- You are working to restore an endangered breed
- You otherwise wish to hatch more chicks than broody hens can handle

Incubators come in a broad range of sizes, styles, and prices. The smallest incubator, designed for classroom and home school projects, holds no more than three eggs. At the other extreme, large room-size commercial incubators hatch thousands of eggs at a time.

Practical incubators for home use hold anywhere from a few dozen eggs to a few hundred. In determining what size you need, remember that not all the eggs you put into the incubator will hatch; a good hatch is considered to be 85 percent. If you get a 48-egg incubator — and assuming all the eggs you put into it are fertile and everything goes right — you can expect no more than about 40 chicks per hatch. With experience, you may coax your incubator to a higher hatch rate; with less attention to detail, you'll get a lower rate.

A tabletop model requires less space than a cabinet model that sits on the floor, but it hatches fewer eggs. Both tabletop and cabinet units range from hands-on models that require frequent attention to digital units that handle everything electronically. Some models in both styles have an observation window that lets you watch the hatch without opening the incubator, which could disrupt the hatch.

TOP-OF-THE-LINE INCUBATORS

Tabletop

The most efficient incubators come with a full line of features including automatic temperature control, an automatic turning device, and a fan for good air circulation.

Electronic Cabinet

Incubators for home use come in a range of sizes from small tabletop models that will hatch a mere 3 eggs to large cabinet models that will hatch up to 300 eggs at a time.

In deciding which incubator is right for you, consider these five features:

- Turning device
- Airflow
- Temperature control
- Humidity control
- Ease of cleaning

Turning Device

By fidgeting in her nest and adjusting eggs with her beak, a setting hen periodically turns each egg beneath her. Constantly adjusting her eggs may make the hen more comfortable or perhaps offers some psychological benefit. The hen can't possibly know that by turning each egg she keeps its yolk centered within the white and that if an egg isn't turned, its yolk will eventually float away from the center and stick to the shell lining.

To imitate the hen's activities, some incubators have automatic turning devices that tilt the eggs from side to side at regular intervals, every hour or every four hours. Whole shelves may tilt from one side to the other, or rows of eggs within racks may tilt back and forth.

The first time you use a tabletop, you may find that as the eggs turn some of them get crushed by hitting the fan housing or the switching mechanism in the cover. Mark the positions of those trouble spots in the racks so you can avoid using them in the future. Or get an expansion ring, if one is available for your model, to place between the base and lid to give the eggs a little more room.

Some incubators, especially budget still-air units, don't have a turning device. Price may be an important consideration in deciding whether or not to get an incubator without a turning device, but your time may be of equal

Pipping — the appearance of the first hole in the shell — is a sign that a chick is about to hatch.

concern. Without a turner, you'll need to be on hand at least three times a day, every day, to turn the eggs yourself.

For manual turning, if your incubator comes equipped with a rack, simply flip the eggs from one angle to another. If the eggs lie flat on a tray, to make sure you've turned each egg place an "X" on one side and an "O" on the opposite side with a grease pencil or china marker. Some folks prefer a soft (number 2) pencil, but I stopped using a pencil after I pierced a few shells.

Whether the eggs are turned automatically or manually, they aren't turned end for end, but side to side. Throughout incubation, the pointed end of the egg should never be oriented upward. Otherwise, according to research at the University of Georgia, fewer chicks will hatch, and the ones that do hatch will be of lower quality.

Signs of improper turning are early embryo deaths due to stuck yolks, and full-term chicks that fail to pip, or break through the shell. Eggs need not be turned after the 14th day of incubation and should not be turned during the last three days before the hatch, when chicks need time to get oriented and begin breaking out of their shells.

Airflow

Developing embryos use up oxygen rather rapidly, while at the same time generating carbon dioxide. An incubator therefore needs a good airflow to constantly replenish oxygen and remove the carbon dioxide. All incubators have vents to admit fresh air and expel the stale air.

Some small incubators rely on gravity to let warm, humid air escape through vents in the cover, which in turn causes fresh air to be drawn in through vents in the bottom. Other units have a built-in fan to circulate the air. Although a fan-ventilated (or forced-air) incubator costs more than a gravity-ventilated (still-air or natural-draft) incubator, the steady circulation of air helps maintain a more uniform temperature throughout the incubator, resulting in a better hatch rate. In tabletop models the fan is usually in the cover, so when you remove the cover, you move the fan away from the eggs. Cabinet models usually have the fan at the top or back of the cabinet and should have a switch that lets you turn off the fan before you open the door. A model that lacks such a switch must be powered down before the door is opened to prevent the fan from blasting cool air across the warm eggs.

Except for its blades, the fan of a commercially made incubator is sealed so it can't get gummed up by fluff and other hatching debris. Nevertheless, whenever you clean your incubator, brush off the fan with a soft-bristle paintbrush, then gently pass a vacuum hose over it to suck out fluff.

Temperature Control

The normal body temperature of a hen is about 103°F (39.5°C). You'd think that would be the temperature at which an incubator should be run, but other factors figure in. For one thing,

the hen's body does not surround the eggs beneath her, resulting in a slightly cooler temperature for the eggs. Furthermore, an incubator's thermometer is not necessarily positioned to read the temperature at the position of the eggs. As a result, operating temperatures recommended by incubator manufacturers vary slightly from one model to the next.

A typical operating temperature for a forced-air incubator is 99.5°F (37.5°C); for a still-air incubator, 102°F (39°C). Lethal temperatures are 103°F (39.5°C) in a forced-air incubator, 107°F (41.7°C) in a still-air incubator. The recommendations given in the incubator's operating manual should be considered baseline; you can improve your future success by making minor future adjustments based on past hatching records.

An incubator with an electronic thermostat is preset by the manufacturer and should need little or no tweaking. Some digital incubators include a readout that will tell you if the room temperature outside the incubator is too hot or cold for the incubator to maintain proper internal temperature.

Even a high-tech electronic incubator may have as a backup an old-fashioned *ether wafer thermostat*, a metal disk filled with ether. When the incubator heats up, the ether expands, causing the wafer to swell until it makes contact with a button switch that turns the heat off. As the incubator cools down, the ether contracts until the wafer loses contact with the switch, causing the heat to go on again.

Compared to an electronic thermostat, the cycling of a wafer against the thermostatic switch allows greater temperature fluctuations — as much as two degrees above or below the desired incubating temperature. A wafer thermostat also typically takes longer to bring the temperature back up after the incubator has been opened.

Regulating the heat in an electronic incubator is done with touch pads or touch buttons. Regulating a wafer involves turning a control bolt to adjust the temperature at which the wafer makes contact with the switch. Most wafer-controlled incubators have an indicator light that goes on when the heat is on and off when the heat is off. After a while, you develop a sixth sense that makes you feel uneasy if the light stays on or off for too long.

Sometimes a wafer goes out, or ceases to function properly, because it springs a leak and the ether escapes. In that case the switch remains in the on position, causing the temperature to stay too high for too long. Unless you catch it right away, the eggs will cook. If your incubator uses a wafer, always keep at least one spare wafer on hand.

A failure of the thermostat, whether it is electronic or wafer controlled, is not the only thing that can cause the incubator's temperature to soar. The incubator could be too close to a heater, for example, or sunlight could fall on the incubator during part of the day. Even just a short-term rise in temperature does greater damage than a drop, which may occur if the incubator's cover or door is left ajar, hatching fluff jams the wafer switch, a child or pet pulls the plug, or the power goes out.

The more often you check the temperature, the more likely you are to catch and correct problems such as a pulled plug or a wafer gone bad. During my early years using an incubator, I kept a piece of paper nearby on which I

POWER OUTAGES

The more valuable your hatching eggs, the more you can count on the power going out at a critical time during incubation. After losing too many hatches to extended power outages resulting from spring storms, we got a portable standby generator. If your incubator is electronic, you should also protect it with a surge suppressor. Another option where electricity is iffy or unavailable is an Amish-made kerosene incubator, such as those offered by Lehman's heritage-hardware store.

If you don't have a generator and the power stays off for any length of time, open the incubator and let the eggs cool until the power goes back on. Trying to keep the eggs warm can cause abnormal embryo development. Furthermore, if you close the vents or wrap the incubator with blankets in your attempt to keep eggs warm, a greater danger than temperature loss is oxygen deprivation, and in a prolonged outage the oxygen level could fall below that necessary to keep the embryos alive.

As soon as the power goes back on, close the incubator and continue normal operation. The effect of the outage on your hatch will depend on how long the eggs had been incubated before the outage and how long the power was out. A power failure of 18 hours will delay the hatch by a few days and significantly reduce the success rate. An outage of up to 12 hours may not significantly affect the hatch — except to delay it somewhat — especially if the outage occurred during early incubation, when cooled embryos naturally go dormant. Embryos that are close to hatching may generate enough heat to carry them through a short-term outage.

recorded the time and temperature every time I passed the incubator. After a while checking got to be a habit, and I no longer needed the paper reminder. But keeping one nearby is a good idea if two or more people share responsibility for maintaining the incubator.

Some cabinet incubators have an alarm that sounds or lights up when the temperature drops or rises past a certain range. If you're handy with gadgets, you could easily rig up something similar for an incubator lacking such an alarm.

When you're ready to start hatching, set up your incubator and let it warm up for at least half a day to stabilize before you put in any eggs. To keep the temperature steady, locate the incubator away from drafts and where the room temperature remains fairly constant. A room temperature between 75 and 80°F (24 and 27°C) is ideal, although between 55 and 90°F (13 and 32°C) is acceptable.

The better the incubator is insulated, the less it will be affected by fluctuations in room temperature, such as might be caused by sun coming through a window, heat or air conditioning turned off during the night, or heavy storms that cause drastic changes in barometric pressure. Occasional minor fluctuations are normal, so once the incubator's temperature has been set, don't keep fiddling with it.

A fully electronic incubator shows the temperature readout in a digital display. Others use a thermometer, which may itself be electronic. Make sure your thermometer is true

ELECTRONIC OR DIGITAL INCUBATOR

An electronic incubator has a digital display and is regulated with convenient touch pads.

by checking it against a second or even a third thermometer that you trust for accuracy.

Some thermometers are designed to be placed inside the incubator at the level of the eggs but are so small you can't easily read them without opening the incubator, which of course causes the temperature to drop before you get a reading. A stem thermometer solves that problem; it has a probe you insert into a hole in the incubator while the readout dial remains on the outside.

If the temperature runs slightly high or slightly low during incubation, the eggs will not hatch in exactly 21 days. Some pipped eggs may not hatch at all, and in general the hatch rate among fertile eggs will be lower than if the temperature had been just right.

At a temperature that is 0.5 to 1°F (0.3 to 0.5°C) low, chicks will take longer than the normal 21 days to hatch. They will tend to be big and soft with unhealed navels, crooked toes, and thin legs. They may develop slowly or may never learn to eat and drink and therefore will die.

If the temperature runs 0.5 to 1°F high, chicks will hatch before the allotted 21 days. They will tend to have splayed legs and can't walk properly.

If the eggs hatch in exactly 21 days but not all pips hatch and the hatch rate of fertile eggs is generally poor, the problem is not the temperature. Rather, most likely the humidity is off.

Humidity Control

For a successful hatch, moisture must evaporate from the eggs at just the right rate. Overly rapid evaporation can inhibit the chicks' ability to get out of their shells at hatching time. Overly slow evaporation can lead to mushy chick disease (*omphalitis*), in which the yolk sac isn't completely absorbed so the navel can't heal properly; as a result, bacteria invade through the navel, causing chicks to die at hatching time and for up to two weeks afterward.

Evaporation is regulated by the amount of moisture in the air — the more moisture-laden the air, the slower moisture evaporates from the eggs and vice versa. To reduce the rate of evaporation from eggs, every incubator has some sort of device that gradually releases moisture into the air.

The water-holding device might be a simple pan, a divided pan, or grooves molded into the bottom of the incubator. Humidity is regulated by adjusting the surface area available for evaporation. Using a pan with a larger surface area increases humidity; a smaller pan decreases it. Filling more divisions or grooves with water increases humidity, and filling fewer decreases it. Increasing surface area with sponges or humidity pads increases humidity; partially covering the water pan with foil decreases humidity.

Some devices must be filled manually. Since cool water draws heat from the incubator, always use warm water. Some incubators may be fitted with an external water container that automatically feeds into the incubator, which must be checked occasionally and refilled as needed so it never runs dry.

Humidity control may be fine-tuned by adjusting the incubator's vents. Some vents must remain open at all times for good oxygen flow; others have either removable plugs or sliding covers that allow you to adjust the size of the openings. Closing vents increases humidity by trapping more moisture-laden air within the incubator. Opening vents decreases humidity by allowing more moisture-laden air to escape.

An accumulation of moisture on the incubator's observation window during the hatch is an obvious indication of excess humidity. Low humidity tends to be the greater problem with small incubators and those that have to be opened to turn the eggs manually. If humidity drops inexplicably during a hatch, the water pan may need to be cleaned. Fluff released by newly hatched chicks will coat the surface of the water, preventing evaporation and causing the humidity to plummet.

Because the eggs themselves contribute to humidity by evaporating through the shell, an incubator that is not filled to capacity may have more difficulty maintaining proper humidity than a full incubator. The larger the incubator, the greater the problem. At the end of the season, when my cabinet incubator has few eggs left in it, I boost the humidity and ensure a good final hatch by adding a cake pan full of water during the hatch.

Most incubators call for about 60 percent relative humidity, except during the last three days prior to the hatch, when it should be increased to 70 percent. An electronic *hygrometer*, or an incubator with an electronic control system, shows the humidity level in a digital display.

Measuring Humidity. Humidity in a nonelectronic incubator is measured by a thermometer in what's known as *wet-bulb degrees*, in contrast to *dry-bulb degrees*, in which the same thermometer measures heat. A wet-bulb reading is more accurate in a forced-air incubator than in a still-air incubator.

To obtain a wet-bulb reading, slip a piece of cotton, called a wick or sock, over the bulb or stem end of the thermometer with the tail end of the wick hanging in water. You can buy wicks or make them by cutting up a white cotton tennis shoelace. You'll need spare wicks to replace those that get crusty with mineral solids and lose their absorbency. If your tap water tends to be hard, use distilled water, and your wicks will last longer.

You can make a temporary hygrometer by wrapping a piece of cheesecloth or gauze bandage around the end of the thermometer and

TEMPERATURE AND HUMIDITY

Optimum incubation temperature and humidity are interrelated. As the temperature goes up, relative humidity must go down to maintain the same hatching rate.

When operating an incubator for the first time, follow the manufacturer's recommendations. Based on the hatch rate, you'll likely need to make future adjustments to suit your specific conditions and the kind of eggs you hatch. For example, small eggs evaporate more rapidly than large eggs, so they hatch better at a lower temperature and higher humidity.

As you make adjustments, keep accurate records of the temperature and humidity for each hatch and of your success rate (percentage of fertile eggs hatched). After a few hatches, you'll hone in on the optimum combination that gives you the best possible hatch for your circumstances.

wetting the gauze with warm water. Place the thermometer in your incubator where you can see it when the cover is on or the door is shut.

As water evaporates from the cotton, the thermometer gives a lower reading than it would without the wick. After a few minutes the wet-bulb temperature will stabilize, and you can take a humidity reading. A typical wet-bulb reading is 86 to 88°F (30 to 31°C) during incubation and 88 to 91°F (30 to 33°C) during the hatch. The table on page 282 shows the relationship between wet-bulb degrees and relative humidity percentage at common incubation temperatures.

In place of a hygrometer, or in conjunction with one, a good indication of humidity is the changing air-cell size inside the developing eggs, as determined by candling. Moisture evaporating from an egg causes its contents to shrink, which increases the size of the air cell. If air cells are proportionately larger than the one shown in the sketch below, increase humidity; if they're smaller, decrease humidity.

As moisture from within an egg evaporates during incubation, the egg's weight decreases. Under proper humidity, an egg will lose 12 to 14 percent of its weight during the 21 days of incubation. So another way to measure humidity is to monitor weight loss as incubation progresses. At the start of incubation, weigh a sample of eggs — say, half a dozen — and weigh the same sample every five days, then compare their actual weight to their expected weight based on average loss. If the actual weight is less than the expected weight, increase the incubator's humidity; if the actual weight is greater than expected, decrease the humidity.

The "Expected Weight Loss during Incubation" table on page 282 shows the expected drops in weight during incubation of a sample of six eggs of different sizes having typical starting weights for their size. Because of the small numbers involved, a scale that measures in the metric unit of grams is more accurate than a scale measuring ounces.

Maintaining adequate incubation humidity is more difficult in really dry weather than in humid weather. Those of us who live where the weather is extremely dry one half of the year and extremely humid the other half of the year must compensate accordingly.

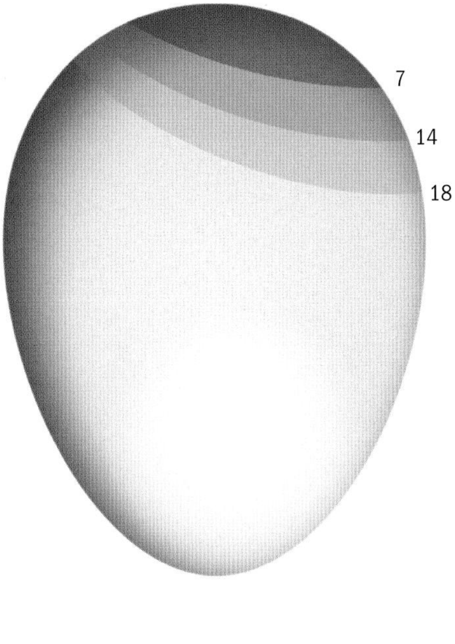

7
14
18

As moisture evaporates, the air cell within the egg expands. Shown here are the relative air cell sizes on the 7th, 14th, and 18th days of incubation.

HUMIDITY AT COMMON INCUBATION TEMPERATURES

Forced-Air		Still-Air				Relative Humidity
99.5°F/37.5°C		100°F/37.8°C		101°F/38.5°C		102°F/39°C	
Wet-Bulb Reading						
80.8	27.1	81.3	27.4	82.2	27.9	83.0	28.3	45%
82.8	28.2	83.3	28.5	84.2	29.0	85.0	29.5	50%
84.7	29.3	85.3	29.6	86.2	30.1	87.0	30.5	55%
86.7	30.4	87.3	30.7	88.2	31.2	89.0	31.7	60%
88.5	31.4	89.0	31.7	90.0	32.2	91.0	32.7	65%
90.3	32.4	90.7	32.6	91.7	33.2	92.7	33.7	70%
91.9	33.3	92.5	33.6	93.6	34.2	94.5	34.7	75%
93.6	34.2	94.1	34.5	95.2	35.1	96.1	35.6	80%
95.2	35.1	95.7	35.4	96.8	36.0	97.7	36.5	85%

EXPECTED WEIGHT LOSS DURING INCUBATION

Day	0	5	10	15	21
Ounces per Sample of Six Eggs					
Bantam	7.5	7.3	7.1–7.0	6.9–6.7	6.6–6.4
Small	9.0	8.7	8.5–8.4	8.2–8.1	7.9–7.7
Medium	10.5	10.2	9.9–9.8	9.6–9.4	9.2–9.0
Large	12.0	11.6	11.3–11.2	11.0–10.8	10.6–10.3
X-Large	13.5	13.1	12.7–12.6	12.3–12.1	11.9–11.6
Jumbo	15.0	14.5	14.1–14.0	13.7–13.5	13.2–12.9
Grams per Sample of Six Eggs					
Bantam	216	210–209	204–202	198–194	190–186
Small	258	251–249	243–241	236–232	227–222
Medium	300	291–290	283–280	274–270	264–258
Large	342	332–331	323–319	313–308	301–294
X-Large	384	373–371	362–358	351–345	339–331
Jumbo	420	408–406	396–392	384–378	370–361

To determine if humidity level is correct, compare actual weight to the expected weight: if the actual weight is greater than expected, humidity is high; if the actual weight is less than expected, humidity is low. Expected weight on any given day lies between W-(0.0057×D×W) and W-(0.0067×D×W), where W = starting weight and D = days of incubation.

Ease of Cleaning

The heat and humidity within an incubator offer ideal conditions for germs to flourish, and hatching produces plenty of organic wastes for the germs to thrive on. An incubator must therefore be cleaned of debris after every hatch and periodically disinfected. No incubator is truly easy to clean and sanitize, but some are easier than others.

A Styrofoam incubator is the most difficult type to clean. My experience, and that of others, is that after a few consecutive hatches the success rate drops sharply. Although a Styrofoam sanitizer is on the market, some incubator manufacturers recommend cleaning Styrofoam only with plain water. Easily sanitized plastic liners are available for some models. An alternative is to press aluminum foil against the bottom, shiny side down, and poke holes for the necessary air vents. If you opt for a Styrofoam incubator, plan on one hatch, or only two or three consecutive hatches, then thoroughly clean the incubator and let it sit for several months before using it again.

Illness in freshly hatched chicks comes from one of two sources: poor sanitation in the incubator or around the brooder and disease in the breeder flock that's transmitted through eggs. Some disease organisms are inside the egg; others get on the shell when it is laid or when it is improperly washed.

Diseases that may be transmitted via the egg spread from infected chicks to healthy chicks in the incubator (often inhaled in fluff) or in the brooder (usually through ingested droppings in feed or water). The most common of these diseases is omphalitis caused by the incubation of dirty eggs, operation of an unsanitary incubator, or improper temperature or humidity during the incubation period.

Good incubator sanitation not only improves hatching success and gives chicks a healthy start in life but also helps break the natural disease cycle in a flock. Because hatching itself is a major source of contamination, take time to clean your incubator thoroughly after each hatch.

The best sanitary measure is to incubate one setting of eggs, then clean and disinfect the incubator before starting another setting. Some incubators are designed for continuous hatching. They typically have three trays to hold incubating eggs and a fourth to hold the hatch, allowing you to fill one tray with eggs every week and hatch a tray every week. Continuous hatching is a significant source of contamination.

To minimize contamination, move the hatch to the lowest tray, where no eggs are below for hatching debris to fall on. If the hatching tray has a screen bottom, place a layer of paper toweling or foil under the tray to simplify cleanup. After each hatch, remove the hatching tray, and scrub and disinfect it as well as the area below it. Clean and disinfect the water pan, too.

A better sanitary measure for continuous hatching is to use a separate incubator as a hatcher. Ideally, it will be a forced-air incubator, but it doesn't need a turning device. After heating up the hatcher, move the about-to-hatch eggs from the incubator to the hatcher. After the hatch is complete, scrub and disinfect the hatcher in preparation for the next hatch. Not only does this plan improve sanitation, but it allows you to leave the incubator at the optimum temperature and humidity for incubation, while adjusting the hatcher for optimum hatching temperature (0.5 to 1°F cooler [0.3 to 0.5°C]) and humidity (6 to 10 percent higher) for an even better rate of hatch.

Between hatches and at the end of every hatching season, thoroughly clean out the incubator or hatcher. When no eggs are in the incubator, unplug the unit, remove and clean all removable parts, vacuum up loose fluff, sponge out hatching debris, and scrub the incubator with detergent and hot water. Take care not to wet down or spray the heater or any electrical parts — instead brush them off with a soft-bristle paintbrush and move the vacuum hose over them gently to pick up the loosened fluff.

When the incubator and all its parts look thoroughly clean, wipe or spray the nonelectrical parts with a disinfectant such as Germex, Tek-Trol, Oxine, Vanodine, or chlorine bleach (¼ cup bleach per gallon of hot water; 15 mL/L). Leave the incubator open until it is thoroughly dry, preferably in sunlight, then store it in a clean place.

If mice, paper wasps, or mud daubers have access to your incubator storage area, prevent them from setting up housekeeping in the incubator by plugging vents or covering them with duct tape; remember to unseal the vents next time you operate the incubator. I fastened pieces of window screening over the vents of my cabinet incubator; the screens keep out critters during storage and don't impede airflow during incubation.

Candling the Eggs

To monitor the progress of each hatch so you can remove spoiled and nondeveloping eggs, you'll need a candling device. A handheld candler looks like a small flashlight with a plug-in cord. In fact, a small flashlight works at least as well as, if not better than, the appliance designed for the job.

So before you go out and spend money on a candling device, look for something around the house that might work equally well. All you need is bright light that comes through an opening smaller than the diameter of the eggs you want to candle. If you have a flashlight that is too big, cut a hole in a piece of cardboard and place it over the business end of the flashlight, or tape a short piece of empty toilet-paper or paper-towel tube on the end, so only the light comes through the narrowed opening.

Use your candling device in a dark room. Hold the egg at a slight angle, large end to the light and pointed end downward. Making sure your fingers don't block the light, turn the egg until either you see something or you're certain there's nothing to see.

Candling Procedure

White-shelled eggs are easier to candle than eggs with colored shells, which is why white eggs have become the industry standard. Similarly, plain-shelled eggs are easier to candle than eggs with spotted shells. A good way to gain practice candling is to hatch white-shelled eggs in your first setting. Don't buy white eggs from the grocery; they're not fertile, but even if they were, they wouldn't hatch well because they have been refrigerated.

Examine eggs after one week of incubation. You will likely find one of three things inside the shell:

- A web of vessels surrounding a dark spot — the embryo is developing properly

- A thin ring within the egg or around the short circumference — the embryo has died
- Nothing (or a vague yolk shadow) — either the egg is infertile or the embryo has died

Assuming you remove infertile eggs and those with blood rings during the first candling, examine the incubating eggs again after two weeks. You'll likely find one of two things:

- A dark shadow except in the air cell at the large end (you may see movement against the air-cell membrane) — the embryo is developing properly
- Murky or muddled contents that move freely and/or a jagged-edged air cell — the embryo has died

The majority of embryos that die do so at two peaks. The first embryo death loss occurs within a few days of the beginning of incubation. The second, larger death loss occurs just before the hatch. The most common cause of embryo deaths clustered during early incubation is improper egg handling or storage. Embryo deaths during mid- and late-stage incubation may result from an inadequate breeder-flock diet.

Embryos that grow to term but fail to hatch may be a result of bacterial contamination: in other words, unclean eggs were placed in the incubator. The multiplication of bacteria within a contaminated egg may cause the egg to rot and stink. Left in the incubator, a rotting egg may explode, spreading bacteria throughout the incubator and contaminating other eggs.

Any time you open your incubator and smell an unpleasant odor, use your nose to ferret out the offending egg. It may or may not exude darkish fluid that beads on the shell or (due to the pressure of gases within the shell) may be cracked and leaking. By candling on days 7 and 14 and removing nondeveloping eggs, you can avoid the unpleasant experience of a rotten-egg explosion.

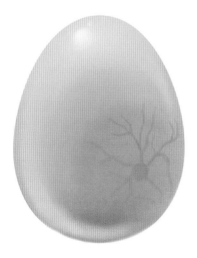

fertile

dead

After one week of incubation, a properly developing embryo (left) appears as a web of vessels surrounding a dark spot. A thin, irregular ring (right) means the embryo has died.

Egg Culling

The first few times you candle eggs, break open the culls to verify your findings; take the eggs outside, in case the contents smell bad. Even after you gain confidence in your candling abilities, continue breaking and examining culls; you might learn something that will help you improve your future hatching success.

Assuming you incubate only properly stored, clean eggs, the majority of your fertiles should hatch. The average hatching rate for artificial incubation is 75 to 85 percent of all fertile eggs.

To optimize incubator space and improve your overall hatching rate, incubate a full setting for 15 hours, then candle and remove infertile eggs and those with germinal discs (the dark spots) that are smaller than the majority. Put the remaining eggs back into storage, where they will safely go dormant, and repeat the process with a second setting. Combine the best of both settings and incubate as usual. This technique requires a more powerful candling device; the best one I ever used was a slide projector.

A wonderful book that shows photographically all that goes on inside a properly developing egg day by day is *A Chick Hatches* by Joanna Cole and Jerome Wexler. This fine book is out of print but is well worth looking for at your local library, borrowing through interlibrary loan, or searching for in used bookstores or on the Internet.

The Hatch

A normal hatch is complete within 24 hours of the first pip. A dragged-out hatch, or one that occurs earlier or later than expected, may be caused by improper incubation temperature or humidity. Another cause for so-called draggy hatch is storing eggs for a long time while collecting enough to fill the incubator — the eggs stored longer will take longer to hatch. Yet another cause is combining eggs of various sizes in one setting. Larger eggs take longer to hatch than smaller eggs, so if you combine eggs from bantams and large fowl, or from light and heavy breeds, the bantam eggs or those from the lighter breeds will usually hatch first.

When the hatch is short and quick, move the chicks to a brooder when most of them (95 percent) have dried and fluffed out, leaving the wet ones in the incubator a little longer. Moving chicks while they're still wet can cause them to chill. If the hatch is draggy, remove the dried chicks every six to eight hours so the incubator's airflow won't cause them to dehydrate. Work quickly, since opening the incubator causes the temperature and humidity to drop, reducing the percentage of remaining eggs that will hatch.

Chicks that can't get free of the shell without assistance are called *help-outs* and are often the result of either an air cell or a chick that is poorly positioned for hatching. Just prior to hatching, a chick should have its head under its right wing and be in a position to break through the air cell at the large end of the egg.

KEEP INCUBATED EGGSHELLS OFF THE MENU

Never feed your hens shells from incubated eggs — they're loaded with bacteria and can spread disease.

WHAT WENT WRONG?

Problem	Possible Cause	Possible Cause
Clear eggs	Deficient breeder ration	Excessive showing of breeders
	Too many or too few cocks	Breeders too crowded
	Hens too fat or too thin	Eggs chilled or overheated
	Cocks too fat or too old	Eggs stored too long or improperly
	Cocks have foot or leg injuries	
Blood ring	Breeder ration low in vitamins	Eggs chilled or overheated
	Unhealthy breeder flock	Eggs stored too long or improperly
	Eggs dirty or contaminated	Irregular incubation temperature
Dead embryos, 1–7 days	Inbreeding	Eggs stored too long or improperly
	Unhealthy breeder flock	Incubation temperature too high or too low
	Deficient breeder ration	Insufficient ventilation
	Eggs dirty or contaminated	Improper turning
Dead embryos, 12–18 days (uncommon)	Breeder ration low in vitamins	Incubation temperature too high or too low
	Eggs dirty or contaminated	Insufficient ventilation
Chicks fail to pip	Eggs dirty or contaminated	Insufficient ventilation
	Incubation humidity too low	Improper turning
	Hatching humidity too low or too high	Hereditary hatching weakness
	Temperature too high (or too low)	
Pips fail to hatch	Eggs dirty or contaminated	Temporary temperature surge
	Humidity too low	Insufficient ventilation
	Temperature too low	
Chicks can't get free of shell	Hereditary weakness	Humidity too low
	Eggs incubated or hatched large end down	Insufficient ventilation
	Improper turning	Irregular temperature
Shells cling to chicks	Hatching humidity too low	
Sticky chicks	Hatching humidity too high	
	Hatching temperature too low	
Splayed legs	Temperature too high	
	Hatching tray too smooth	
Crooked toes	Temperature too low	
	Too few eggs in hatching tray	
Prolonged hatch	Eggs stored improperly	Warm and cool spots in incubator
	Large and small eggs hatched together	Insufficient ventilation

A chick with its head under its left wing or between its legs is not properly positioned to pip. Even when a chick is positioned correctly, if the air cell is not at the large end of the egg, the chick won't be able to break free.

If the eggs were incubated and set into the hatching tray with the large ends up and the temperature and humidity were properly controlled, help-outs could well be a hereditary issue. Chicks that need help breaking free of the shell, assuming they live, will mature to produce more chicks that have difficulty hatching. For this reason, as tempting as assisting help-outs might be, it is a decidedly bad idea.

Among newly hatched chicks, two common problems are crooked toes and splayed legs. Crooked toes are often caused by a too-low incubation or hatching temperature. But they may also result from placing too few eggs in the hatching tray, thus leaving newly hatched chicks too much room in which to move around too soon. When the number of eggs you're incubating won't fill the hatching tray, use pedigree baskets (described on page 290) to more closely confine the newly hatched chicks.

Splayed or spraddled legs may be caused by a too-high incubation or hatching temperature. They may also result from a hatching tray that is too smooth, causing chicks to slip and slide before they get strong enough to stand and walk properly. If your incubator has a hatching tray with a smooth bottom, line it with paper toweling or an excelsior pad before filling it with eggs.

Splayed legs (left) may be caused by a too-high temperature or a slippery surface in the hatching tray or brooder. Crooked toes (right) may be caused by a too-low hatching temperature or from the chick's enjoying too much activity too soon after the hatch.

Chick Identification

Pedigree breeding requires having some way to track offspring. Chicks may be identified at hatch in one of two ways: embryo dyeing and pedigree baskets. They may then be tracked by means of wing bands or leg bands.

Toe punching is another method of identification, involving a coded pattern of removing the skin webs from between the toes. Aside from the fact that chicks don't like it — *Ow, that hurts!* — the loss of toe webbing can cause foot problems later in life. For these reasons, I quit toe punching years ago.

Embryo Dyeing

Embryo dyeing works best on chicks with white or light-colored down and wears off as soon as the chicks grow their first feathers. By injecting dye into eggs before they hatch, you can color-identify chicks of different matings. Dyeing in no way affects a chick's health or growth rate, provided you handle the eggs carefully and use only clean materials.

To dye embryos, you will need a 20-gauge, 1-inch-long (25 mm) hypodermic needle, a sharp sewing needle of the same size or a little bit bigger, and a set of food dyes in 2 or 3 percent concentration. Among the primary colors, red, green, and blue show up best. Purple, made by combining red and blue, also works well. Yellow and orange don't show up well at all, since chick down is often yellowish. Embryos dye best during the 11th to 14th day of incubation. To avoid chilling the eggs, remove no more eggs from the incubator than you can dye within 30 minutes.

About ½ inch (13 mm) from the tip of the pointed end, disinfect an area about the size of a quarter by wiping it with 95 percent rubbing alcohol or povidone-iodine (such as Betadine).

Dip the sewing needle into the alcohol or iodine. Cushioning the egg in one hand, make a tiny hole in the center of the disinfected area by pressing against it with the needle, twisting the needle back and forth until it just penetrates the shell and membranes. Take care to make only a tiny hole that does not go deeper than necessary to pierce the inside membrane (no more than ⅛ inch [3 mm]).

Dip the hypodermic needle into the alcohol or iodine and fill it with ½ cc of dye. Insert it into the hole so the tip is just beneath the inner shell membrane. Slowly depress the plunger to release the dye without letting it overflow. To avoid inadvertently mixing two colors, use a clean needle when you change dyes.

DYEING AN EMBRYO

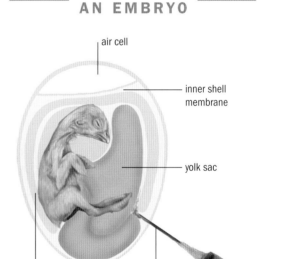

Dyeing embryos is a sure way to identify chicks as they hatch.

Seal the hole with a drop of melted wax or a tiny piece of Band-Aid — a sheer strip sticks best. Return the eggs to the incubator, and continue as normal. The chicks will hatch having down of the various colors you gave them.

Pedigree Baskets

Pedigree baskets are placed in the hatching tray to separate eggs from different matings. They might be nothing more than pint-size plastic baskets set upside down over groups of eggs, although make sure that chicks can't slip underneath and get mixed together.

Small plastic baskets might be found at an office-supply, school-supply, toy, drugstore, or dollar-type store. Before you go shopping, measure the height of your hatcher so you will get baskets that aren't too tall to fit.

We make our baskets from ½-inch (12.5 mm) hardware cloth, bent to form a box 2 inches (5 cm) high. We chose a height of 2 inches because it fits nicely in our lidded hatching tray, which is a standard height.

We have some baskets that are 5 inches (12.7 cm) square. That size starts out as a 9-inch (22.9 cm) square of hardware cloth with

A pedigree basket is used to separate eggs from different matings and may be used to restrict activity of hatching chicks to prevent crooked toes.

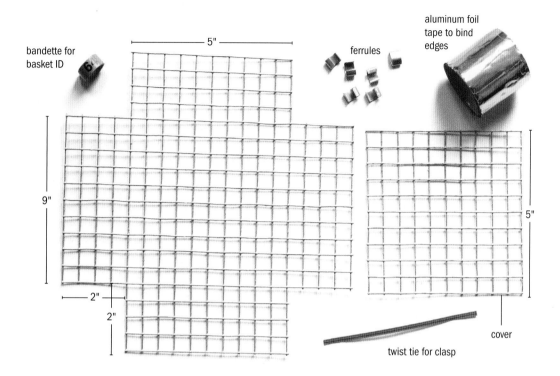

bandette for basket ID

ferrules

aluminum foil tape to bind edges

5"

9"

2"

2"

5"

cover

twist tie for clasp

This pedigree basket, big enough for five large eggs, will be fastened together with one ferrule at each corner and two ferrules to hinge the cover; zip ties (cable ties) or twist ties may be used in place of ferrules.

a 2-inch square cut out of each corner. These baskets hold about five large eggs or eight bantam eggs, more or less, depending on the exact size of the eggs. Our larger baskets are 7½ inches (19 cm) square and hold considerably more eggs. One of those is made from an 11½-inch (29-cm) square of hardware cloth with a 2-inch square cut from each corner.

A reasonable size is whatever will hold the number of eggs you wish to put in it — typically one week's worth of eggs from one hen — which you can determine by placing that number of eggs in a square on a flat surface

and measuring how much space they take. To give the newly hatched chicks sufficient wiggle room, make the basket big enough to hold two or three more eggs than you intend to put in it.

Cut a hardware cloth cover to fit, and hinge it to the basket on one side. We use cage-making ferrules to fasten together the sides of the basket and to hinge the lid —one ferrule at each corner and two to hinge the lid, but cable ties (zip ties) work just as well. Twist ties work, too, but don't last as long and must be periodically replaced. A twist tie makes a handy clasp for fastening down the lid after eggs have been placed in the basket.

To prevent a chick or yourself from getting scratched by the cut-off wire ends, bind off the edges of the basket and the lid. We use aluminum foil tape because it's designed to withstand moisture and therefore won't peel off in the incubator's high humidity.

If you have two matings to keep track of, you need only one basket to separate one mating's eggs from the rest in the hatcher. If you use two or more identical baskets, you'll need to identify each one so you can keep track of which basket holds which eggs. A numbered bandette fastened to the top of each basket offers a handy way to keep track of which is which. Alternatively, baskets fastened together with zip ties might be color-coded by the color of the ties used for each, or spiral leg bands may be used for color coding.

On the 18th day of incubation, group the eggs in the baskets by the mating that produced them, noting in your records which mating is in which basket. After the chicks hatch, remove one group at a time, wing-band or leg-band all the chicks in the group, and again note in your records which mating produced which chick.

Wing Banding

Wing banding involves applying a numbered band to the wing web — the wide flap of skin between the chick's wing and its body. Wing bands may be used for both large breeds and bantams, although most show judges

APPLYING A WING BAND

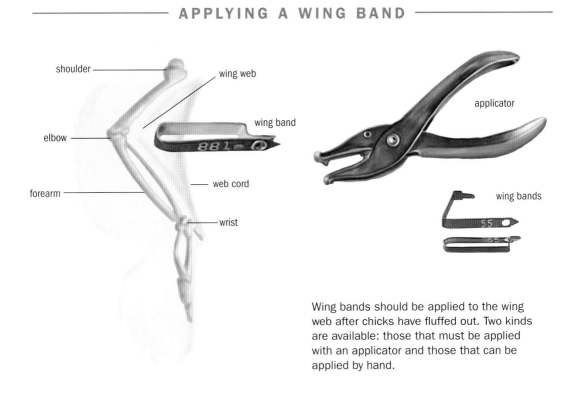

Wing bands should be applied to the wing web after chicks have fluffed out. Two kinds are available: those that must be applied with an applicator and those that can be applied by hand.

would frown on a small bantam wearing a visible wing band.

To avoid tearing the wing web, apply the band after chicks have completely fluffed out and have begun to toughen, usually at 1 day old. Check week-old chicks and adjust any band that might have slipped over a tiny wing, and continue checking until the chicks are big enough for the bands to stay in place and not restrict wing development.

Except in tightly feathered breeds, a wing band is clearly visible only on close examination. It provides a permanent means of identification, unless you deliberately choose to remove it.

Leg Banding

Leg banding chicks requires careful attention to changing the plastic bands several times as little legs grow. Furthermore, bands must initially be applied with care to avoid breaking a delicate leg. Apply bands as soon as you remove chicks from the hatcher.

For chicks of the large breeds, start with #4 bands; for bantams you'll need #3 or maybe #2 for the really little guys. You have the option of using either numbered bandettes to identify individuals or spiral rings to identify matings by color code. Spiral rings are thinner and therefore easier to apply on little legs than the wider numbered bands.

Both kinds tend to break eventually, making leg banding less dependable than wing banding. A good idea as the chicks grow is to use more than one band, in case one gets lost. When using color-coded combinations, put the same combination of bands on both legs.

Some breeders find that applying zip ties (cable ties) to the legs of baby chicks is easier and safer than applying leg bands. Be sure to cut off the excess to keep chicks from tightening a tie by picking on the end. Zip ties are even less flexible than leg bands and therefore must be watched more closely. A tie that gets too tight is difficult to remove without injuring the bird.

Replace bands or ties as the birds' legs grow and the bands or ties become too tight to slide up and down easily. Never leave a band or tie on so long that it binds. If either a tie or band is left too long on a growing bird, the leg will grow around it and the band will become imbedded in the leg, resulting in serious problems. Similarly, as a cock's spurs begin to grow, make sure the leg band remains above the spur where it won't bind.

Record Keeping

Hatching records are essential not only for fine-tuning your incubation procedures but also to help you pinpoint breeder-flock management problems that affect the hatch. The more detailed your records, the easier you'll be able to spot patterns that reveal room for improvement. Leave space at the bottom of each page to note such transient events as the dates and times of power outages or changes in your incubation procedure that you'd like to try in the future.

If you mark your chicks with individual identification, set up a second sheet with a line for each chick and space to note any data you find important, such as the chick's down color and (later, when it becomes apparent) its gender. Under notes you might also keep track of such things as deaths, reasons for culling, or changes in plumage color.

HATCHING RECORD

Breed/pen: _____   Cock: _____

Date set: _____   Time set: _____

Date candled: _____   Results: _____

Date candled: _____   Results: _____

Hatch should start on: _____

Hatch started on: _____   Time: _____ a.m./p.m.

Hatch ended on: _____   Time: _____ a.m./p.m.

Band code: _____

RESULTS:

A. Number of eggs set   _____   G. Culled, too weak   _____

B. Not fertile   _____   H. Culled, deformed   _____

C. Died 1st week   _____   I. Other   _____

D. Died 2nd week   _____   J. Total lost (add lines B through I)   _____

E. Full-term, failed to pip   _____   K. Total hatched (line A minus line J)   _____

F. Pipped, failed to hatch   _____   L. Percentage of fertile eggs hatched
(line K divided by [line A minus line B])   _____%

NOTES: _____

CHICK RECORD

Number	Sire	Dam	Hatch Date	Sex	Color	Notes

11

CHICK CARE

Newly hatched chicks are not entirely helpless, but until they grow a full complement of feathers, you'll need to keep them warm and dry and protect them from dogs, cats, and other animals. Like any other babies, they must also be kept clean and well fed.

Some chicken breeders feel safer raising chicks in a mechanical brooder. Others find just the opposite, preferring to have a hen brood chicks naturally. The difference is in management style and facilities setup.

Natural Brooding

If you're away from home much of the time, you'll likely find it less trouble to let a hen rear, or brood, the chicks for you. Until the chicks are 10 to 12 days old, they will typically stick pretty close to the mother hen. At about 2 weeks old, they start getting adventuresome and wander off to feed on their own but frequently return to the hen for warmth, comfort, and to sleep. As time passes, depending on their environment, the chicks will eventually go their separate ways or remain together as a social group.

While the chicks grow, a brooding pen offers the hen and chicks safety from the elements and from predators. A hen won't always find a safe place on her own, however. In my early years of keeping chickens, I let a broody make her nest in the layer house. As the chicks left the nest, they fell into the droppings pit and had to be rescued.

Later, another hen made her nest in the hayloft. One morning I found chicks wandering around the barn peeping miserably while the oblivious mother hen sat happily on her nest overhead in the hay. To prevent such mishaps, move the hen to a safe brooding pen either from the time she starts setting or as soon as the hatch is complete.

Brooding Pen

A *brooding pen* is simply a small-size chicken coop just big enough for a hen and her growing brood. No specific design is better than any other. You might adapt a small doghouse, a cleaned storage drum, or an unused camper shell. If you make the coop from scratch, it needn't be more than about 3 feet (1 m) square (3 feet × 3 feet [1 m × 1 m]). Be sure the floor, walls, and roof are rat-proof and snakeproof, and install screened vents near the top for good airflow. Add a door that opens onto a small, enclosed run. If the setup is outdoors, be sure to provide shade.

Chicks are amazingly active and self-sufficient from the moment of hatch, but still require basic care that includes warmth, protection, clean water, and properly formulated chick feed.

BROODING COOP

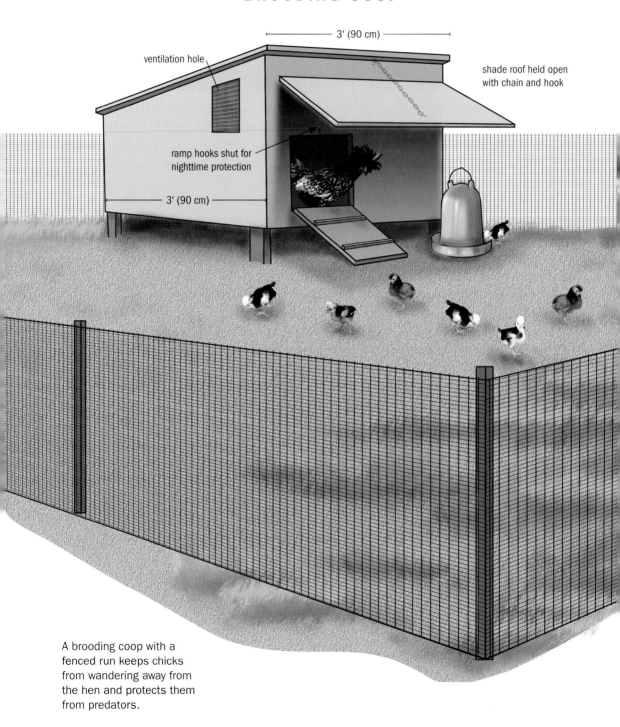

3' (90 cm)

ventilation hole

shade roof held open
with chain and hook

ramp hooks shut for
nighttime protection

3' (90 cm)

A brooding coop with a
fenced run keeps chicks
from wandering away from
the hen and protects them
from predators.

Never let a hen and her brood just wander around an open yard, or the chicks may fall prey to house cats, hawks, and other predators. They might get chilled in damp grass or get lost and not be able to find their way back. Once a chick is out of earshot, a hen has no way of knowing the little one is missing.

Feed the chicks starter ration, and feed the hen lay ration. Place the hopper of lay ration high enough up so the chicks can't reach it — the large amount of calcium required by hens will damage a chick's tender kidneys. On the other hand, if you have trouble keeping the chicks out of the layer ration, no harm will be done by letting the hen share their starter ration.

To keep the chicks and hen from scratching in the starter, place it in a chick feeder with holes too small for the chicks to climb through (although bantams and other small chicks might weasel into the feeder for a few days until they get too big). To protect the chick-size feeder from being knocked over by a too-exuberant hen teaching her chicks to scratch in the nearby ground, put up a barrier of 2-inch (5 cm) wire mesh that the chicks can pop through to reach their feeder, or place the feeder inside a cage with wire openings large enough for the chicks to get through. A quick and inexpensive option is to turn a sturdy cardboard box upside down over the feeder, with chick-size holes cut into the sides and a heavy board on top to keep the hen from knocking over the box. Where the hen shares the chicks' ration, fastening the bottom of the feeder to a short piece of 2 by 6 lumber provides a sturdy base.

You'll also need a water container the hen can't knock over. A 1-gallon (4 L) waterer is ideal. If you're brooding bantams, for the first few days put marbles or pieces of clean gravel in the rim to prevent a tiny chick from drowning.

Foster Moms

A setting hen can often be tricked into raising any chicks she finds under her — even those of another species. Mixing different bird species usually doesn't work well, but it's worth a try in a pinch. I have had hens raise both guinea keets and turkey poults along with their own chicks. I also had a chicken hen raise ducklings, although she got mighty perturbed when her kiddies started swimming.

A hen is most likely to accept foster chicks if she's been seriously setting for a couple of weeks but not necessarily the entire 21 days. By the same token, the chicks must be not much more than a day old and still receptive to accepting (or *imprinting*) a new mom.

To increase the chances both parties will accept each other, slip the chicks under the hen at night. In the event things don't work out as you had planned, be prepared to gather up the chicks and raise them yourself.

You can sometimes induce a hen to brood even if she hasn't been setting. Put the hen and baby chicks together in a dimly lit brooding pen, and watch for the four typical signs the hen is willing to care for the chicks. Not necessarily in this order, a hen willing to accept chicks will

- Spread her wings to cover the chicks
- Tidbit — sound the food call and pick up and drop bits of food
- Rush to the assistance of a chick making sounds of distress, such as if you pick it up
- Cluck loudly and continuously

Under normal conditions, a hen and her brood make initial contact through sound rather than sight, so a foster broody is more likely to accept chicks slipped under her at night and have the entire night to listen to them peep. By morning, both parties ideally will have developed an amiable relationship through sound, even if the chicks might look like little aliens to the hen.

Since a hen can successfully mother as many as three times the number of chicks she hatches, you might wish to artificially incubate additional eggs, with the intent of having your hen brood the resulting chicks along with her own. This ploy works best if all the chicks hatch at the same time. Again, slip the freshly hatched incubated chicks under the hen at night when she's sleeping. If you're moving the hen and her brood from the nesting site to a brooding pen, place all the chicks in the pen first so they'll get mixed together before you introduce the hen.

Mechanical Brooders

A mechanical brooder replaces a brooding hen as a place where chicks are temporarily raised until they have enough feathers to keep themselves warm. A brooder should provide

- Adequate space
- Protection from predators
- Protection from moisture
- A reliable heat source
- Freedom from drafts
- Good ventilation

Brooders come in many sizes and styles. The style that's best for you will depend on how often you plan to raise chicks, how many you wish to brood at once, and how much money you want to spend. Ready-made brooders are available in many sizes from small plastic ones that hold a dozen or so chicks to large metal ones that hold hundreds. Or you can make your own of any size or style that suits your needs.

Whatever style you choose, locate your brooder where you won't be bothered by the dust chicks invariably stir up with their constant activity. Brooding chicks in the living room does not sit well with the person who cleans the house.

Box and Battery Brooders

A typical commercially available *box brooder* is made of metal or metal and plastic. It has a built-in heat source and a low-level light to attract chicks to the heated area. A wire-mesh floor has a removable tray beneath it to catch waste. Built-in feed and water troughs are attached to the outside along three sides to furnish plenty of feeding and drinking space while keeping the chicks from fouling the feed or water.

A barred gate, adjustable as the chicks grow, protects each trough from droppings and prevents chicks from popping through and drowning or falling to the floor. For bantams, the bars start out closer together than for larger chicks. For all chicks, you have to watch the adjustment of the gates to make sure the birds haven't grown so big they can no longer reach through to eat or drink.

A *battery brooder* consists of a series of box brooders stacked one on top of the other. The removable droppings trays prevent waste from one tier from falling on chicks in the tier below. This style of brooder is what you typically see at a traditional farm store, where batteries are favored because they house lots of chicks in a small space, and the tiers offer a convenient way to separate birds of different breeds, species, and ages.

A battery brooder lets you keep lots of chickens in a small space.

Some batteries have a rack that holds a specific number of tiers; others come as individual units that may be stacked as high as you can reach. Each level has its own heater, which allows you to save electricity by turning off the heat of an unoccupied tier.

The number of chicks you can brood in a battery depends on the dimensions and the number of tiers; a typical box holds about four dozen chicks, and a typical battery has three to five tiers. Common practice is to keep chicks in the battery for about a week, or two at most, then move them to roomier housing. In any case, birds must be moved from a battery brooder before they grow so tall their heads rub against the top or so big they can't get their heads through the gates at the widest adjustment.

Battery brooders are available through farm stores and poultry-supply catalogs. They're pretty expensive, but you might get lucky and find a good deal on a used one. To operate a battery, you'll need a draft-free but well-ventilated outbuilding where the temperature is warm and fairly steady.

Area Brooder

An area brooder has the advantage over a box or battery in being easily expandable so you can brood more chicks at a time and they don't outgrow the housing as quickly, if they outgrow it at all. This brooder is the style traditionally used by farmers who start chicks and grow them to maturity in a single chicken coop, barn, or other outbuilding.

The heat source, called a *hover*, is hung from the ceiling so it hovers over the chicks. The birds huddle beneath the hover to be warmed by its electric or gas heater and move away from the heat to eat and drink. The hover may be round or rectangular and may have curtains hanging around the edges to keep in heat and keep out drafts.

An area brooder is basically a heater that hovers over the chicks and is ideal for brooding birds in the same facility where they will grow to maturity.

A hover is an economical way to brood a hundred chicks or more. The chicks are housed on a litter-covered floor in a small predator-proof outbuilding or a draft-free stall or corner of a larger building. Feeders and waterers are spaced around the outer edges of the hover where chicks can easily find them.

When you use an area brooder, you need some way to keep the chicks from wandering far from the heat source and getting chilled. A brooder guard confines chicks close to the heat, reduces floor drafts, and eliminates corners where chicks tend to pile up and smother. Ready-made brooder guards are available, but you can easily make your own by cutting up a large cardboard box, because a brooder guard is nothing more than a 12- to 18-inch-high (30 to 45 cm) circular fence of corrugated cardboard.

Run the fence at a distance of 2 to 3 feet (60 to 90 cm) around the heat source and fasten the ends together with duct tape. In 7 to 10 days, after the chicks have become fully familiar with the locations of heat, feed, and water, either remove the brooder guard or, if it's still needed to reduce floor drafts, expand it to give the birds room to grow.

Homemade Brooders

A brooder need not be an expensive commercially built affair and need not even be a permanent fixture. Many a chick has been brooded in a sturdy cardboard box, which has the distinct advantage that it may be disposed of and replaced, instead of having to be cleaned.

Another ready-made option is a galvanized tank designed for watering livestock, which is easy to hose out between uses. A light bulb or heat lamp furnishes warmth, and a piece of wire mesh secured over the top provides ventilation while keeping out cats and other chick eaters. If necessary to eliminate drafts, lay a piece of cardboard or plywood across part of the top.

A large plastic tote, with openings cut into the lid for ventilation, is a popular brooder option, but requires careful monitoring to avoid crowding, overheating, and excessive moisture.

An extra-large plastic tote or bin makes an easy-to-clean brooder for a small number of chicks, and the snap-on lid secures them from predators. To provide heat and ventilation you'll need to cut into the lid, which is not an easy job but may be done neatly with a utility knife and plenty of elbow grease. Hardware cloth fastened to the lid will keep out cats and other predators. A plastic tote can easily get too hot, so keep an eye on the chicks' comfort level and adjust the heat accordingly.

Moisture also tends to collect in the bedding because it has no place to go; clean out and replace the bedding at least once a week to keep it from getting moldy.

An infrared heat lamp with an aluminum reflector and a porcelain socket, hung over a stock tank lined with litter, makes a dandy brooder.

Brooder Features

As long as you maintain the principles of security and warmth, the possibilities for brooding chicks are limited only by your imagination. Any brooder must be designed to minimize stress, since stress drastically reduces the chicks' immunity, making them susceptible to diseases they might otherwise resist. Stress is minimized by making sure the chicks are neither too cool nor too warm; have a clean, safe environment; are provided sufficient space for their numbers; and can always find feed and water.

Stress may also be reduced by approaching the chicks from the side, rather than from the top. Commercial box and battery brooders are designed with this feature in mind. Most other brooders are designed for the convenience of the chicken keeper, who scares the living daylights out of chicks by approaching them from above — after all, most predators swoop down on baby chicks. Whenever you approach chicks from the top, the polite thing to do is talk or hum to let them know you're coming.

Brooder Heat

A chick's body has little in the way of temperature control, although a group of chicks can keep themselves warm by huddling together in a small space — which is why a box full of newly hatched chicks may be shipped by mail. When given sufficient space to exercise, eat, and drink, chicks need an external source of warmth while their down gives way to feathers, starting at about 20 days of age.

Chicks tend to feather out more quickly in cooler weather, but if the air temperature is quite low, they need auxiliary heat longer than chicks brooded in warmer weather. For this reason, chicks hatched in winter or early spring typically require brooder warmth longer than chicks hatched in late spring or early summer.

Start the brooding temperature at approximately 95°F (35°C) and then reduce it by about 5°F (3°C) each week until the brooder temperature is the same as ambient temperature. Within the chicks' comfort zone, the more quickly you reduce the heat level, the more quickly the chicks will feather out.

Commercial box brooders, batteries, and hovers operate by adjustable thermostat. Most homemade setups provide heat with either incandescent lightbulbs or infrared heaters that have no thermostat.

An incandescent lightbulb is the least expensive heat source for batches of 25 to 50 chicks. If your brooding area is large enough to handle the extra heat, you're better off using two bulbs, in case one burns out when you're not around. Screw each bulb into a fixture with a reflector and hang it over the brooder. The heat may be adjusted two ways: by raising or lowering the fixture and by decreasing or increasing the bulb's wattage. Start with 100- or 60-watt bulbs, depending on the size of the brooder and the number of chicks.

One 250-watt infrared heat lamp provides sufficient heat for 25 to 100 chicks. Infrared lamps with either red or clear bulbs are available at farm stores, electrical-supply outlets, and some hardware stores. A red lamp is more expensive than a white lamp but won't burn out as quickly, and the red glow discourages picking; as long as everything looks red, truly red things won't attract attention.

An infrared lamp gets quite hot, so use a porcelain rather than a plastic socket, because the plastic might melt. A standard brooder lamp holder has a porcelain socket, as well as a couple of stout wires bent across the front so the lamp can't come into direct contact with bedding — for instance, if the lamp falls — or other flammables and create a fire hazard.

Hang the heat lamp by an adjustable chain, starting about 18 inches (45 cm) above the chicks. As the chicks grow, raise the lamp to reduce the heat. A general rule is to raise the lamp about 3 inches (7.5 cm) each week.

Be especially watchful with chicks confined in a small brooder, since an infrared lamp can get pretty hot and you don't want the chicks to be cooked alive. As they get older and require less heat, give them more room so they can move away from the heat, or switch from infrared lamps to incandescent bulbs.

A safer infrared option is an Infratherm heating panel, which is more expensive than an infrared heat lamp but uses so much less electricity that in the long run the panel turns out to be considerably cheaper. A panel directs heat only beneath itself, making it easier for chicks in a small area to move away to maintain their comfort level. Panels come in various lengths and, unlike light bulb and heat lamp fixtures, are entirely sealed, making them much easier to clean and sanitize.

Theoretically, brooder temperature is measured with a thermometer placed 2 inches (5 cm) above the brooder floor (and at the outer edge of a hover), but you shouldn't need a thermometer. Just watch the chicks, and adjust the temperature according to their body language.

Chicks that aren't warm enough — due either to insufficient heat or to draftiness — crowd near the heat source, peep shrilly, and may have sticky bottoms or outright diarrhea. In an effort to get warm while they sleep, the chicks will pile up and smother each other. Piling is most likely to occur at night when the ambient temperature drops, so in cold weather check your chicks before you go to bed, and if necessary, increase the heat overnight.

Chicks that are too warm move away from the heat, spend less time eating, and as a result grow more slowly. They pant and try to get away from the heat source by crowding to the brooder's outer edges, perhaps smothering one another. If the brooder is hot enough to raise their body temperature above 117°F (47°C), chicks die.

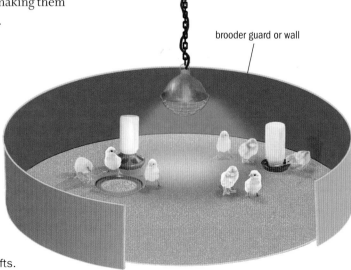

brooder guard or wall

Floor hover provides heat;
a brooder guard eliminates drafts.

Happy chicks that are warm and cozy wander freely throughout the brooding area, emit musical sounds of contentment, and sleep sprawled side by side to create the appearance of a plush down carpet. The sight can be dramatic to someone who has never seen chicks resting comfortably. An overnight guest once woke my husband and me early one morning, in a panic because "the chicks are all dead." Meanwhile, his commotion had awakened the chicks, and by the time we rushed back to check on them, to the astonishment of our guest, the chicks were busy having breakfast.

Brooder Light

Chicks are attracted more to light than to heat, which is why commercial brooders have a small light, appropriately called an attraction light, near the heat source. One 25-watt bulb will adequately light about 10 square feet (1 sq m). To help chicks find feed and water, light the brooder continuously for the first 48 hours. If the brooder gets natural daylight, after the first two days you can turn off the light during the day. Windows on the south side furnish the best sunlight.

Even if the light is also your source of heat, turn it off for half an hour during each 24-hour period — but obviously not during the coolest hours — so the chicks learn not to panic later when the lights go out at night or in the event of a power failure. Putting the light on a timer will save you the trouble of remembering to turn it off and on each day.

Light affects the growth rate of chicks, so never keep them in the dark. Even if you have to dim the lights to control cannibalism, the light should still be bright enough for you to see what's going on in the brooder. A rule of thumb is that dimmed lighting should be barely bright enough for reading a newspaper.

Brooder Flooring

Raising newborn chicks on wire-mesh flooring has both advantages and disadvantages. The biggest advantage is ease of cleaning, since droppings and other debris fall through the brooder floor and are caught on a droppings tray or a layer of newspaper.

A disadvantage is the lack of gradual exposure that allows floor-reared chicks to develop immunity to coccidiosis, a major chickhood disease. Unless the birds spend the rest of their lives in wire cages, those brooded on wire are likely to suffer an outbreak of coccidiosis when moved to open housing. Brooding

---

SEPARATING COCKERELS FROM PULLETS

At 3 to 8 weeks of age, depending on their breed, chicks start developing reddened combs and wattles. The cockerels' combs and wattles will become larger and more brightly colored than the pullets'. Unless they're Sebrights or Campines, which are hen-feathered breeds, the males will soon develop pointed back and saddle feathers and long tail sickles, in contrast to the more rounded back and saddle feathers and shorter tails of hens. At about the same time, peck-order fighting will get serious and sexual activity will start. If you haven't already done so, it's time to separate the cockerels from the pullets or at least to pare down the number of cockerels to a reasonable ratio for the number of pullets.

chicks on litter from the start gives them the gradual exposure to coccidia they need to develop immunity.

Another disadvantage of brooding on wire is that the chicks are more likely to become cannibalistic than chicks raised on litter. No one is exactly sure why, but the theory is that, since pecking is normal behavior, chicks that have nothing else to peck at will peck each other.

Bedding also helps keep chicks dry, insulates the floor for added warmth, and absorbs droppings. Ideal kinds of litter are peat moss, wood shavings (pine, not hardwood), crushed corncobs, crushed cane, finely shredded paper, vermiculite, and coarse sand. Chopped straw retains more moisture than most other types of bedding, and manure tends to cake on the surface, but it is usable if that's all you have or can obtain cheaply. Avoid whole straw, since chicks have trouble walking on it and it quickly mats down.

When brooding chicks on litter, stir up the litter every day to keep it from packing down, and add fresh bedding as often as necessary to keep it fluffy and absorbent. Remove and replace moist bedding around waterers, since damp litter turns moldy and can cause *aspergillosis*, known as brooder pneumonia. This brooder disease and others are easily prevented by proper litter management.

If you area-brood chicks on dirt, first cover the dirt with well-anchored ½-inch (12.5 mm)-mesh aviary netting or hardware cloth to keep out burrowing predators. Then spread a layer of litter at least 3 inches (7.5 cm) deep on top.

Whether you brood on wire or litter, for the first week, cover the floor with paper. In my smaller brooders, I first put down a few layers of newspaper; the big brooder I line with opened-out feed sacks. Freshly hatched chicks can't be brooded directly on newspaper or other smooth paper, because little guys damage their legs slipping around on it. So on top of the paper liner, I put a layer of paper toweling, adding a fresh layer at least daily or as often as necessary to keep it clean.

Covering a wire-bottom brooder with paper prevents chicks from getting their little hocks jammed through the wire, and also eliminates drafts. Covering litter bedding with paper encourages the chicks to peck at feed instead of filling up on litter. Once the chicks are eating and walking well, all the paper may be removed easily by rolling up the newspaper or feed sacks.

Brooding Space

The rate at which chicks grow varies from breed to breed. Bantams and most of the heritage breeds grow more slowly than others. Hybrid broilers are bred for excessively rapid growth. Separating the pullets from the cockerels as soon as you can identify them helps both parties grow more steadily.

For the first few days, all chicks spend a lot of time sleeping and therefore don't need much room to roam. But as they grow and become more active, they need increasingly more space for sanitary reasons and to prevent the boredom that leads to picking at each other. On the other hand, more isn't necessarily better — chicks given too much space during cold weather have trouble staying warm.

If you start chicks in a box or other closely confined brooder, giving them more space as they grow means either dividing them up or periodically moving the entire batch to larger quarters. If you start chicks in an area brooder, giving them more room is simply a matter of expanding the draft barrier until it is no longer needed.

The rule of thumb is to begin with about 6 square inches (40 sq cm) of space per chick. Bantams and light breeds can get by with a little less; broilers and the really big breeds need a little more. Base the size of the brooding and growing space you provide more on common sense than on a meticulous measuring of the floor space. Common sense tells you chicks are overdue for expanded living quarters if

- They have little room to move and exercise or to spread out for sleep.
- They dirty the brooder floor faster than you can keep it reasonably clean — droppings pack on the floor, manure balls stick to feet, you can smell ammonia.
- They run out of feed or water between feedings, indicating the need for a larger area to accommodate more or larger feeders and waterers.

A move to unfamiliar housing is a frightening experience for chicks, and chicks that are frightened tend to pile together and smother one another. For the first few nights after moving chicks, provide dim lighting and check often to make sure they are okay. Moving their old feeders and waterers to the new location also helps by bringing along something familiar.

If you move chicks or growing birds to housing that has held chickens in the past, the facility must be thoroughly cleaned, swept free of dust and cobwebs, and washed down. First clean it with warm water and detergent, then with warm water and chlorine bleach (¼ cup bleach per gallon of hot water; 15 mL/L) or other disinfectant approved for use with poultry. Leave doors and windows open until the housing is completely dry before bringing in the new birds.

When light breeds reach about 4 weeks of age and heavy breeds about 6 weeks, they're ready to roost on low perches. Don't provide roosts for broilers or they'll develop breast blisters and crooked keels. For others, nighttime roosting is a natural and healthy habit, and hopping on and off perches during the day is good entertainment, as well as good exercise. Allow 4 inches (10 cm) of roosting space per chick. Provide low and high perches, or start the perches close to the floor and move them up as the majority of birds learn to use them.

Before about 1950, most chicks were reared on range, which has once again become popular. A portable range shelter on skids is especially attractive for raising a batch of meat birds, since they don't require permanent year-round housing. If the weather is warm enough — and presumably weather that's warm enough for pasture growth is warm enough for chicks — broilers may be put out when they are as young as 2 weeks of age. If the weather is still cool, wait until the chicks are fully feathered before putting them on pasture.

Raising pullets on pasture, away from older birds, gives them time to develop immunities through gradual exposure to the diseases in their environment. Even in a colder climate where the pullets must later be moved to winter housing, range rearing during spring and summer offers sunshine, fresh air, and green feed that combine to keep the birds healthy.

When moving chicks to a range shelter, for the first few days confine them to the shelter or to a small enclosed yard while they can get oriented before you give them freedom to roam. If it rains, especially a long, hard rain at night, go out and check your birds to make

From an early age, growing chicks enjoy having a place to perch.

sure they aren't huddled outside or squatting in a puddle inside the shelter.

Even if you don't have pasture or range, when the chicks are at least 2 weeks old and the weather is warm and sunny, you can put them in a pen or wire-bottom cage on the lawn for a few hours each day, provided the lawn has not been sprayed with toxins. They'll need shade and water, a ½-inch-mesh (12.5 mm) wire guard around the perimeter so they can't wander away, and wire mesh over the top to keep out hawks and cats. For good sanitation and to provide fresh forage, put the pen in a new spot each day. If you put a pen of growing chicks in your garden, spade under the soil after each move.

Feed and Water

Newly hatched chicks need chick-size feeders and waterers to start. As they mature, they'll need different feeders and waterers that provide better access for larger birds and can hold the larger amounts of feed and water consumed by growing birds.

Watering Chicks

Chicks can go without water for 48 hours after they hatch, but the sooner they drink, the less stressed they will be and the better they'll grow. A chick's body needs water for all life processes, including digestion, metabolism, and respiration. Water helps regulate body temperature by taking up and giving off heat, and it also carries away body wastes. A chick that loses 10 percent of its body water through dehydration and excretion will experience serious physical disorders. If the loss reaches 20 percent, the chick will die.

Chicks that can't find water won't grow at a normal rate, will develop bluish beaks, and will stop peeping. Deaths start occurring on the fourth day and continue until the sixth day. If water was available, but some chicks couldn't find it, those surviving after the sixth day are the ones that did find it.

To make sure your chicks know where the water is, take advantage of the fact that a chick starts drinking as soon as its beak is wet. As you place each chick in the brooder, dip its beak into the waterer and watch that it swallows some water before releasing the bird. Even though they may not drink right away, most of them will get the idea. Once several chicks start drinking, the rest will learn to imitate them.

Some chicks start life quite active and curious, while others initially aren't adventuresome at all, so for the first day or two, keep waterers fairly close to the heat source, where all the chicks can easily find them — but not directly under the heat, where they are likely to sleep crowded against the waterer and drown. As soon as the chicks are drinking well, move waterers farther out to give the chicks more space to rest near the source of heat.

Various health boosters consisting of vitamins and electrolytes may be added to drinking water to give chicks the best start in life. If the chicks have been shipped, a booster solution for the first three weeks will help them overcome shipping stress. Even if they haven't been shipped, a booster solution for at least one week will help them cope with the stress of transforming from embryo to chick.

In the old days, before chick-booster blends were readily available, old-timers put sugar in the drinking water of newly hatched chicks to give them extra energy. If the chicks were shipped by mail or just looked droopy and in need of a spurt of energy, as much as 1 pound (0.5 kg) of sugar was stirred into each gallon (4 L) of water (or 0.1 kg per liter), or about a half cup per quart (120 mL per 1 L).

Chick Waterers

Chicks must have access to fresh, clean water at all times. The easiest way to provide water to newly hatched chicks is to use a 1-quart (1 L) canning jar fitted with a metal or plastic watering base, available from most feed stores and poultry-supply catalogs. A plastic base will crack over time, and a metal one will break

CHICK WATER REQUIREMENTS

A chick's need for water increases as the bird grows. To determine approximately how much water a batch of chicks needs, divide the chicks' age in weeks by 2 to determine how many gallons you should provide per one hundred chicks per day. For example, 4-week-old chicks need about 2 gallons of water per day per hundred chicks, or 1 gallon per 50 chicks. (In metric, multiply 2 times the age in weeks to determine how many liters of water one hundred chicks need per day; at 4 weeks old, one hundred chicks need about 8 liters of water per day.)

away from the portion that screws onto the jar, so keep a few extra ones on hand.

If you can't find a chick-watering base, or yours springs a leak, make an emergency waterer from an empty can and an aluminum pie tin that's 2 inches (5 cm) larger in diameter than the can. Bore two small holes on opposite sides of the can, ¾-inch (2 cm) from the open end. Fill the can with water, invert the pie tin on top, and turn the assembly upside down so the pan becomes a water-filled basin the chicks can drink from.

Don't be tempted to cut corners by furnishing water in an open dish or saucer. Chicks will walk in it, tracking litter and droppings that spread disease. They'll tend to get wet and chilled, and the stress will open the way to disease. Some chicks may drown.

Drowning is generally not an issue when chicks have a proper waterer unless they are so crowded some chicks end up falling asleep with their heads in water. Tiny bantams may also have trouble with drinkers designed for standard-size chicks, in which case put marbles or clean pebbles in the water for the first few days until the birds get big enough to avoid drowning.

Within a week or so, the chicks will outgrow the 1-quart watering jars. You'll know the time has come when you find yourself filling drinkers more and more often to keep your chicks in water at all times. Another sign that a waterer is too small is finding chicks perched on top, a habit that gets droppings in the water. You'll know the time for changing drinkers has long since passed when you find the quart jar overturned and spilled due to boisterous play in the brooder.

When it's time to change to larger waterers, you have several choices. Regardless of the design, a good waterer has these features:

- It is the correct size for the flock's size and age — chicks should neither use up the available water quickly nor be able to tip over the fount.
- The basin is the correct height — a chick drinks more and spills less when the water level is between its eye and the height of its back.
- Chicks can't roost over or step into the water — droppings plus drinking water make a sure formula for disease.
- The drinker is easy to clean — a fount that's hard to clean won't be sanitized as often as it should be.
- The waterer does not leak — leaky drinkers not only run out fast but create damp conditions that promote disease.

One style of waterer is similar to the quart-jar setup, only in larger versions. Depending on the number of chicks you're brooding and the amount of space they have, you might first switch from quart jars to 1-gallon (4 L) plastic drinkers, then move up to 3-gallon (12.5 L) or 5-gallon (20 L) metal founts.

Another style is a trough-type waterer. You'll need 20 linear inches (50 cm) of trough per 100 chicks up to 2 weeks old, and 30 linear inches (75 cm) thereafter. A trough that allows chicks to drink from both sides offers twice the watering space of a one-sided trough.

Automatic waterers are a great time saver, are the most sanitary of all designs, and ensure that chicks won't run out of water when no one is around to refill founts. Anytime you use indoor metal waterers connected directly to water lines, if chicks refuse to drink, stick a finger into the water. If you feel a buzz, a bad electrical connection such as poor grounding is causing chicks to get a shock each time they try to drink. Fix the problem if you can or call in an electrician.

Clean waterers daily with warm water and vinegar or other poultry-approved sanitizer. Initially place drinkers no more than 24 inches (60 cm) from the heat source. Later, as you move the chicks to expanded housing, make sure they never have to travel more than 10 feet (3 m) to get a drink. Whenever you change waterers, leave the old ones in place for a few days until the chicks get used to the new ones. This stress-reducing measure applies whether or not the chicks are moved at the same time the drinkers are upgraded.

To keep chicks from picking in damp litter around waterers, after the first week place each fount over a platform created from a wooden frame covered with ½-inch-mesh (12.5 mm) hardware cloth. Drips will fall below the mesh where chicks can't walk or peck. During the first days, do not place waterers on a platform where chicks can't reach them, although you might use something thin and flat underneath to steady the waterer. Some folks use a small square of plywood; I prefer easily sanitized ceramic tiles left over from remodeling or purchased as seconds at discount stores.

To minimize stress, chicks should drink soon after they hatch and eat within five hours.

Feeding Chicks

Chicks experience less stress if they eat on the first day after they hatch, even though they can survive for another day or two without eating. Yolk reserves provide nutrients that in nature allow early-hatched chicks to remain in the nest until all the stragglers have hatched. In modern times, the yolk reserves sustain newly hatched chicks when they're shipped by mail.

Since chicks have an inherent instinct to peck and may peck at feed almost immediately on being placed in the brooder, make sure they are drinking before they start eating. They seem to experience less of a problem with sticky bottoms if they get a good dose of water before they get a belly full of feed, especially when the feed is commercially formulated chick starter.

Starter ration for chicks contains a mixture of grains, protein, vitamins, and minerals. It is higher in protein and lower in calories than rations designed for older chickens. Never feed layer ration to chicks, even as an emergency measure if you run out of starter. The high calcium content of layer ration can seriously damage a chick's kidneys.

If you run out of starter, or you forget to pick some up and you have chicks to feed, you can make an emergency starter ration by cracking scratch grains in the blender or, if you have no scratch, by running a little uncooked oatmeal through the blender and mixing it 50/50 with cornmeal. Grains are high in calories and low in the protein, vitamins, and minerals a chick needs for good growth and health, so don't use this mixture any longer than necessary.

If you have extra eggs on hand, mashed hard-boiled egg makes an excellent starter ration. In the old days before commercial rations were available, farmers typically started their chicks on mashed boiled eggs. After the first few days, the chicks were fed oatmeal and cracked grains until they got big enough to forage for themselves. Chicks that are fed cracked grains also need grit. If your local farm store doesn't carry chick grit, use cage-bird grit from a pet store.

Some brands of commercial starter are medicated with a coccidiostat to prevent chicks from getting coccidiosis. Whether or not you need medicated starter depends in good part on your management style. Use medicated starter if:

- You brood chicks in warm, humid weather.
- You brood large quantities of chicks at a time.
- You keep chicks in the same brooder for more than three weeks.
- You brood one batch of chicks after another.
- Your sanitation isn't up to snuff.

You shouldn't need medicated starter if you brood in late winter or early spring (before warm weather allows coccidia and

CHICK FEED REQUIREMENTS

As a general rule, each chick will eat approximately 10 pounds (4.5 kg) of starter ration during its first 10 weeks of life.

other pathogens to thrive), you brood on a non-commercial scale, you keep your chicks on dry litter, and your chicks always have fresh, clean drinking water. On the other hand, if you're raising your first-ever chicks, using medicated starter gives you one less thing to worry about while you work through your learning curve.

In areas where chickens are big business, farm stores may offer a starter ration, a separate grower or finisher ration for broilers, and a grower or developer ration for layers and breeders. Switch from one ration to another as indicated on the labels, gradually making the switch by combining the old ration with greater and greater amounts of the new ration to avoid problems related to digestive upset.

In many parts of the country, farm stores carry only one all-purpose starter or starter-grower ration. If you can find only a single chick ration, continue using it until broilers reach butchering age and layers/breeders are ready for the switch to lay ration.

Chick Feeders

Soon after they hatch, chicks start looking for things to peck on the ground. If they don't see anything else on the ground, they'll peck their own feet. To help them find feed, sprinkle a little starter ration on a paper towel or paper plate. As soon as most chicks are pecking freely, remove the feed-covered paper before it begins to hold moisture that attracts mold.

For the remainder of the first week, put the starter in a shallow lid or tray, such as a shoebox lid or anything with sides of a similar height. When the chicks start scratching out the feed, switch to a regular chick feeder, available from farm stores and through poultry-supply catalogs.

Like chick waterers, chick feeders come in several styles. A good feeder

- Prevents chicks from roosting over or scratching in feed
- Has a lip to prevent billing out
- May be raised to the height of the birds' backs as they grow
- Is easy to clean

One style of chick feeder is a base, similar to a drinker base, that screws onto a feed-filled quart (1 L) jar, and has little openings through which the chicks can peck. Because this style has a small footprint, it's ideal for use in a brooder where space is limited.

Another style is a trough that comes in various lengths, the shortest of which is about a foot (30 cm). Those designed for baby chicks have a lid with individual openings into which the chicks peck. The styles designed for growing chicks have either a reel to discourage birds from perching over the feed or a grill to keep them from scratching in it.

If you use a round or trough feeder with individual openings, allow one slot per chick. If you use an open trough, allow 1 linear inch (2.5 cm) per chick to 3 weeks of age, 2 linear inches (5 cm) to 6 weeks of age, and 3 linear inches (7.5 cm) to 12 weeks. Count both sides in your measurement if chicks can eat from either side.

Like grown-up chickens, chicks waste feed through billing out — the habit of scratching out feed with their beaks. To minimize wastage, fill trough feeders only two-thirds full. An inwardly rolled lip discourages billing out, as does raising the feeder so it's always the same height as the birds' backs.

By about their second week, the chicks will be ready for a bigger feeder. You'll know it's time to switch when chicks that climb through slots to eat get too fat to pop back out. The time to switch has passed when chicks are strong enough to overturn the feeder or they've grown too large to get their beaks into it easily. Whenever you change to a different feeder, leave the old one in place for a few days until you're sure all the chicks are eating from the new one.

If the brooder is roomy enough, a hanging feeder is ideal because it holds a lot of feed, so chicks are less likely to run out during the day; it minimizes feed wastage because chicks can't scratch in it and are less likely to bill out feed if the feeder is maintained at the proper height; and it is easy to raise on the hanger to the height of the chicks' backs as they grow. Hanging feeders come in various sizes, from small plastic ones to large metal ones. To determine how many chicks aged 6 weeks or younger may be fed from one tube feeder, multiply the feeder's base diameter in inches (or centimeters) by 3.14 and divide by 2 (in metric divide by 5).

A chick feeder that screws onto a quart (about a liter) jar has a small footprint, making it ideal where brooder space is limited.

BENEFICIAL GUT BACTERIA

Old-time poultry keepers spiked their chicks' water with a tablespoon of apple cider vinegar per gallon (3.75 L). In the days before soda pop, people regularly drank water with vinegar in it (they called it "an acid" or "a shrub"), so why not give it to the chickens? Chickens like it, and the poultry keepers saw positive effects. Could they have known that the beneficial bacteria and yeasts naturally colonizing a chick's intestines prefer acidic conditions? I doubt it. The science of probiotics is all pretty new. But we know now some reasons why it was/is beneficial. Encouraging the growth of beneficial gut flora fends off harmful organisms through a process called *competitive exclusion*.

Chicks raised in an incubator acquire beneficial gut flora more slowly than chicks raised under a hen. To enhance their immunity, probiotics are available that are either dissolved in water or sprinkled on feed to give the chicks an early dose of the same gut flora that will eventually colonize their intestines. A handy substitute is live-culture yogurt, but a little goes a long way — giving chicks too much yogurt will cause diarrhea.

Fill feeders in the morning, and let the chicks empty them before filling them again. Leaving feeders empty for long invites picking, but letting stale or dirty feed accumulate is unhealthful, so strike a happy balance. Clean and scrub feeders at least once a week.

Chick Problems and Solutions

Healthy, well-formed chicks that are housed and fed properly generally thrive. Disease is rarely a problem in newly hatched chicks unless they pick up an egg-transmitted illness during incubation, the incubator or brooder isn't kept clean, or the chicks are otherwise exposed to a high concentration of typical poultry microbes before they have a chance to develop natural immunity through gradual exposure.

Because chicks are especially susceptible to adult illnesses such as Marek's disease and leukosis during their first 6 to 8 weeks of life, do not brood them in unclean housing where adult birds have been raised in the past or within 300 feet (90 m) of an adult flock. When you do chores, tend to your chicks before visiting your older birds.

Stress due to chilling, overheating, dehydration, or starvation can drastically reduce the immunity of newly hatched chicks, making them susceptible to diseases they might otherwise resist. Routine activities such as vaccination or being moved also cause stress. To minimize the effects of stress, avoid exposing chicks to more than one stressor at a time. For example, avoid moving and vaccinating chicks at the same time.

Different chickhood diseases occur in different parts of the country. Your county Extension agent or state poultry specialist can tell you whether or not your chicks should be vaccinated and for which diseases.

Signs of Disease

Lots of people successfully brood chicks without losing any. People who brood large numbers of chicks expect a normal death rate of about 2.5 percent during the first two weeks, increasing to up to 5 percent by the seventh week. As distressing as a death may be, the occasional loss of a chick is not a cause for any panic.

A pattern of increased deaths, on the other hand, is a sure sign of disease. Other signs to watch for include listlessness, weakness, poor growth, drooping wings, decreased appetite, increased thirst, huddling near heat with heads down and eyes closed or swelling in one or both eyes, and diarrhea. The most common cause of diarrhea in chicks is coccidiosis.

A listless chick with droopy head and wings isn't feeling well.

Coccidiosis

Coccidiosis is a common intestinal disease among chicks. It results from the protozoa that naturally colonize a chick's intestines multiplying and getting out of hand, which can happen if the chick picks up protozoa too fast by eating droppings in feed, water, or litter. Litter picking occurs when feed runs out or chicks are so crowded they either get bored (because they have too little room to move around) or can't get enough time at the feeder (because not enough feeder space has been provided for the number of chicks).

Coccidiosis is most likely to occur in chicks 3 to 6 weeks of age, with the worst cases appearing at 4 to 5 weeks. It may be prevented by medications, including the use of medicated starter ration, but the better option is to keep brooder litter clean and dry and to clean feeders and drinkers often. Wash waterers each time you refill them. If you find chick droppings in the water, raise the drinker by elevating it on a low platform, or switch to one more suitably designed for chicks. The best measure of all is to get chicks out of the brooder and onto clean pasture as soon as the weather permits.

When you purchase chicks from a hatchery, you will be offered the option of having them vaccinated against coccidiosis. Vaccination stimulates a natural immunity that produces lifetime protection against coccidiosis, but you must take care to never feed medicated rations, which would neutralize the vaccine.

Chicks that have coccidiosis can't be cured with medicated feed, and besides, sick chicks typically stop eating. This disease is serious — now is not the time to experiment with concoctions described on the Internet. Chicks with coccidiosis must be treated with an anticoccidial dissolved in their water, since they will continue to drink even after they stop eating. Treatment works quickly and effectively if started at the first signs of disease.

See page 158 for more on coccidiosis.

Sticky Bottoms

Sticky bottoms, also known as *pasting*, is a common condition in newly hatched chicks. Although it may be caused by disease — typically in chicks older than 1 week — it is more likely to be caused by chilling, overheating, or improper feeding. Soft droppings that stick to a chick's vent will harden and seal the vent shut, eventually causing death.

The hardened droppings must be removed before they plug up the works. Run a little warm water over the chick's bottom to soften the mess, then gently pick it off, taking care not to tear the chick's tender skin. Depending on how thick the pasting is, you may have to pick off a little at a time and then apply more warm water.

When all the droppings have been cleared away, dry the chick's bottom by gently dabbing it with a piece of paper towel, and apply a little Neosporin or Vaseline to protect the affected area and prevent more poop from sticking.

Pasting is less apt to occur if chicks are drinking well before they start eating. Another preventive measure is to combine starter with chick scratch for the first two or three days. If your farm store doesn't carry chick scratch, run regular chicken scratch through the blender or crush uncooked oatmeal and combine it with an equal amount of cornmeal.

Sometimes the simplest solution is best: change to a different brand of starter. Our chicks persisted in pasting year after year until we switched feed brands. We haven't had a single sticky bottom since.

CHICK TOXINS

Although chicks are more susceptible to toxins than mature birds, brooding them in a properly managed environment will protect them from poisoning. Toxins to watch out for that chicks are especially susceptible to include these:

- **Carbon monoxide** — when chicks are transported in the poorly ventilated trunk of a car; chicks die

- **Disinfectants** — from overuse, particularly of varieties containing phenol, and especially in a poorly ventilated brooder; chicks huddle with ruffled feathers

- **Fungicides** — on coated seeds intended for planting; chicks rest on their hocks or walk stiff-legged

- **Pesticides** — when used to control insects in or around the brooder; chicks die

- **Rose chafers** *(Macrodactylus subspinosus)* — a type of beetle found in late spring and early summer in eastern and central North America; from eating them, chicks become drowsy, weak, and prostrate, go into convulsions, and die or recover within 24 hours

- **Coccidiostats** (nicarbazin, monensin, sulfaquinoxaline) — if added to water in warm weather, when chicks drink more than usual; they may obtain a deadly dosage

Starve-Out

Chicks don't need to eat the moment they hatch. For their first two days of life, they can survive on residual yolk. If, however, they don't eat within two to three days of hatch, they become too weak to actively seek food and will die of starvation, a condition known as *starve-out*.

Starve-out also may be caused by feeders placed where chicks can't find them or feeders set so high the chicks can't reach them. Other causes are placing feeders directly under the heat and bedding newly hatched chicks with sand or sawdust that they fill up on instead of feed. When using loose litter, cover it with paper towels until you know your chicks are eating properly.

To get chicks started pecking, sprinkle a little feed on the paper towels or in a paper plate, shoebox lid, or other low-sided container they can stand in while pecking. Yes, they will waste some feed by pooping in it, but at this stage it's more important to get them started pecking for feed. Place feed no more than 24 inches (60 cm) from the heat source but not directly under it. As soon as all the chicks are eating well, switch to normal chick feeders.

Toe Picking

Once chicks get started pecking, they'll look for things on the ground to peck. Among the things they'll find are toes — their own toes and the toes of other chicks. A common cause of toe picking is running out of feed to peck

The end cut from a tissue box makes a handy first feeder to encourage baby chicks to peck for food.

at or not having enough feeder space for the number of chicks.

Other causes include a too-warm or too-crowded brooder, a too-bright brooder light, or a starter ration that's too low in protein. Obvious solutions include increasing the number of feeders, increasing the available living space, regularly decreasing the brooder temperature by 5°F (3°C) per week, using low-intensity or red lights, and feeding a high-protein chick starter ration.

Make sure the chicks haven't outgrown their baby feeders or that replacement feeders are not too high for them to reach. Keep the chicks active by moving feeders far enough from the heat source to encourage exploration.

Crooked toes in a newly hatched chick may be straightened with Band-Aids or first-aid tape.

Feather Picking

Feather picking and toe picking share common causes, and both are early forms of cannibalism. Feather picking usually starts when chicks begin feathering out. The newly forming feathers don't look like the original down. Chicks get curious about what a feather is and check it out the only way they know how — by pecking it.

Feather picking has many causes, often working in combination. Conditions that can trigger it include the following:

- Crowding, especially in fast-growing chicks that quickly fill the available space and can't get away from each other
- Bright lights left on 24 hours a day
- Inadequate ventilation
- Too few feeders and drinkers
- Feed and water stations left too close together as chicks grow
- Diet too low in protein

Emerging feather quills are filled with blood, and once chicks get a taste of it, they want more. If feather picking draws blood, clean the bloodied quills and change the brooder light to a red bulb, which makes blood less attractive because everything else looks red, too.

In addition to correcting any conditions that might have led to feather picking, give the chicks something to peck on, such as a shiny aluminum pie tin hanging from a chain or string. Installing perches offers chicks another way to entertain themselves, as well as giving them more places to get away from each other.

Manure Balls

Balls of manure that cling to the chicks' toes invite toe picking and can result in crippling.

To remove manure balls, set the chick's feet in warm, not hot, water until the hardened manure has softened, then gently pick it off. If the toes bleed after they've been cleaned, dry them and coat them with povidone-iodine (Betadine).

Be sure to correct the condition that caused the problem in the first place: droppings accumulating in the brooder because of either crowding or improper bedding management. Manure balls may also result from crooked toes that make a chick walk unnaturally on the sides of its feet.

Crooked Toes

Crooked, or curled, toes may occur during incubation, during the hatch, or after hatching. Brooder conditions associated with crooked toes are overcrowding and a too-smooth floor. Other causes include nutritional deficiency, injury, and heredity.

Crooked toes may be straightened in a newly hatched chick while its bones are still soft. Use first-aid tape or a Band-Aid cut to size and wrapped around the toes to hold them in normal position. If the toes are going to straighten out, they should do so within a day or two as the bones harden.

If you can't identify the cause, don't include crooked-toe chickens in your breeder flock, or they're likely to produce future crooked-toe generations. And a chicken with crooked toes is not suitable for show. On the other hand, although crooked toes may not look great, a chicken will get along fine in all other respects despite its curled toes.

Splayed Legs

Like crooked toes, splayed legs may occur during incubation, during the hatch, or after hatch. The usual cause during or after the hatch is a hatcher or brooder floor that's too smooth for a chick to walk on, so its legs slide out to the side. As a result, the leg muscles can't develop, the chick cannot walk properly, and, if the condition is not corrected, the chick will die. Lining a smooth-bottom hatcher with paper toweling and covering the brooder floor for the first few days will prevent legs from splaying.

A chick with one or both legs splayed must have its legs hobbled, or mechanically brought together, to help the chick stand until its muscles develop sufficiently to let it stand and walk on its own. The hobble should be made of something soft and flexible that will not cut into the leg or bind, as string or yarn would. First-aid tape works. A Band-Aid is handier, and the gauze pad may be used as a spacing guide to get the legs the appropriate distance apart.

Use a smaller Band-Aid for bantams than you use for larger chicks. In both cases, cut the Band-Aid in half lengthwise to make it the right width to fit the chick's shank. To avoid getting the tape or Band-Aid too tight and cutting off circulation to the foot, protect the legs with a bit of double-sided foam mounting tape wrapped almost all the way around except for the inner sides of each leg. Then cover the foam with first aid tape or a Band-Aid, making sure the sticky side is completely covered so it won't stick to down when the chick squats to rest. Keep an eye on the color of the feet to make sure the hobble isn't cutting off circulation.

Depending on how long the chick has been sliding around with its legs out from under, it may have a hard time getting used to having its legs properly positioned underneath. If the hobbled chick falls over, space the legs a little farther apart than they normally would be, and each day retape them a little closer together.

During recovery, the chick must have a nonslip surface to walk on. Training a hobbled chick to walk may take several days and several reapplications of the hobble. The sooner a splay-legged chick is hobbled while its bones and muscles are still flexible, the more quickly it will learn to walk properly.

Splayed legs may be strengthened with a hobble made from double-sided foam mounting tape wrapped with first-aid tape or a Band-Aid.

12

EXHIBITING YOUR CHICKENS

How well your bird places in an exhibition depends on its physical condition, its disposition, how closely it conforms to the standard description for its breed and variety, how it compares with other birds in its class at the show, and the adeptness of the judge who reviews the class. How well you do as an exhibitor — and how much enjoyment you get out of it — depends on your reasons for showing and how well prepared you are for each event.

Why Show?

People who show poultry are called *fanciers* because they fancy, or are fond of, their chickens. They enjoy showing poultry for many reasons:

To have an affordable hobby. Everybody needs something relaxing to do in his or her spare time. Conditioning and showing chickens as a hobby is fun and inexpensive, involves a never-ending learning process, includes people of all ages and all walks of life, and is an ideal activity for the entire family.

As a learning experience. A poultry show is a good place to meet other people who keep your chosen breed or variety, who know where to obtain the best foundation stock and equipment, who can answer any questions you may have, and who are eager to share information you might otherwise never learn.

To get feedback from judges and other exhibitors. Serious breeders enjoy the challenge of constantly trying to improve their breeding stock and their showing skills. They like to compare the results of their breeding and conditioning programs with other entries, and they enjoy the opportunity to talk shop with other experienced breeders at a show.

To win awards. Winning may be the result of experience and skill at breeding and conditioning birds or the result of acquiring the best birds money can buy. Those with experience and skill tend to win consistently. Those who habitually purchase the birds they show tend to be inconsistent winners and poor losers.

Entering the county fair and other competitions is a fun hobby for chicken keepers of all ages.

To sell birds. Not all exhibitors are interested in selling chickens, and not all shows allow selling on the premises. Where selling is not allowed, simply winning blue ribbons is great advertising. Where selling is allowed, serious breeders may sell an occasional bird to help someone get started in their favorite hobby or to help defray the cost of showing. Some exhibitors enter shows solely to advertise and sell large quantities of birds, which tend to go for lower prices than chickens sold by serious breeders because they are generally of inferior quality.

To promote the cause. A few exhibitors enter shows to serve some higher purpose, such as to encourage interest in a breed or variety they fear may be nearing extinction or to educate the public about the benefits and joys of keeping chickens. Such people take pleasure in developing elaborate displays about their chosen breed or some other aspect of keeping and showing chickens.

For the camaraderie. Some exhibitors simply enjoy chickens and like to visit with others who also enjoy chickens. People who attend shows for social reasons tend to be less concerned about winning than with having a good time, and indeed some of them don't exhibit any birds at all. But most people who regularly attend shows eventually get caught up in the action and become serious about showing their chickens.

Before you start showing your own chickens, a good idea is to attend a few shows and become familiar with the procedures. Once you have an idea how things are handled, you'll be better prepared to get involved. And by making friends with some of the people who regularly exhibit, you'll be more relaxed and have more fun than you would by entering your first event unfamiliar with the process and dealing with complete strangers.

Selecting Birds for Show

Breeds and varieties of ornamental chickens represent so many sizes, shapes, colors, and personalities that settling on just one can be daunting. As a result, many novices try to show too many different breeds or varieties, and don't do well with any of them. These newcomers have the mistaken idea that the more birds you enter, the better chance you have of winning. But showing chickens isn't a game of chance — it's a game of skill. The first secret to success is to specialize.

Exhibitors who win consistently are breeders with an in-depth knowledge of the genetics of their chosen breed and variety. They know that showing starts in the breeding pen: making carefully thought-out matings; watching young birds grow; and culling any with deformities, incorrect type or color, or disqualifications. To make these decisions, every exhibitor needs a copy of the American Poultry Association's *American Standard of Perfection*, the American Bantam Association's *Bantam Standard*, or both.

At most shows, bantams outnumber the larger breeds. They are easier to house, cheaper to feed, and easier to transport in the family car. They generally respond well to confinement and handling and are easier than the bigger breeds for a youngster to manage. Bantams are a good choice for anyone just getting started in the world of exhibition.

Defects and Disqualifications

No chicken is perfect, meaning all chickens have defects of one sort or another. The *Standard of Perfection* establishes a general scale of points that assigns a value to every feature — size, condition and vigor, comb, crest, beak, skull and face, eyes, wattles, beard, earlobes, neck, back, tail, wings, breast, body and fluff, legs and toes, symmetry, and carriage or station.

A perfect score would be 100. In a process called *cutting*, points are deducted for specific defects. The severity of the defect determines the number of points deducted. A list of general defects applying to all breeds sets forth a value for each cut in relation to the total value established in the general scale of points. Additional defects apply to individual breeds or varieties. These guidelines are invaluable in evaluating your potential show entries.

A serious defect constitutes a disqualification, meaning the bird cannot win an award at show. Disqualifications for all breeds and varieties include a lack of appropriate characteristics for the breed, any indication of disease, and various deformities of the body, beak, comb, tail, legs, and feet. Common deformities include crossed beak, side sprigs (projections growing from the side of a single comb), humped back, crooked feet, crooked breastbone, and wry tail (a tail flopping over to one side).

Other disqualifications pertain to each specific breed or variety, and a feature that is standard for one breed or variety may be a disqualification for another. For instance, a feather-legged breed is defined as having feathers on the outer sides of the shanks and on the outer (or outer and middle) toes; a clean-legged breed with feathers appearing on the shanks or toes would be disqualified. Most disqualifications are inherited; therefore, any chicken with a disqualification not only cannot be shown but should not be used in breeding for future show birds.

TYPICAL DISQUALIFICATIONS

Side sprig (left), in which the comb has an extra appendage, and wry tail (right), in which the tail feathers lean to one side, will get a bird disqualified.

Type and Color

Since type defines breed, type comes before color in selecting a show bird. Type includes not only a chicken's overall weight and size but also the shape of its head, the slope of its back, the carriage of its tail, the breadth of its stance, the quality of its feathering, and a myriad of other details that characterize each breed, as described in the relevant standard — the exhibitors' bible.

The relevant standard also describes the proper color that defines each variety. Color includes not just the appearance of visible plumage but the color of the underfluff and skin underlying the plumage, as well as the comb, shanks, earlobe, eye, and skin surrounding the eye.

In selecting your potential show winners, you can often sort out probable losers from among young growing birds, but you cannot make a fully informed final selection for

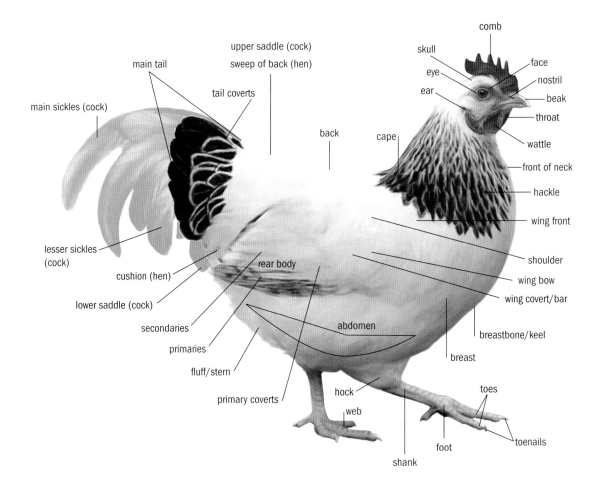

Discussing the finer points of an exhibition chicken, including any defects or disqualifications, requires the ability to identify all its various parts.

exhibition or breeding until the chickens are mature and acquire their adult plumage. Since preparing a bird for show takes lots of time, train and condition only your best birds. Select chickens for exhibition that

- Have the fewest defects
- Lack any disqualification
- Are vigorous and free of disease
- Are mature and well developed
- Have the best body type for the breed
- Are near standard weight for the breed
- Have a full complement of intact plumage
- Are of a uniform color appropriate to the variety

Rules of the Game

Before deciding which of your chickens to show, and mapping out their training and conditioning program, obtain copies of the premium lists for the shows you wish to attend. The premium list outlines the classes and varieties that will be accepted for each particular show and should indicate which breed standard (American Poultry Association or American Bantam Association) will be used, as the two don't always agree.

Not all birds are eligible for all shows. Some shows allow only large breeds or only bantams; others allow only certain breeds or varieties (due perhaps to space limitations or to specific interests of the show's sponsors). If you have any doubts, call the show superintendent or secretary.

Other information you'll find in the premium list includes

- The show's time, date, and location
- The deadline for sending in your entry form
- Entry fees involved
- Health requirements (such as vaccinations, blood tests, and health certificates)
- The time and date when birds must be signed in and removed
- Acceptable methods for identifying birds
- Prizes or premiums being offered

The value of the prizes often determines the amount of competition you can expect in your class. Competition will likely be stiff if your class involves a *sanctioned meet*, at which prizes are offered by a specialty club for its members. To be eligible for such prizes, you must be a member of the group. Joining the specialty club for your breed is a good idea, anyway. By reading the group's newsletter, attending meetings, and chatting with other members, you will learn about issues that are peculiar to your breed.

AGE CLASSES

Most shows have four classes for each variety: cock, hen, cockerel, pullet. The terms cock and hen generally refer to birds 1 year of age or older, but some shows define them as having hatched before January 1st of the year of the show. The terms cockerel and pullet refer to birds that are younger than 1 year of age, although some shows define them as having hatched during the calendar year the show is held. If you will be showing young chickens, check the definitions in the premium list to make sure you enter your birds in the right classes.

Coop Training

While genetics determine a chicken's championship potential, training and conditioning determine whether or not that potential is realized. Many a fine-looking bird has been entered into a show, only to make a bad appearance because it hadn't been properly trained. Few chickens naturally show well. Most exhibition birds need to be coop trained, which teaches them what to expect and what's expected of them at a show.

Training helps a bird show to its best advantage by getting it used to being close to people and being handled, and it ensures the bird won't be disoriented by unfamiliar surroundings. Training also minimizes stress, thus reducing the chance the bird will experience health problems as a result of being shown.

Some exhibitors continually train their show birds; others give them a brushup course just before each show. The younger a bird is when you start coop training, the more quickly it will respond. But occasionally you'll run across a chicken that does not care to be handled, no matter how patient you are or how much time you take. When a chicken does not take kindly to being cooped at home, entering it in a show is a waste of time, energy, and money. If you have the patience, go ahead and work with the bird until the next show. Otherwise concentrate your efforts on your more cooperative individuals.

Exhibiting chickens is a great way to meet people of all ages and all walks of life who share a common interest.

Show Coops

No less than one week before a show, put each bird in an individual cage, or coop, similar to one that will be used at show. Serious exhibitors purchase *show coops* for training purposes. These collapsible coops are quite costly, but you might reduce the price by purchasing used coops from a fair or club that's upgrading or by combining your order with that of others in a group. As an alternative, use a wire cage of similar shape, size, and style.

Not all show coops are exactly the same; they vary slightly with each manufacturer. A typical coop for bantams is 18 inches high and 18 inches (or a little less) square (45 cm high, 45 cm square). A typical coop for larger chickens is 27 inches high and 25 inches (or a little less) square (70 cm high, 65 cm square). The floor is usually smooth, not wire, and at most shows is covered with shavings. A double partition between sections prevents adjoining cocks from fighting and injuring each other. The door generally slides upward from the bottom, so the chicken is put into and taken straight out of the coop rather than being lifted upward. Most show coops have built-in cup holders for water and feed.

These features appear pretty basic to us humans but can be scary to a chicken encountering them for the first time. A bird that is not prepared in advance won't show well if it stands or walks awkwardly because of an unfamiliar coop floor or bedding, or it crouches down or flutters around because it's frightened by the strange surroundings. A chicken that tries to get away by pressing against the side of the coop may crush feathers or bend its tail into an unnatural position. A chicken unused to close confinement may get restless and flighty, especially with so many strangers wandering by or stopping to stare at it.

Setting Up

While your show chickens are housed in their training coops, work with each separately, one, two, or three times a day. Keep training sessions short so the bird won't get tired and lose interest. Begin by approaching the coop slowly and calmly; avoid rapid or jerky movements, which can frighten even a well-trained bird. If the chicken appears the slightest bit frightened, stand quietly until it calms down. Begin your training session only when the bird seems calm.

First teach the bird to set up, or strike a pose suitable for its breed. Some breeds (such as Cochin) show best in a compact or horizontal stance. The other extreme is an upright or vertical stance, called *station* and defined in the *Standard* as "the ideal pose and symmetrical appearance including height and reach." In the general scale of points, station is assigned 10 points for Modern Games, Malays, Aseels, and Shamos.

A chicken may be taught to set up properly by tapping it in appropriate places with a judge's stick, which is simply a telescoping pointer available from any office-supply store. To get a bird to lower its tail, for example, tap above the tail. To raise the tail, tap below the tail.

Teach the bird to stretch upward or bend downward, as appropriate, by getting it to reach for a tasty tidbit at the end of the judge's stick. Old-timers used raw hamburger, which is no longer suitable due to the danger of *Salmonella* poisoning. Instead, use tiny morsels of high-quality canned cat food. This treat not only rewards the bird but also gives it a little protein boost at a time of stress.

This Modern Game hen demonstrates the breed's elegant vertical pose, called *station*.

Handling

After spending a few minutes on developing the bird's pose, next work on getting it used to being handled. At show, the judge will remove each bird from its coop to examine it in detail. A chicken that's not prepared to be handled will flap and struggle and may get loose in the showroom, all of which does not make a good impression with the judge.

The proper handling of a show chicken is designed to not only keep the bird calm but also to prevent breaking or ruffling of its feathers as the bird is removed from and returned to the coop. Begin by calmly opening the coop door and gently maneuvering the bird until it stands sideways, with one wing facing you.

Reach across the bird's back, and place one hand over the far wing at the shoulder. Get a firm grip to keep the wing from flapping, and rotate the bird to face you.

Slide your other hand, palm side up, under the bird's breast, with one of its legs between your thumb and index finger and the other leg between your second and third finger. Your index finger and second finger should be between the bird's legs, and the bird's breastbone should rest against your palm.

Keeping a firm grip on its legs, gently lift the bird out of the coop head first. Always remove and replace a bird head first, reducing the chance it will catch its wings in the doorway and damage the feathers.

DUBBING AND CROPPING

If you plan to exhibit Modern or Old English Game cocks, they must be dubbed and cropped to qualify for most shows. *Dubbing* involves surgically removing the comb; *cropping* is removal of the wattles. The earlobes might also be trimmed at the same time.

If you are among the many exhibitors who don't believe in dubbing and cropping or who live in a state where doing so is illegal, show only your cockerels younger than 6 months of age, as well as your pullets and hens, and reserve your intact cocks for breeding.

Intact cock

Dubbed and cropped
Old English cock

With the bird outside the coop, hold it quietly for a moment, then remove your hand from its wing and let the bird rest in your hand another moment to ascertain that you are in full control. Imitate a judge's actions by turning the bird to examine its comb and wattles and by opening each wing to examine the long feathers. At some shows, especially junior shows, each bird might be set on a table for examination, so a further step is to carry the bird to a table and teach it to stand calmly.

Don't try to do all these things in the first training session, but work up to the full routine a little at a time, doing only as much in each session as the bird can comfortably handle. Start out by reaching in and taking hold of the chicken. Notice how it tenses and relaxes. Then lift it off its feet and put it back down. When the bird appears comfortable that far, remove it from the coop, hold it a moment, and put it back in.

At the end of each training session, gently return the bird to its coop, head first. If you're training a Cornish or other heavy breed, teach it to expect a judge to determine its body balance and set of leg by dropping the bird 6 inches (15 cm) to the coop floor. A quick recovery indicates good balance and leg placement. For other breeds, gently place the chicken on its feet and let go. A well-trained bird will stand quietly when released.

Conditioning

Conditioning is the process of bringing a show bird to the peak of cleanliness and good health. A properly conditioned bird has that undefinable quality known as bloom. No matter how much time you spend with a chicken, you cannot keep it in a constant state of bloom. No

bird is suitable for showing, for example, while going through a molt.

Constant showing causes stress, and stress is not conducive to bloom (or to good fertility in the breeding pen). If you plan to attend a large number of shows, rotate the birds you condition rather than expecting all your potential winners to compete at all the shows.

Plumage

No less than two months before the show, treat each potential winner for lice infestation and move it to its own small house with a grassy run. Although chickens need some sunshine for good health, show birds should not spend hour after hour in direct sunlight. Sunshine can fade the plumage of solid buff or red breeds and those of red background color and give a brassy yellow metallic hue to white or blue plumage.

While you're moving each bird to individual housing, examine its plumage and remove the occasional broken or off-color feather. To avoid injuring the feather follicles, never yank out a wing or tail feather. Instead, cut off the broken feather, leaving a 2-inch (5 cm) stub, then cut the shaft down the middle toward the skin. After a few days the feather will loosen, and you can pull it out easily.

Unfortunately, incorrect coloring often shows up in flight feathers of the wing or sickle feathers of the tail, and these feathers take a long time to grow back, so allow at least two months. The feather that grows back may be the right color or may not be; in either case the bird should be fully feathered by the time of the show. Because points will be deducted if the feather is the wrong color, the bird may need to be removed from the show string if the color is too far off.

Feeding

Sooner or later, all experienced exhibitors develop a custom diet for their show birds, taking into consideration, among other things, the desired weight for their breed and the effects of certain feeds on plumage color. A good starting place is basic breeder ration.

Do not feed yellow corn, especially to white varieties or those with white earlobes. Corn tends to make chickens run to fat, and its pigment gives white birds a brassy hue. Feed no additional grain other than whole oats, offered free choice in a hopper separate from the breeder ration. Whole oats improve feather quality without making a bird fat. Be sure to offer granite grit as well, so the bird can digest the oats.

To stimulate natural oil production and give your bird a radiant glow, or bloom, include a small amount of oil-rich feed, such as safflower seed, sunflower seed, flax, or linseed meal. To further enhance the gloss of varieties with black or red plumage, feed a tiny amount of quality canned cat food daily, preferably offered as a treat during coop training.

Washing

A chicken's plumage color can be enhanced, and the fluffiness of loosely feathered breeds emphasized, by washing. Light-colored birds should always be washed before a show. Dark-colored, tightly feathered varieties may need a thorough washing only if their plumage is soiled.

Wash a chicken no less than 48 hours before a show, giving feathers time to get back some of their natural oil and giving you time to fix such goofs as streaking or improper shaping. A good plan if you've never washed a chicken before is to wash one or two practice chickens ahead of time, just to get the hang of it, so you don't run the risk of ruining your potential winner just before a show. Washing a chicken takes 15 or 20 minutes.

I like to wash chickens in the laundry room. It's a warm, draft-free place to work and has a deep basin with plenty of warm running water. Others prefer to work in the garage or, if the weather is nice, outdoors, using three tubs filled with clean water — one for washing, two for rinsing.

If possible, use soft water, which cleans birds better than hard water. Begin with a basin full of warm water (90°F [32°C]). The temperature is right if you can hold your elbow in it without discomfort. Water that's too hot can cause a bird to faint. If a bird should happen to faint, revive it by pouring a little cold water over its head. Place one hand against each of the chicken's wings so it can't flap and give you the bath, and slowly immerse the bird to its neck. If it struggles, dip its head briefly under water. Most birds relax as soon as they realize they're in for a warm, soothing bubble bath.

Add enough shampoo or mild liquid dish soap to the water to whip up a good head of suds. Don't use a harsh detergent, which makes feathers brittle. I've had great results with flea and tick shampoo for household pets, which not only gets birds shiny clean but also zaps any lice or mites that may have gone unnoticed.

Thoroughly soak the bird by raising and lowering it and drawing it back and forth through the water. With a sponge, soak the feathers through to the skin. To avoid breaking feathers, make sure they are thoroughly soaked before doing any rubbing, and work only in the direction they grow. Rub in extra lather around the tail, where feathers tend to be stained by the oil gland, and around the

Most chickens learn to enjoy a warm, relaxing bubble bath.

vent. When the chicken is thoroughly clean, lift it from the bath and press out soapy water with your hands, working from head to tail.

If you're washing a crested bird, hold it upside down by its legs, and dip the crest into the soapy water, keeping the bird's beak and eyes above water. Work suds into the topknot, or if the crest feathers are particularly dirty, apply a drop of shampoo directly to the top-knot. Rinse the crest under running water to remove all traces of soap, taking great care not to get any into the bird's eyes or nostrils. Washing a crest is easier if two people work together, one holding the bird while the other washes the head feathers.

With the plumage thoroughly washed, rinse the whole bird in fresh warm water that's slightly cooler than the wash water. Let the bird soak for a few minutes, until its feathers fan out or float, then move it back and forth in the water to work out remaining soap. Lift the chicken from the rinse, and press out excess water.

If any soap remains, the feathers will look dull and faded when they dry and the plumage won't fluff out properly. So rinse the bird once more. This time add a little vinegar or lemon juice to remove any remaining soapy residue or feather oil.

If you're washing a white bird, brighten its plumage by adding two drops of liquid laundry bluing to this final rinse. To prevent streaking, thoroughly stir the bluing into the full basin of water before rinsing the bird. And remember, more is not better — add more than two drops and your white bird will turn blue.

Squeeze out excess water from the feathers, and gently towel off the bird. Wrap the bird in a fresh towel, and blot it to soak up remaining water.

After the bird's shanks and feet have been soaked, dirt and scales will be soft and easy to clean. Even if you don't wash the whole bird, at least soak its shanks and feet in warm water. Leg scales molt annually, just as feathers do, and you can easily remove dead, semitransparent brittle scales by popping them off with a nail file, a toothpick, or your thumbnail. If any dirt is clinging beneath the scales or toenails, gently and carefully remove it with a toothpick. Use a toothbrush and soapy water to scrub the shanks and toes. A little scouring powder may be used on rough spots, but take care not to rub too hard and cause bleeding.

Grooming

To groom a chicken after giving it a bath, tightly wrap it in a dry towel with only its head showing, taking care not to bend any feathers with the towel. Cover the feet with one corner of the towel so you can open the flap to work on one leg and foot at a time.

Clean the bird's comb and wattles with a little rubbing alcohol mixed with an equal amount of water, being careful not to get any into the bird's eyes. Dry the comb and rub it with a bit of baby oil, vitamin E oil, or a mixture of equal parts alcohol and olive oil (but not petroleum jelly, which gets feathers greasy and gathers dust). Gently buff the comb until all the oil has been worked in, being careful not to get oil on any feathers.

Use a toothpick to clean around the bird's nostrils. If the upper beak is a little too long, trim it back to proper length (see page 140) for beak and nail-trimming instructions). When grooming a breed with white earlobes, such as Leghorn, Minorca, or Rosecomb, coat the washed and dried earlobes with a little baby powder to keep them clean.

Now turn your attention to the legs and feet. Trim long nails with clippers or nail

A light dusting of baby powder on white earlobes, such as on this Leghorn cock, helps keep the birds' earlobes clean.

scissors. Excessively long toenails must be trimmed in several stages, allowing a few days between each trimming for the quick to recede; if you wait until the last minute before a show, you'll run out of trimming time. Rub the cleaned feet and shanks with the same oil used on the comb, again taking care not to get any on the feathers.

Drying

Release the cleaned, groomed bird into a clean training coop or a roomy pet carrier strewn with fresh shavings. The drying coop must be scrupulously clean, as any dirt that touches damp feathers may cling. To avoid soiled or damaged feathers, put only one bird in each coop.

Dry a crested variety in a container high enough that the topknot won't rub against the top. A feather-legged variety needs a roomy drying coop to avoid breaking foot feathers against the side. A cardboard box works better than a wire cage, or place a rim of heavy cardboard around the bottom few inches of a wire cage to protect leg feathers from the wire.

Top: After a bath, a loosely feathered breed like this Silkie may fluff up nicely with a blow dryer set on warm.

Bottom: A tight-feathered breed like this Old English bantam looks best when allowed to dry naturally in a warm area away from any drafts.

You can dry birds outdoors, provided the temperature is at least 70°F (21°C) and either the weather isn't breezy or the birds have good wind protection. If you're drying loosely feathered birds like Cochins or Silkies, you can speed things up, as well as nicely puff out the feathers, with a hair dryer. A blow dryer also works well to fluff up a crest. Always use the warm, not the hot, setting.

Most breeds, though, look better if they dry naturally. Fluffing plumage with a hair dryer can be downright disastrous in tight-feathered breeds like Cornish, Modern Game, or Old English Game. In cold weather that may chill your damp birds, place the drying coops in a warm room well away from the heater, so they won't dry so quickly that the feathers curl. In an unheated room, hang a heat lamp no closer than 2 feet (60 cm) above each bird.

As the birds dry, arrange their feathers for proper shaping, especially around the base of the tail. Complete drying takes 12 to 18 hours, depending on the density of the feathers.

To further whiten a white bird and give it a nice sheen, sponge it off with hydrogen peroxide when the plumage is half dry. When the white bird is completely dry, a light dusting of cornstarch will help keep it clean.

Faking

A fine line exists between grooming and faking. Grooming is the process of making a bird look its best. Faking goes beyond grooming to alter a bird's appearance for the purpose of hiding natural defects. Arranging feathers while they dry is grooming; bending, breaking, trimming, or crimping feathers to change their natural angle is faking.

Faking has two purposes: to fool judges and to deceive potential buyers. No one knows how common the practice is, since expert jobs are hard to detect and even more difficult to prove. The deceitful practice evolved into a high art in the days of the old-time *stringmen*, who traveled the show circuit exhibiting and selling their string of birds far from home. *The Art of Faking Exhibition Poultry*, first published in 1934 in England, describes their methods in detail and shows what to look for if you have suspicions.

Even today, someone at a show may pull you aside to reveal "tricks of the trade" that are not only unethical but will get your bird disqualified if you have the bad luck to show under a judge who's knowledgeable enough to spot the fake job. Many judges, however, are reluctant to disqualify a bird that shows signs of having been tampered with, mainly because faking is so difficult to prove. A cock, for example, may have a scarred comb because it got into a fight or because a side sprig was deliberately snipped off or the comb otherwise reshaped.

One faking practice that's both common and impossible to detect (unless you see it done) is the removal of stubs, or downy feathers on the legs or between the toes of a clean-legged bird. Some clean-legged breeds, including Wyandottes, readily sprout stubs. I once saw a fine Cornish cock disqualified at a show for showing stubs while its well-known and embarrassed owner stood by swearing the bird had shown no sign of stubs the day before. That may well have been true: oiling the legs prior to judging likely loosened the scales enough to let the stubs slip out. The irony is that if this fellow had surreptitiously plucked the stubs, his bird would not have been disqualified.

Other clear cases of faking include using chemical solutions of various sorts to loosen or tighten plumage; starching tail feathers to make them stand up better; stitching a wry tail to straighten it; applying a coloring substance to the beak or earlobes (some substances, like lipstick and rouge, are easy to detect because they rub onto plumage); and rubbing a caustic chemical on undesirable white earlobes to make them blister, scab over, and turn red.

Borderline cases of faking most often involve feather color. White is the most likely color to be faked, since white plumage commonly looks brassy. Brassiness may occur for environmental reasons (such as too much sunshine or pigmented feed) or may be hereditary. Altering feather color by means of "softening, deepening, intensifying, or otherwise changing the natural color" is faking. Bringing out natural color by washing a bird, controlling its diet, or keeping it out of the sun is conditioning, not faking. Is rinsing a bird with bluing or rubbing it down with hydrogen peroxide faking? Not if your intent is to bring out the natural whiteness of its plumage. But bleaching a bird with harsh chemicals to whiten naturally brassy feathers is faking. Just as bleaching human hair causes the hair to become brittle, bleaching a bird's plumage makes its feathers brittle. One criterion used to determine faking in white plumage is feather brittleness.

Feather pulling and beak trimming are additional practices that may be either grooming or faking. If you pull an off-color feather in plenty of time for a new one to grow back, hoping its color will be correct, that's grooming. If you remove an off-color feather on the sly 20 minutes before the judging starts, that's faking. Trimming the upper beak of a bird that's been housed where it couldn't keep its own beak worn down is grooming;

trimming a crossed beak to hide the genetic defect is faking.

The intent of faking is to make a bird look like something it is not. An unscrupulous exhibitor uses faking, rather than selective breeding, to make his birds appear superior to others at the show. An unscrupulous seller uses faking to peddle genetically defective culls to unsuspecting buyers. Either way, faking is fraud.

At the Show

To get your well-conditioned and groomed birds to the show, you'll need an appropriate carrier. Prevent dirtied or damaged feathers by putting no more than one bird in each carrier, and line the bottom with clean straw or shavings.

Don't use wire carriers for feather-legged and crested birds. A good carrier for transporting show birds is

- **Clean** — disinfected between shows
- **Safe** — no protruding wires or nails
- **Suitable in size** — not so small a crest or comb rubs the top or foot feathers rub both sides at once; not so large the bird has room to launch into panicked flight
- **Well ventilated** — more so if the carrier will travel inside a car rather than in an open, windy pickup bed; if you carry chickens in the back of a truck, either cover the bed with a camper top or place each carrier inside a large cardboard box to protect the birds from drafts

Paint your name prominently on each carrier so that, in the scramble to leave and go home when the show ends, no one will mistakenly grab your box. The more expensive your carrier is, the more important some form of permanent identification becomes.

Stress Reduction

Your birds will show best, and remain healthiest, if you make every attempt to reduce the stress they naturally experience as a result of being transported and shown. The biggest stress-reducing measures you can take are avoiding drafts and long periods without water or feed. Bring along the water and feed your chickens are used to; at the show, a bird may not drink water that tastes strange or eat rations with unfamiliar texture or composition.

Electrolytes and vitamin/mineral supplements help reduce stress when offered to birds for several days before and after a show, but supplements should not be used during the show. Their taste could cause a bird in unfamiliar surroundings to go off feed or water, increasing its stress level. On the other hand, adding a little vinegar to strange-tasting water at the rate of 1 tablespoon per gallon (15 mL per 4 L) will make the water more palatable to the chicken and encourage it to drink.

Chickens at show should have water before them at all times, and you don't always know in advance if the management will provide containers, so bring along drinkers that can't be dumped. Save small food cans, making sure none have sharp edges. In the side of each, near the top, punch a hole and thread a piece of wire through. Use the wire to tie the can inside the show cage. If water containers are already provided, make sure they can't be knocked over by a frightened bird. Stabilize a loose waterer with snugly tied string, a paper clip, a rubber band, or a length of wire.

Bring along extra cans for feed. Birds at show are generally fed scratch to keep their droppings solid. Many shows arrange to have someone travel from coop to coop making sure all birds have feed and water. Sometimes, however, that person is not reliable or has too

FEEDERS AND WATERERS

Whether store-bought galvanized or plastic (left) or homemade from a small can (right), cup feeders and waterers should be clipped securely to the show coop.

many birds to tend or the birds themselves dump their water or feed in the excitement. A few shows require individual exhibitors to be responsible for watering and feeding their own birds. The premium list should indicate a show's feeding and watering policy.

Some shows do not allow exhibitors to feed, water, or handle their birds once they're on display, since show officials have no way of knowing whether you are tending to your own birds or interfering with the competition. If you see that someone else's bird needs feed or water or has any other kind of problem, find the owner and let him or her know. Some exhibitors don't like their chickens to be fed on the morning of the day their class is judged, because they want to avoid bulging crops. Besides, tending to or handling someone else's bird, no matter how well intended, may be mistaken for interfering.

Last-Minute Touch-Up

When you arrive at the showroom, you will notice two types of exhibitors. One type will be running around tossing birds into their assigned coops. The other will be off in a corner or sitting in the back of a truck or van, calmly examining, grooming, and reassuring each bird before placing it in its show coop. Keep an eye on those quiet fellows — they're your stiffest competition. They're also your best teachers.

Take a tip from them, and give each of your chickens a last-minute going over. Polish the skin around the eyes and the comb, wattles, earlobes, leg, and toes of each bird with a touch of baby oil, mineral oil, vitamin E oil, or a good-quality hand lotion on a soft cloth. Use a mentholated rub on the combs and wattles to open up the chicken's airways and give your

bird a healthy bloom. Be especially careful at this time not to get oil on plumage.

Arrange any feathers that may have gotten out of place during travel. To remove road dust and shine up the feathers, rub your clean bare hands or a piece of silk or wool over the plumage, working in the direction in which the feathers grow. When your chicken is competing against another bird of similar quality, the cleaner bird always places higher.

Cooping In

Every show has a time by which all birds must be in place in the showroom. Some shows are more tolerant of latecomers than others. Most exhibitors arrive at the last minute, which makes things pretty hectic for a while. Being a little early lets you take your time with each bird, and then you can relax, watch the action, and spend time chatting with others who have finished cooping in.

When you arrive, you'll be faced with row after row of show coops. *Cooping in* is the process of putting your bird in its assigned show coop. The birds are organized according to classification, so all the birds to be judged against one another are displayed in adjacent coops along one row (or in adjoining rows, if the class is large).

Attached to each coop is a coop card identifying both the bird assigned to that coop and the bird's owner. The card has room for the judge to note the bird's placement in relation to others. A thoughtful judge may also jot down some comment about the bird's outstanding good or bad feature, such as "nice lacing" or "too long in the back."

An *open show* is one in which the coop cards are left open during the judging. A *closed show* is one in which either the cards are folded or the exhibitors are identified by numbers, so the judge can't see the name of the bird's owner and be influenced by the owner's reputation. Novices believe they have a better chance of winning at a closed show, but experienced exhibitors know that a skilled judge can identify chickens coming from strains owned by top exhibitors simply by the birds' unique and uniform appearance. Some judges feel that closed shows insult their integrity.

And the Winner Is ...

A licensed judge is supposed to appraise birds according to the scale of points published in the *Standard*. A specific number of points is allowed for each trait (comb, tail, back, symmetry, weight, condition, and so forth), with deductions (or cuts) made for such things as incorrect weight, missing tail feathers, off-color eyes, and other defects. In the old days, coop cards had this scale printed right on them, but rare is the judge today who refers to the point system at all. Judging by points and cuts takes too long.

Judging these days tends to be much more subjective. Birds are often ranked according to the judge's personal likes and dislikes, which may significantly diverge from the relevant standard. To do well at a show, you have to know who the judge will be so you can show the kind of birds that judge likes to see in your breed and variety. As in all things, some judges are fairer than others. The fairest judges tend to be willing to discuss the placings when the judging is over. The least fair judges don't like to explain their reasoning because they know they'll end up in an argument.

Differences in judging may be accounted for not only by differences in the judges' taste but also by differences in their conscientiousness,

interpretation, powers of observation, experience, and age. Chickens, like other things in life, are subject to fads. Refinements in color or body type vary in desirability from year to year, so what was popular a generation ago may be unfashionable today and vice versa.

Successful exhibitors ignore these fads and work as closely as possible with the standard descriptions for their breed and variety. To avoid discouragement, however, keep in mind that the birds depicted and described in the *Standard* are ideals to strive for, as determined by interested breeders and specialty clubs; such birds do not exist in real life.

Chickens, like judges, have their good days and bad days. A bird may display itself differently from one day to the next because of changes in its physical condition, health, training, or stress level. Even the fairest of judges may place the same group of birds in one order on one day and in a different order on another day. If a bird doesn't place high in its class, don't be hasty to cull it until you find out why the bird did poorly. The best bird in the world won't win if it lacks maturity, is out of condition, or is out of sorts.

Ultimately, all judges base their decisions on two things:

- How each bird compares to the ideal (or standard) for its breed and variety
- How each bird compares in type, condition, and training to the others in its class

In ranking the birds in a class, most judges first compare two birds, then compare the better of the two with a third bird, and so on until all the birds in the class have been ranked. A good way for novice exhibitors to become more observant is to practice doing the same: Rate

JUDGING SYSTEMS

Two different systems of judging are used in the United States: American and Danish. Adult shows and open shows generally use the American system, in which the birds in each class are ranked against each other according to the standard for their breed and variety; awards are generally given to the top three or five, although some shows go all the way down the line. Youth shows may use the Danish system, in which birds are not compared to one another but instead each bird is judged on its own merit according to how well it meets the standard description for its breed and variety. Rather than pitting competitors against one another, the Danish system helps exhibitors gauge their individual progress. The 4-H uses a modified Danish system, in which the standard is adjusted according to the exhibitor's age and years of experience.

each entry in the class according to its description in the relevant standard, then rank them against each other. When the judging is over, compare your ideas against the judge's. If your ranking is far off, ask a knowledgeable breeder to show you why.

Regardless of your opinion, avoid getting into an argument with other exhibitors or with the judge. Even if every exhibitor in a class disagrees with the judge, the judge has the final say.

Clerking for Experience

An excellent way to learn the ins and outs of showing is to volunteer to help out. One of the best jobs is that of the judge's clerk, who gets to see firsthand how the judge determines the placing of each bird. If you're interested in a specific class or in working with a specific judge — assuming the show is big enough to have more than one judge — say so at the time you volunteer to clerk, but don't be upset if the assignment you want is already taken.

If you let the judge know ahead of time that you're there to learn, he or she may be willing to think out loud while judging. Avoid the temptation to pester the judge with questions, though. You'll only slow things down at a time when the judge needs to concentrate. One of the functions of a clerk, in fact, is to keep bystanders from kibitzing with the judge.

At most shows, while judging is taking place, the aisle is blocked off to give the judge a clear view of the class and prevent interference from passersby. Yet a crowd invariably gathers at the barriers to watch the judge in action. To satisfy their curiosity, a good judge will discuss the placings after finishing each class.

A good clerk makes sure the judge has time for this important educational aspect of showing by keeping things running smoothly and efficiently. In doing so, the clerk locates each class to be judged, lets the judge know how many birds are in each class so none will be missed if the class continues around the corner or across the aisle, keeps track of the judge's breed standards guide and other paraphernalia that might otherwise be misplaced, records the judge's placings, and returns forms to the show secretary so ribbons and other awards may be distributed without delay.

After the Show

While a class you are interested in is being judged, note who in the crowd is especially attentive. Those folks likely have a bird in the class. After the judging is over, a good way to get tips about the breed is to seek out some of those folks and ask questions, such as "Which bird is yours?" and "What do you think of the bird that won?" If theirs is the winning bird, you might ask why it placed higher than the runner-up. The responses will help you learn to notice the finer details of the breed and variety. If you have a bird in the same class, you might also ask what they think of your bird, but please don't be offended by a frank response.

Cooping Out

Cooping out refers to the removal of your bird from the showroom after all the prizes have been awarded and the show is over. Cooping out occurs in anticipation of going home, unless you are asked to coop out early because one of your birds appears to be ill. Having a bird disqualified is not sufficient reason for early coop-out. Neither is selling a bird to someone who's eager to take it home. Some shows have strong rules against people who coop out early, such as losing eligibility to enter the next show.

People who attend a poultry show come to see chickens, not empty cages. So if you don't win, avoid the temptation to pick up your birds and go home. It's not sportsmanlike. Whether you win or lose, stick around to discuss the placings with other exhibitors. You might be amazed at what you learn.

The appropriate time for cooping out will be given in the premium list. Birds must be removed in a timely manner before the building is closed and locked up or, typically at a county fair, cleaned out and readied for another show. Coop-out time is usually a mad scramble, with everyone in a hurry to start for home. It's a good time to keep a close watch on your birds, carriers, and other equipment. If you have more stuff than you can carry to your vehicle in one trip, try to buddy up with someone so one of you can watch things inside while the other carries things to the parking lot.

Theft is so uncommon at shows that the disappearance of a superior bird or an expensive carrier causes a big stir among exhibitors. But even ordinary birds and simple cardboard boxes are sometimes inadvertently grabbed in the rush, especially by novice exhibitors or those who show large numbers of birds and don't take time to count them at coop-out time. Although such losses are not likely to be economically important, they're disheartening if not downright inconvenient. So in addition to guarding your own belongings, take care in the fracas not to pack up something that's not yours.

Bringing Home More Than You Bargained For

Some shows require exhibitors to get a health certificate from a veterinarian before entering chickens. Unfortunately, even a health certificate doesn't ensure that a bird is healthy. A chicken may be fully capable of spreading a disease without appearing sick to the vet who signs the certificate. Indeed, some vets handle certification this way: "Have you had any problems with this bird? No? Well, it looks fine to me. Here's your certificate. And here's a bill for my services."

Even if you are certain the birds you exhibit are healthy, you can't count on the same from everyone else. Your chickens are less likely to catch a disease from other birds at a show if you take every precaution to reduce stress.

Putting medications into the birds' drinking water is definitely a bad idea. Medications may cause a bird to drink less than usual, increasing stress and the likelihood of disease. Besides, if you don't know what disease your chickens are exposed to, if any, how would you know which medication to use?

When chickens do catch something at a show, it is likely to be a cold. Colds spread among birds the same way they spread among people —through coughing and sneezing. As you coop in, check the entries in adjoining coops. If a bird doesn't look healthy or coughs and sneezes, notify show officials so the bird can be removed.

Upon return from a show, clean and disinfect all carriers, waterers, feeders, and other equipment you used at the show. Isolate returning birds for at least two weeks, and watch them for signs of disease. Feed the isolated birds after attending to the needs of your other chickens to avoid potentially contaminating the birds that stayed at home during the exhibition.

Despite these necessary precautions, take heart from the fact that a chicken rarely picks up disease at a show. Most people who exhibit their birds take great pride in keeping them healthy.

Become a Master Exhibitor

If you enter enough shows and bring home enough ribbons, eventually you may become a Master Exhibitor. Both the American Bantam Association and the American Poultry Association have Master Exhibitor programs, although they use different systems for establishing the honor.

The ABA offers additional awards for Master Breeder and Citation of Merit, while the APA offers Grand Master Exhibitor and a Hall of Fame award. Some of these honors encourage specialization in a single variety, and qualifying for any of them requires years of dedication.

A bird exhibited in such fine condition as this Sumatra cock is a testament to the skill and dedication of its owner.

13

MANAGING MEAT BIRDS

If the idea of raising your feathered friends for meat doesn't appeal to you, read no further. But if you are among the many folks for whom an important reason to keep chickens is access to clean, healthful, and delicious meat, read on. Of all the different forms of livestock, chickens can put meat on your table with the least amount of time and effort. In a matter of weeks, your chicken-keeping chores are over, and your freezer is full of poultry that's tastier and better for you than anything you could buy at the store.

Meat Breeds

When it comes to raising chickens for meat, you have three basic choices. You can produce a commercially developed Cornish-cross strain with either white plumage or colored plumage, or you can raise one of the old-fashioned heavy or dual-purpose utility breeds.

White Cornish hybrids are the most efficient chickens to raise for meat. They are an industrial creation developed by combining white Cornish and white Plymouth Rock genetics. The resulting broilers — the type sold at the supermarket — share these characteristics:

- They grow and feather rapidly.
- They efficiently convert feed into meat.
- They are broad breasted.

- Chicks of the same age and sex grow at the same rate.
- They reach target weight in six to seven weeks.
- They have white feathers for clean picking.
- The edible meat is approximately 75 percent of live weight.

Since these broilers do little more than eat, they grow fast and tender. Under careful management, they consume approximately 2 pounds of feed for every pound of weight gained. They must be butchered as soon as they reach target weight, before they develop bone ailments or die of heart failure resulting from their excessively rapid growth.

And because they were developed to be raised in climate-controlled housing, they don't actively forage and won't do at all well outdoors when the weather is extremely hot or cold. Managing these hybrids therefore requires careful attention, but at least the homegrown result is tastier than the store-bought version.

Industrial Cornish crosses have fewer feathers to pluck and no underlying hairs to singe, making them easier and faster to clean than other broilers. I like them for roasting with the skin intact. Whenever I raise other kinds of broilers, I skin them.

Chickens raised for meat are plumper and faster growing than chickens kept strictly for eggs.

Colored Cornish hybrids are another industrial creation, developed for France's famous Label Rouge organic free-range chickens and adopted by America's pastured broiler industry. Common trade names include: Black Broiler, Color Yield, Colored Range, Freedom Ranger, Kosher King, Redbro, Red Broiler, Red Meat Maker, Redpac, Rosambro, and Silver Cross.

Most strains have red plumage, but they also come in black, gray, or barred — anything but white. Their colored feathers make them less attractive to predators, especially hawks, but more difficult to pluck clean.

These broilers grow more slowly than white hybrids, taking at least 11 weeks to reach target weight, and don't necessarily grow at a uniform rate. They eat about 3 pounds of feed per pound of weight gained, in part because of their longer growth period and in part because of the calories they use while foraging. Because of their slower growth, their meat is more flavorful than that of faster-growing hybrids. The edible portion (excluding excess fat, intestines, feathers, heads, feet, and blood) is approximately 70 percent of live weight.

Old-fashioned meat or utility breeds are hardier than commercial-strain broilers, but they grow much slower. Where a commercial broiler reaches 5 pounds (2.25 kg) in six to seven weeks, a purebred meat breed takes 9 or 10 weeks to reach the same weight, and a utility breed takes 12 to 16 weeks. The old-fashioned breed offers more flexibility in butchering age and, unlike the Cornish cross, does not grow uniformly, so not all the birds of the same age will be ready to butcher at the same time. They are not as efficient as hybrids at converting feed to meat, consuming at least 4 pounds of ration per pound of weight gained.

And the edible portion is about 65 percent of live weight.

Compared to a Cornish hybrid strains, purebred strains have thinner breasts and more dark meat. Because of their greater activity, their meat is lower in fat and firmer in texture. And because of their older age at harvest, the meat has a stronger chicken flavor. According to some reports, the longer growing period also makes the meat more nutritious, because it has more time to develop complex amino acids.

Some people describe nonhybrid meat as tough and suitable only for slow, moist cooking. I couldn't agree less. After years of enjoying the meat of dual-purpose birds cooked in all the same ways as store-bought meat, I find the latter has a bland taste and an unnatural mushy texture. Old-time American utility breeds with the greatest potential for meat production are these heritage breeds: Delaware, New Hampshire, Plymouth Rock, and Wyandotte. Naked Necks don't qualify as a heritage breed, but do make good meat birds and have sparse plumage that's an advantage at plucking time.

We raise meat birds as a by-product of keeping layers. Each spring we hatch a batch of dual-purpose chicks to get replacement pullets. When the pullets are big enough to go out on range, we separate them and confine the surplus cockerels until they reach fryer size. Later, any pullet that doesn't measure up is culled as a roaster. When the pullets start laying, our old layers become stewing hens.

The idea of raising meat birds as a by-product of the laying flock is far from new. It is, in fact, how today's broiler industry got started in the first place. In the 1920s, many housewives like Mrs. Wilmer Steele of Ocean View, Delaware, purchased chicks every year to raise as layer replacements. One year, Mrs. Steele mistakenly received 500 chicks instead of the 50 she had ordered, so she raised the surplus for meat and sold them at a dandy profit. The next year, she bought 1,000 chicks, and again the money rolled in. When Mrs. Steele had worked her way up to 25,000 chicks a year, Mr. Steele retired from the Coast Guard to stay home and help. For years thereafter, Delaware was the center of the broiler industry and development of efficient meat strains.

Meat Classes

The class of poultry meat you prefer may influence your choice between hybrids and nonhybrids. Meat birds are divided into these basic classes:

Rock-Cornish game hen: Not a game bird at all and not necessarily a hen, but a Cornish, Rock-Cornish, or any Cornish-cross bird weighing between 1 and 2 pounds (0.5 to 1 kg). To get plump, round game hens, you must raise Cornish hybrids, although I've grown surplus bantam cockerels into respectable single-serving birds.

Broiler-fryer: A young, tender bird of either sex weighing between 2½ and 4½ pounds (1 and 2 kg) dressed. The skin is soft, pliable, and smooth textured; the breastbone is flexible; and the meat is tender enough for any method of cooking. Hybrids and nonhybrids alike make good broiler-fryers, although a hybrid reaches target weight in about half the time of a nonhybrid. The United States Department of Agriculture (USDA) defines a broiler-fryer as being "about 7 weeks old."

Roaster: A young tender bird of either sex, usually weighing 5 to 7 pounds (2.25 to 3.2 kg) dressed. The skin is soft, pliable, and smooth textured, but the breastbone is less flexible than a broiler-fryer's. This bird is usually roasted whole. Because of the hybrid broiler's voracious appetite beyond the fryer stage, and the ensuing health issues, nonhybrids are more economical to raise as roasters but will be a little smaller, in the 4- to 6-pound (2 to 2.5 kg) range, and may take longer to get there. The USDA defines a roaster as being "about 3 to 5 months old."

Stewing or baking hen: A mature (10 months or older) female chicken with less-tender meat than a broiler or roaster and a nonflexible breastbone, requiring use of moist cooking methods such as stewing, braising, or pressure cooking. Stewing hens are generally older laying hens that are no longer economically productive.

Cock or rooster: Any male chicken that has entered the stag stage, when the comb and spurs develop, the skin becomes coarse, and the meat turns dark, tough, strong tasting, and generally not fit to eat, although with long, moist cooking, it may be rendered tender enough to chew.

Capons

Before highly specialized, fast-growing broiler hybrids were developed, meat birds were nothing more than cockerels from a batch of straight-run chicks raised to get pullets for laying. Those of us who keep dual-purpose flocks still raise meat birds the old-fashioned way. Due to the slow growth of nonhybrids, we have to take care to butcher cockerels before they reach the stag stage, when the meat gets strong tasting and tough.

In the old days, before the development of modern hybrids, such cockerels were caponized, meaning their testicles were surgically removed to channel their energy into

continued growth rather than sexual maturity. A capon grows to the size of a small turkey and was once considered an alternative to the holiday gobbler, only much easier to grow — turkeys being notorious for sitting around thinking up ways to die.

The hackle, saddle, and tail feathers of a capon grow longer than those of a cockerel, but instead of developing a large comb, the capon keeps the cockerel's small, pale head. The capon has a calmer disposition than a cockerel and rarely crows. He gains weight at about the same rate as a cockerel to about 18 weeks of age; then his growth rate surpasses that of a cockerel.

Compared to an intact cockerel (above), capons (below) remain calmer, grow larger, and rarely crow.

The capon is more expensive to raise than a cockerel, since he isn't properly grown and finished until he's at least 20 weeks old, and the heavier capon is more susceptible to weak legs and breast blister (a blister caused by the pressure of resting all that weight against the breastbone). For these and other reasons, few people bother caponizing these days.

Color Preferences

Many fine meat breeds never caught on in the United States simply because of the color of their skin. The skin color of a meat bird is a matter of preference, and consumers usually prefer what they've been taught to like. As a general rule, European consumers favor white-skinned breeds, Asians like black-skinned breeds, and Americans prefer yellow-skinned breeds. The hybrids developed for meat production in this country have yellow skin.

Even within a single breed, skin color varies with the chickens' diet. Marigolds, for instance, are sometimes fed to hens to make their egg yolks a richer yellow and will also deepen the color of the skin. A yellow-skinned bird that isn't feeling well, or for some other reason isn't eating well, will have paler skin than its flockmates. A young chicken with little fat may have bluish-looking skin.

In North America, birds with white plumage are preferred for meat because they look cleaner when plucked than dark-feathered birds. Regardless of feather or skin color, the taste is pretty much the same. Since many people now remove the skin before cooking or serving chicken, preferences in feather and skin color have become less important than they once were.

Management Methods

The breed you raise will, to some extent, determine how you manage your meat birds. Or your chicken-managing practices will determine the most suitable breed for your purpose. Methods for managing broilers fall into three basic categories:

- Indoor confinement
- Pasture confinement
- Range feeding

The last two methods are sometimes grouped together as pasturing, a method favored by those of us who prefer our food to be produced as naturally as possible. Since the first two methods involve confinement and the last two involve pasturing, the middle method, pasture confinement, bridges the gap between indoor confinement and the freedom to openly forage.

SKIN COLOR BY BREED

Yellow		
Aseel	Delaware	Naked Neck
Barnevelder	Dominique	New Hampshire
Brahma	Holland	Orloff
Buckeye	Java	Plymouth Rock
Chantecler	Jersey Giant	Rhode Island
Cochin	Langshan	Welsumer
Cornish	Malay	Wyandotte
White		
Ameraucana	Houdan	Orpington
Australorp	La Fleche	Penedesenca
Dorking	Maran	Redcap
Favarolle	Minorca	Sussex
Pinkish	**Black**	
Catalana	Silkie	

Indoor Confinement

Indoor confinement is the preference of large commercial growers because it lets them maximize capacity in terms of capital investment and facilities. It is also favored by small-flock owners who don't have much space for raising meat birds. It involves housing chickens on litter and taking them everything they eat.

Indoor confinement requires less land than the other two methods (all you need is a building) and less time (set up your facilities properly, and you should spend no more than a few minutes each day feeding, watering, and checking your birds). The goal is to get the most meat for the least cost by efficiently converting feed into meat.

The standard feed conversion ratio is 2:1— each bird averages 2 pounds (1 kg) of feed for every 1 pound (0.5 kg) of weight it gains. To get a feed conversion ratio that high, you must raise a commercial strain developed for its distinct ability to eat and grow. Even then, you likely would not achieve the same results without the controlled environment and specialized rations used in industrial production.

Even if your management is meticulous, you can't raise broilers indoors for the same low cost you would pay at the grocery store. For starters, the cost of chicks by the dozen is much higher than the cost of buying by the thousands. The same holds true for buying feed by the bagful, rather than periodically having a truck roll in to fill your silo.

You won't get the same high feed-conversion ratio, either, unless your facilities are designed to encourage feed consumption, right down to conveyor belts that keep feed moving to attract birds to peck. On a small scale, the best you can do to pique the flock's interest in eating is to feed your birds often to stimulate their appetite.

Efficient feed conversion also means keeping housing at a temperature between 65 and 85°F (18 and 29°C), which entails providing supplemental heat or crowding birds enough for them to keep each other warm. If you give your meat birds more room than the minimums shown in the chart (opposite), be prepared to either heat their housing or feed them longer to get them up to weight.

Another aspect of efficient feed conversion is controlled lighting. Compared to artificial light, natural light causes birds to be more active. As a result, they use up more calories and grow more slowly. Because birds in confinement have little else to do, when they're not eating, they get bored and resort to feather picking. If you let your confined birds enjoy natural light through windows or screened doorways, closing the openings to limit natural light to no more than 10 hours a day will result in fewer picking problems.

During the rest of the time, provide just enough light to let the birds find the feeders but not enough to inspire them to engage in other activities. Get chicks started eating and drinking under 60-watt bulbs, placed in reflectors 7 to 8 feet (2 to 2.5 m) above the floor. After two weeks, switch to 15-watt bulbs. Allow one bulb-watt per 8 square feet (0.75 sq m) of living space.

The total number of light hours meat birds should get per day is a matter of debate. The trend in commercial production is to shorten daylight hours for chicks 2 to 14 days old. Shorter days give them less time to eat, slowing their growth rate and thereby reducing leg problems and other complications resulting from too-rapid growth. After the birds reach 2 weeks of age, light hours are increased to encourage them to eat and grow.

MELODIOUS GROWTH BOOSTER

Fowls favor euphonious sounds — a phenomenon discovered by animal physiologist Gadi Gvaryahu, who found a flock of chickens crowded around a radio inadvertently left running. As Gvaryahu discovered, you can boost the growth rate of meat birds by letting them listen to classical music. The less-soothing strains of pop music don't work as well, so stick with classical FM.

Raising broilers under continuous light is a bad idea, in any case, since they may panic, pile up, and smother if the power fails. To get the birds used to lights-out, turn lights off at least one hour during the night. Some growers contend that as little as 14 hours of light per day is enough for efficient growth. During hot weather, when chickens get lethargic, turning lights on in the morning and evening encourages them to eat while the temperature is cooler. Putting lights on a timer will save you the trouble of having to remember to flick the switch.

Since rapid growth characteristically causes weak legs, don't provide roosts for your meat birds. Leg injuries can occur when heavy birds jump down from roosts. Perching can also cause blistered breasts and crooked breastbones. Injuries and blisters may also occur when heavy birds are housed on wire; wood; bare concrete; or packed, damp litter. Avoid these problems by housing confined meat birds on a soft bed of deep, dry litter.

MINIMUM SPACE FOR CONFINED MEAT BIRDS

Age	Floor Space per Chick
0–2 weeks	0.5 sq ft / 465 sq cm
2–8 weeks	1 sq ft / 929 sq cm
8+ weeks	2–3 sq ft / 1,858–2,787 sq cm

Pasture Confinement

Like indoor confinement, pasture confinement involves keeping broilers in a building, but this building is portable, is kept on range, and is moved daily. Pasture confinement is suitable for hybrid and purebred strains alike, although (as with indoor confinement) the former will grow more quickly than the latter. Furthermore, since Cornish broilers were developed for climate-controlled confinement, they won't do well on pasture if the weather is much cooler than 65°F (18°C) or much warmer than 85°F (29°C); other breeds have a much wider range of temperature tolerance.

The upside of pasture confinement is a slight reduction in feed costs, especially if you move the shelter first thing each day to encourage hungry birds to forage for an hour before feeding them their morning ration. On the downside, you need enough good pasture (or unsprayed lawn) to move the shelter daily, and without fail you must do it each day. As they reach harvest size, birds will graze plants faster and deposit a greater concentration of droppings, so they'll have to be moved more often — at least twice a day — for the health of the broilers and to avoid burning the pasture with too much nitrogen-rich manure.

A good plan is to arrange your shelter rotation so the chickens don't get moved farther into left field the bigger they get. Otherwise, you'll end up hauling greater quantities of feed and water a longer distance, and have to transport all those pudgy broilers back to home base for butchering.

During the first few moves, the birds will be reluctant to follow their shelter, and if you're not careful, they'll pile up, and some may get hurt. After a few times, they learn to walk along when you move their floorless shelter. Don't be tempted to try making the move easier by adding a wire floor so you can lift the birds along with the shelter. For one thing,

adding the birds' weight makes the whole she-bang heavy and unwieldy. Furthermore, while their shelter is in motion, the birds tend to stabilize themselves by curling their toes around the wire, causing toes to get crushed by wire when you set the shelter back down.

The pioneers of modern small-scale commercial pasture confinement are Joel and Theresa Salatin of Swoope, Virginia, who describe their method in detail in their book *Pastured Poultry Profit$* (see Recommended Reading on page 402). The Salatins confine up

Pasture confinement requires enough good range or unsprayed lawn to move the shelter to new ground once or twice a day.

to one hundred broilers per 10-by-12-by-2-foot (3 by 3.5 by 0.6 m) pen made of chicken wire stapled to a pressure-treated wood frame and roofed with corrugated aluminum. Weather permitting, chicks may be moved from the brooder to the pen once they reach 2 weeks of age. For broilers to do well on pasture, they must begin foraging by the age of 25 days.

Using a homemade dolly, the Salatins move each pen daily, requiring a total of 5,000 square feet (465 sq m) of good grazing per pen per 40-day growing period.

The couple raises a commercial strain that reaches a butchering weight of 4 to 4½ pounds (1.8 to 2 kg) in eight weeks, the same as they would if confined indoors. The management differences are basically twofold: feed costs are reduced, and because the birds spend so much time grazing, they have less time to pick at each other. The end result is meat that contains less fat and more omega-3s and other nutrients than that of confinement-fed broilers.

Range Feeding

Range feeding is similar to pasture confinement, except the birds are allowed to come and go freely from their shelter. The extra activity creates darker, firmer, more flavorful meat but also causes birds to eat more total ration because they take longer to reach target weight.

The time-honored method of range feeding chickens, widely practiced in the days before confinement became conventional, is traditionally called *free ranging*. But since the term "free range" has been corrupted by the USDA to mean "the poultry has been allowed access to the outside" (but not necessarily requiring that chickens actually go outside), the latest descriptive phrase is *day ranging*, as described by Andy Lee and Patricia Foreman in their book *Day Range Poultry: Every Chicken Owner's Guide to Grazing Gardens and Improving Pastures*.

Range feeding involves less labor than pasture confinement, because you don't have to move the shelter daily, but requires more labor than indoor confinement, because you do have to move it periodically. Compared to either form of confinement, ranging requires more land — enough for the shelter itself as well as pasture for grazing (and trampling), multiplied several times to allow for periodic fresh forage.

Utility breeds take to grazing quite readily because they retain some of the foraging instincts of their ancestors. Commercial strains don't think too much of getting out

Range feeding involves less labor than pasture confinement because the shelter is moved less often.

and around, but they will roam more than they do in confinement, and the energy used during roaming slows their growth and makes them less susceptible to leg problems. The end result is a trade-off between faster growth and better health.

If you raise straight-run chicks, you'll have to separate the cockerels as soon as they become sexually active; otherwise they'll harass the pullets, and neither will grow well. Sexual harassment is not a problem with confined broilers, since they go into the freezer before they get old enough to notice the opposite sex.

Not everyone is willing to raise broilers an extra few weeks, and not everyone appreciates the full flavor and firm texture of naturally grown chicken. As a result, backyard pasturing is used less often for growing meat birds than for keeping laying hens.

Feeding Meat Birds

A wide variety of meticulously formulated starter, grower, and finisher rations has been developed with one thing in mind — to keep feed costs down. Newly hatched chicks need a lot of protein. As chicks grow, their protein needs go down and their carbohydrate needs go up. Since protein sources (legumes and meat scraps) are more expensive than carbohydrate sources (starchy grains), switching to rations with progressively less protein saves money.

To precisely target the protein-versus-energy needs of meat birds according to their stage of growth, big-time growers formulate their own rations and have them privately milled. We small-flock owners are at the mercy of local feed providers. Depending on where you live, you may have little choice in the available combination of starter and finisher or starter/grower and finisher.

How much that matters to you depends on your method of management. A confinement-fed broiler eats approximately 2 pounds (1 kg) of feed for every pound (0.5 kg) of weight gained. If you raise your birds to 4 pounds (1.8 kg), each one will gobble up at least 8 pounds (3.6 kg) of feed during its lifetime. A purebred strain may eat twice that much, a factor you can somewhat mitigate by letting your birds forage for some of their sustenance.

In any case, if all you have available to you is one all-purpose starter/grower ration, nothing is inherently wrong with using it from start to finish. But don't expect the same rapid growth or low feed-to-meat ratio you would get with a more targeted ration. If you do have a wider choice, follow directions on the label regarding when to switch from one

FEED CONSUMPTION GUIDELINE

The bigger a broiler is, the more it eats. It stands to reason, then, that a flock's feed use steadily increases as the birds grow. To estimate the minimum amount of feed one hundred confinement-fed chicks should eat each day, double their age in weeks. For example, at 4 weeks old, one hundred broilers should eat no less than 8 pounds of feed per day. If feed use levels off or drops below this guideline, look for management or disease problems. (In metric terms, the broilers' age approximates the number of units of feed eaten: in the above example, one hundred broilers at 4 weeks of age would have eaten about 4 kg of feed.)

ration to another. Each manufacturer's feeding schedule is based on the formulations of its particular rations.

Unfortunately, standard commercial rations may not contain sufficient nutrients to sustain the rapid growth rate of Cornish-cross broilers, and as a result, they develop leg problems. If you raise a commercial-broiler strain, supplement the rations with a vitamin/mineral mix, added either to feed or to drinking water according to directions on the label. Some backyard growers withhold feed overnight to limit growth in an attempt to prevent lameness.

The older a chicken is, the less efficient it becomes at converting feed into meat and the costlier it becomes to raise. The conversion ratio starts out at less than 1 in newly hatched chicks and reaches 2:1 at about the fifth or sixth week. During the seventh or eighth week, the cumulative, or average, ratio reaches 2:1— the point of diminishing returns.

From then on the cumulative ratio has nowhere to go but up, and the amount of feed eaten (in terms of cost) can't be justified by the amount of weight gained. Although the most economical meat comes from birds weighing 2½ to 3½ pounds (1 to 1.5 kg), most folks prefer meatier broiler-fryers in the 4- to 4½-pound (1.8 to 2 kg) range. If you want nice plump roasters, be prepared to pay more per pound to raise them.

Two Pasturing Philosophies

Growers of pastured or range-fed broilers fall into two distinct camps: those who feed a high-protein diet and those who favor a high-energy diet. The high-protein group focuses on economics, while the high-energy group focuses on flavor.

Muscle (meat) growth relies on protein. Broilers with access to pasture, and expected to grow as rapidly as confined broilers, may be fed a ration of up to 30 percent protein. Using a ration with as little as half that amount of protein can reduce the growth rate by as much as 50 percent. Broilers taking longer to grow eat more total ration. Therefore, feeding the less expensive low-protein ration may cost more in the long run than feeding a pricier high-protein ration. A diet that is high in nutritional energy, in the form of grains, reduces the growth rate of muscle and increases the development of fat. Grain-fed fryers, therefore, aren't in favor with cholesterol-conscious nutritionists but are trendy in natural-food circles, where the goal is to avoid feed additives by using so-called organic grains. The problem is that unless you grow your own, organically grown grains are hard to come by and sometimes mislabeled.

Chances are good the scratch you feed your chickens is the same stuff used, in ground-up form, to make commercial rations.

The feeding method developed by producers of France's famous Red Label (*Label Rouge*) organic poultry has become a model for American organic-broiler growers. The Red Label ration is of 100 percent vegetable origin, supplemented by insects and plants the chickens find in the pine forest where they actively forage. These broilers are not of the Cornish-cross commercial type but are a cross between the slower-growing heritage Cornish and other old-time breeds. These broilers do not have white plumage like industrial broilers, hence some of the trade names for similar hybrids in the United States include a reference to color: Black Broiler, Color Yield, Colored Range, Freedom Ranger, Kosher King, Red Broiler,

Redbro, Red Meat Maker, Redpac, Rosambro, and Silver Cross, to name a few.

The main element in the diet of the French broilers is corn, produced by the French farmers themselves. The corn is crushed and mixed into a ration consisting of:

- 80 percent corn (for energy)
- 15 percent soy (for protein)
- 3 percent minerals and vitamins
- 2 percent alfalfa

Although an important part of the Red Label diet is natural forage, if you don't have available pasture, you can still enjoy broilers of similar flavor by raising grain-fed fryers. Feeding your birds up to 70 percent of their diet in scratch grains results in a slower growth rate that compares with the growth of range-fed chickens. Like ranged birds, they won't be ready for butchering until about 13 weeks.

If a finisher ration is available, feed chicks commercial rations for the first six weeks, switch entirely to scratch grains (with a vitamin/mineral supplement) until the last two weeks, then let finisher supply 30 percent of their diet. If you have grower ration available or only one starter/grower formula, feed your chicks commercial rations for the first four weeks, then switch to grower (or stay with the starter/grower) for 30 percent of their diet and scratch grains for the remaining 70 percent right to the end. In any case, as soon as you offer your birds scratch, they'll need free-choice granite grit to digest the grains.

Avoiding Drug Residues

The use of drugs to improve the growth rate of chickens has been banned since January 2017. Until then, ingredients found in rations fed to industrially produced chickens included low levels of antibiotics to improve feed conversion. The widespread use of antibiotics in poultry and other livestock feed contributed to the evolution of human disease–causing bacteria that became increasingly more resistant to treatment by drugs.

The use of antibiotics to treat backyard chickens has dropped dramatically, now that a prescription from a veterinarian is required for the same antibiotics that were once readily available over the counter. However, without using drugs you can discourage disease-causing bacteria through competitive exclusion, by reducing the pH in the crop to encourage the number of beneficial bacteria that make it to the intestines. On a commercial scale, intestinal pH is reduced by adjusting the feed formula to be more acidic. An easy way to introduce acidity for the home flock is to add vinegar to

FEEDER SPACE

Provide enough feeder space so broiler chicks can eat at will and so lower ones in the peck order won't get pushed away from feed by dominant birds. The general rule is to provide enough feeder space so at least one-third of your birds can eat at the same time. If you use hanging tube feeders, you'll need one for every three dozen birds. For trough feeders, follow these space recommendations:

Age	Trough Space per Bird
0–2 weeks	1 in (2.5 cm)
3–6 weeks	2 in (5 cm)
7–12 weeks	3 in (7.5 cm)
12+ weeks	4 in (10 cm)

the drinking water, but take care not to add so much the chickens stop drinking.

One tablespoon (15 cc) of vinegar per gallon (4 L) of water is sufficient unless the water is naturally alkaline, in which case double the vinegar. Some growers additionally help along the process of competitive exclusion by feeding broilers beneficial bacteria and yeasts in a probiotic formula designed to stimulate the immune system and fend off disease-causing bacteria.

A medication still found in some rations available for backyard flocks is a coccidiostat to prevent coccidiosis, an intestinal disease that interferes with nutrient absorption and drastically reduces the growth rate of infected broilers. If you raise chicks in a warm, humid climate or at a warm, humid time of year, the only way to avoid using a coccidiostat is either through the meticulous management of confinement housing or by getting the birds into a pasture rotation as quickly as possible, at least by the time they are 3 or 4 weeks old. You must be especially careful to keep litter clean and dry for indoor birds and move range-fed birds frequently to prevent a buildup of droppings. Otherwise, without the use of a coccidiostat, you'll be fighting a losing battle.

Unlike most medication, the common coccidiostat amprolium requires no withdrawal period. Many other medications do require a withdrawal period, which represents the minimum number of days that must pass from the time drug use stops until drug residues in the meat have been reduced to a level deemed acceptable by the U.S. Food and Drug Administration (FDA). If the medication label does not specify a withdrawal period, ask your feed dealer to look it up for you in his spec book. If you simply cannot find out, allow a withdrawal period of no less than 30 days.

At all times, provide drinking water that is clean of droppings and eliminate manure-laden puddles your broilers might sample. Clean water serves more purposes than preventing disease — broilers that don't have free access to fresh water eat less and therefore grow at a slower rate.

Broiler Health Issues

Health issues of homegrown meat birds primarily relate to rapid growth and heavy weight. Common issues for commercial strain broilers are lameness, breast blister, and heart failure. Among utility breeds, breast blister is the most common health concern.

Lameness

Commercial-strain Cornish-cross broilers are developed for such rapid growth that their bodies get too heavy for their little legs to carry them. Difficulty walking is therefore a significant issue among strains developed for industrial production. The faster a bird grows, the greater its risk of going lame.

During the past half century, the rate of industrial-broiler growth has increased from less than 1 ounce (25 g) per day to today's rate of 3½ (100 g) ounces per day. Where a broiler once took 13 weeks to reach 4½ pounds (2 kg), today's commercial broilers reach that weight in as little as five weeks. As a result of the strain on their legs and joints, those fast-growing birds can't get around well, and a small percentage can't walk at all. A study as early as 1972 concluded the "birds might have been bred to grow so fast that they are on the verge of structural collapse."

When a broiler gets so heavy its legs can't support its body, the bird can't get to feed and water, leaving it to get trampled by the more mobile birds and eventually die of either starvation or dehydration. In industry, up to 2 percent of lame broilers must be killed before they reach market weight.

Although it shouldn't take a PhD to see when a chicken is in distress, the broiler industry has devised various lameness scoring systems to determine when a review of management practices might be needed and at what point a lame bird should be humanely put down. The typical progression of lameness is

- **No lameness** — the chicken walks freely, bending the toes of its raised foot as it walks
- **Detectable lameness** — the chicken is mobile but somewhat unsteady, and the raised foot may remain flat with spread toes
- **Abnormal gait** — the chicken is mobile, although one leg takes short, quick, unsteady steps (like a person with a sprained ankle)
- **Severely impaired** — the chicken doesn't want to walk; if you stand it up, it may take a few steps before plopping down, perhaps struggling backward (backpedaling) on its hocks
- **Completely lame** — the chicken can't stand up but may try to shuffle around on its hocks

When a chicken reaches the impaired stage, it may develop a kinky back. Too-rapid growth causes the vertebrae to twist and pinch the spinal cord. Typically, an affected broiler will arch its back, extend its neck, squat with its feet off the ground and its weight on its hocks and tail, and backpedal. The bird may fall over and be unable to get up, become paralyzed, and die from dehydration. The humane thing to do for impaired and completely lame broilers is put them out of their misery.

The best way to minimize leg-health problems in commercial-strain broilers is to reduce their growth rate. Some backyard broiler growers do so by removing feed overnight to decrease eating time during early-morning hours. A study published in 2008 by researchers at England's Bristol University determined that broiler lameness also may be reduced by feeding whole wheat, which has the added benefit of improving digestion; reducing the number of hours the broilers are under light, thereby decreasing the amount of time they spend eating; reducing crowded conditions; feeding a nonpelleted ration, which increases the amount of time required to eat the same amount of feed. Of course, you could avoid the lameness issue altogether by raising a utility breed that doesn't achieve the exaggerated growth rate of industrial broilers.

Breast Blister

Broilers that have trouble standing or walking spend a lot of time resting with their weight against their breastbone, or keel. The pressure of the breastbone against the ground causes irritation and inflammation, resulting in a large blister, also known as *keel cyst*, *keel bursitis*, or *sternal bursitis*.

Housing broilers on wire or on wet or hard-packed litter and providing roosts for heavy birds to perch on increase the chance that breast blisters will develop. Another factor is poor feather development, which results in fewer feathers to protect the breast.

Although a blister is uncomfortable, it does not pose a serious health risk unless it becomes infected. The blister may, however, mar the appearance of the bird's meat.

Avoid blisters in utility breeds by keeping them on soft litter or grassy pasture and by not furnishing roosts. Avoid breast blisters in commercial-strain broilers by reducing their rate of growth using the same methods as would be used to minimize lameness.

Heart Failure

As short as the life of a meat bird is, commercial-strain broilers run the risk of dying prematurely from heart failure. Their little hearts and lungs simply can't keep up with the exaggerated growth rate of their body muscles. The high oxygen demand of rapid muscle growth, coupled with too-little space for blood flow through the capillaries of the lungs, causes *ascites* — the pooling of yellowish or bloody fluids in the abdomen. The condition is thus commonly called *broiler ascites*, *dropsy*, or *water belly*.

Affected broilers grow more slowly, sit around with ruffled feathers and are reluctant to move, and may die suddenly from heart failure. This condition is more likely to occur in commercial-strain broilers pushed for maximum growth than in the same strain sensibly managed in your backyard. Like leg and joint problems, heart failure is not an issue for slower-growing utility meat breeds.

Production and Marketing Choices

Selling meat birds can be profitable. It can also be economically risky. For starters you'll have a hard time competing with low market prices unless you find customers willing to pay a premium for homegrown meat. For another thing, many customers are not prepared for the differences in taste, texture, and appearance between homegrown and commercially grown chicken, although some folks are wising up to industrial practices and starting to seek out healthier alternatives. Still, selling poultry meat may require you to provide a certain amount of consumer education.

To add to your woes, you may need expensive state-approved facilities and a license to sell dressed birds, although in some states small producers enjoy exemptions. Your state poultry specialist can fill you in on the details. You may be able to get around dressed-bird laws by using a little creativity, such as selling live birds and giving buyers the option of plucking their own or having you pick the birds for free.

Marketing possibilities available in some areas include selling live birds to local butchers, custom pickers, or processing plants. The latter generally want huge numbers of birds and operate under various kinds of contracts. At one end of the spectrum, these buyers may require you to purchase all of your chicks and feed from them, with the understanding that you'll sell back the finished birds. At the other end, chicks and feed may be furnished without charge and you'll be paid for your labor and expenses. Be sure to read the fine print, including who's responsible for chickens that die or fail to measure up to market quality.

Unless you're willing to eat your profits, literally, determine who your customers will be before embarking on a meat-bird marketing venture. Educate yourself as well, regarding the descriptive terminology you are legally allowed to use, which changes with frustrating regularity.

Marketing Terminology

Depending on your style of management, you may or may not be allowed to use such terms as free range, natural, or organic. Even the poultry class designations (broiler, roaster, and so forth) are occasionally changed, and you may run into trouble calling your chickens broilers or fryers if they weigh less or take longer to reach market weight than the USDA says they should, based on the growth rate of industrial strains. Here are some basic definitions set by law, subject to change at any time:

Free range — generally means a significant portion of a chicken's life is spent outdoors; federal regulations say the birds must be allowed access to the outside but doesn't say they actually have to go outside

Free roaming — chickens have room to move around (are not caged)

Loose housing — free roaming within the confines of a building

Naturally raised — chickens are raised without growth promotants, fed no animal by-products, and given no antibiotics with the exception of coccidiostats called ionosphores that are chemically similar to antibiotics; this definition does not address such things as housing in confinement, genetic engineering, and the use of pesticides

Natural — the meat contains no artificial ingredient or added color and is only minimally processed; the label must explain the use of the term "natural," such as "No added colorings or artificial ingredients; minimally processed"

Minimally processed — processed by methods traditionally used to make food edible, preserve it, or make it safe to eat (such as freezing) but that do not fundamentally alter the raw product; a whole chicken separated into parts is considered minimally processed

Natural versus synthetic — the National List of Allowed and Prohibited Substances lists synthetic substances that may be used for organic production, as well as natural substances that are prohibited. Any natural substances not on this list are allowed. Items on this list that are commonly used for poultry production include disinfectants, sanitizers, medical treatments, topical treatments, and feed additives. Some ingredients for processing are also listed. The Organic Materials Review Institute (OMRI) keeps a list of generic and brand-name products that are permitted and includes brand-name mineral mixes, botanicals, probiotics, kelp, and homeopathic preparations.

Organic — certified by an agency as, among other things, having been raised in conditions that are less crowded than conventional birds; fed a vegetarian, antibiotic-free diet; and processed according to organic specifications.

Raising Heritage Chickens

In addition to legal definitions of terms that may be used in marketing chickens, The Livestock Conservancy offers the following definition of heritage chickens used for meat production. The breed must

- Have been recognized by the American Poultry Association prior to the mid-twentieth century (as listed in the table on page 365)
- Reproduce through natural mating
- Have the genetic ability to live a long, vigorous life (breeding hens should be productive for five to seven years, cocks for three to five years)

- Thrive outdoors under pasture-based management
- Have a moderate to slow rate of growth (reaching market weight in no less than 16 weeks).

Such terms as heirloom, antique, old-fashioned, and old-timey are considered to be synonymous with heritage. The Livestock Conservancy suggests that when any such terms are used in the marketing of chicken meat (or eggs), the variety and breed name should be indicated on the label.

Growing Organic

Determining what exactly "organic" means has been problematic since the word came into common use decades ago. Those of us who raise chickens to provide our own families with healthful meat and eggs can set our own standards. Those who produce poultry for sale, however, must comply with a myriad of regulations before labeling products as being organic.

Numerous state and private certifying agencies have sprung up throughout the United States, each with its own set of standards regarding what exactly organic means. The USDA has worked out a set of standards called the National Organic Program (NOP), which in typical government fashion includes many unclarified areas, leaving interpretation to the various certifying agencies. Each agency therefore makes judgment calls as to whether a particular poultry producer is in compliance or violation.

Until the day that all agencies agree to a standardized interpretation, if you want to gain organic certification, you must comply with the interpretations of the agency certifying your operation. To give you a hint of what's involved, here are some NOP standards:

Poultry or poultry products must come from chickens that have been under organic management from the second day of life.

All feed, except minerals and vitamins, must be organically produced. Nonsynthetic vitamins are preferred, but synthetic sources are allowed if nonsynthetics are not available. Animal by-products from mammals and poultry, such as meat and bone meal, may not be included in rations. Fish meal and crab meal are not permitted, because of the difficulty of determining if they were organically produced. Synthetic amino acids are not permitted. Nonsynthetic nonagricultural products such as oyster shell and diatomaceous earth are permitted, provided they comply with the Federal Food, Drug, and Cosmetic Act.

The handling of feed ingredients must comply with organic requirements. Solvent-extracted soybean meal, for example, is not permitted, but expelled soybeans are permitted, as are roasted and extruded soybeans. Premixed organic feeds are available but are usually expensive, especially if they must be shipped a long distance. The mill producing such feeds need not be dedicated organic but must implement a thorough cleanout before handling organic feed. Many organic poultry producers prefer to mix their own feeds. Some grow their own ingredients, others purchase ingredients from local certified-organic sources, and still others use a combination of the two.

No antibiotics or other drugs may be used to promote growth. Probiotics are typically used instead.

Physical alterations, such as beak trimming and toe trimming, are permitted if they promote the welfare of the animal and are "appropriately performed and within the context of an overall management system." Caponizing, as surgery by a nonveterinarian, is not considered an "appropriately performed" procedure.

Living conditions must be established and maintained to "accommodate the health and natural behavior of animals, including access to the outdoors, shade, shelter,

exercise areas, fresh air, and direct sunlight." Continuous confinement in cages is not permitted, but temporary confinement is allowed if adequately justified (the justification being up to the certifying agency's judgment), such as for hatchlings and for older birds during bad weather. Pasture confinement is open to interpretation, and some certifiers consider it unfavorable to natural behavior.

Access to the outdoors, according to an NOP policy statement, "simply means that

HERITAGE MEAT (AND LAYING) BREEDS AS DEFINED BY THE LIVESTOCK CONSERVANCY

Breeds deemed likely to benefit from being labeled as "Heritage Chicken"		Breeds deemed unlikely to benefit from being labeled as "Heritage Chicken"
Ancona	La Fleche	Belgian Bearded d'Anvers Bantam
Andalusian	Lakenvelder	Belgian Bearded d'UCCLE Bantam
Australorp	Lamona (likely extinct)	Booted Bantam
Brahma*	Langshan	Cochin Bantam
Buckeye	Leghorn	Frizzle Bantam
Campine	Minorca	Frizzles
Catalana	New Hampshire	Japanese Bantam
Chantecler	Orpington	Leghorn Bantam
Cochin	Plymouth Rock	Malay
Cornish*	Polish	Malay Bantam
Crevecoeur	Redcap	Minorca Bantam
Cubalaya	Rhode Island Red	Modern Game
Delaware	Rhode Island White	Modern Game Bantam
Dominique	Sicilian Buttercup	Old English Game
Dorking	Silkie Bantam	Plymouth Rock Bantam
Faverolle	Sussex	Polish Bantam
Hamburg*	White-Faced Black Spanish Wyandotte*	Rhode Island Red Bantam
Holland		Rosecomb Bantam
Houdan		Sebright Bantam
Java		Sultan
Jersey Giant		Sumatra

*both large size and bantams

a producer must provide livestock with an opportunity to exit any barn or other enclosed structure." No required amount of outdoor access is specified, although some certifiers require a minimum of eight hours per day or natural daylight, whichever is less. NOP standards call for outdoor areas to be free of pesticides for three years. Certifiers may additionally require these areas to be covered with vegetation and managed to prevent them from accumulating standing water or becoming bare.

Dusting areas must be provided.

Clean, dry bedding is required and, if of a type typically consumed by poultry, must meet organic feed requirements.

Chickens are not permitted to come into contact with pressure-treated lumber, including wood used to construct field pens.

Health-care measures must include preventive practices such as good nutrition and sanitation, healthy living conditions, and vaccinations. Antibiotics are not permitted. Synthetic parasiticides, including coccidiostats, are not permitted. Although some natural alternative products are allowed, health problems must be controlled primarily through good management. (You wouldn't think this commonsense notion would require federal guidelines.)

On the other hand, necessary medical treatment should not be withheld simply to preserve a chicken's organic status. The bird should be medicated and, after observing the appropriate withdrawal period, sold in the conventional marketplace.

All manure must be managed in a way that does not contribute to the contamination of crops, soil, or water and that optimizes the recycling of nutrients.

Among areas that remain unclarified is the minimum requirement for floor space. Some certifiers calculate minimum space based on birds per square foot, others base it on the overall weight of broilers—such as a minimum of 1 square foot per 6 pounds (0.1 sq m/2.7 kg), meaning each square foot would accommodate three 2-pound (0.9 kg) birds, two 3-pound (1.4 kg) birds, or one 6-pound (2.7 kg) bird.

Another area open to some interpretation is the use of artificial lights to extend daylight hours. Many certifying agencies require 8 hours of darkness every 24 hours. Some certifiers additionally require the lights to be bright enough to encourage normal activity and windows to provide some sunlight for broilers confined indoors.

The NOP standards continue to evolve. You may review the current set of rules on the Internet or by contacting the United States Government Printing Office. The National Sustainable Agriculture Information Service Appropriate Technology Transfer for Rural Areas (ATTRA) monitors these standards and translates them from governmentese into everyday language. ATTRA also maintains a list of organic-poultry-feed suppliers and provides information on natural methods for dealing with coccidiosis, mites, cannibalism, and other common poultry problems.

Certification

If you're going to sell meat birds, obtaining organic certification may be worth your while. Certification is essentially a marketing tool to reassure customers of the quality of your poultry and justify your higher prices compared to industrially produced chicken.

But the never-ending stream of paperwork required for certification discourages a lot of people right from the start. First you fill out an application, which in itself can be a formidable document. Then you have an interminable amount of record keeping, detailing every step of production. Your operation will be inspected before your application is approved and at least once a year thereafter. Finally, you'll have to pay certain fees, which may include a flat fee, an inspection fee, and a percentage of your sales volume.

Various certification groups work on local, regional, national, and international levels, but not all of them certify animal products. You'll need to find the group that best suits your production needs, then determine whether or not you meet their criteria. Up-to-date details on organic certification and a list of current programs is included in the information packet "Organic Certification" available from ATTRA.

Economic Efficiency

The poultry industry uses several different methods to determine the economic efficiency of producing meat birds. Since the cost of feed accounts for at least 55 percent of the cost of production, most economic indicators factor in the amount of feed used.

Feed-conversion ratio is the total amount of feed in pounds (or kilograms) eaten by the birds, divided by the flock's total live weight in pounds (or kilograms). Commercial broilers raised under efficient methods get by on as little as 1.85 pounds (0.8 kg) of feed per pound (0.5 kg) of weight gained to 4 pounds (1.8 kg) live weight. At home, don't expect a conversion ratio from an industrial strain much better than 2 pounds (0.9 kg) of feed per pound of live weight, or about 3 pounds (1.4 kg) of feed per pound of dressed weight. If your rate is significantly higher, take stock of your management methods.

Feed cost per pound (or kilogram) is the feed-conversion ratio multiplied by the average cost of feed per pound (or kilogram). Determine the average cost of feed per pound by dividing the total pounds of feed used into your total feed cost. The lower this number is, the better you're doing.

Performance-efficiency factor is the average live weight divided by the feed conversion ratio, multiplied by 100. In industry, this index hovers around 200. The higher you get above 200, the better you're doing.

Livability is the total number of birds butchered or sold divided by the total number started. To convert that number to a percentage, multiply by 100. Good livability is 95 percent or better. If your livability is above 90 percent, you're doing as well as most commercial growers.

Average live weight is the total live weight divided by the total number of birds. The industry average for industrial broilers efficiently raised to 8 weeks of age is 4 to 5 pounds (1.8 to 2.25 kg). If your birds aren't even close at the end of eight weeks (13 weeks for nonhybrids), look for ways to improve your management methods. One way to improve your average is to raise cockerels instead of straight-run chicks. As a general rule, finished cockerels weigh 1 pound (0.5 kg) more than pullets at the same age and on the same amount of feed, and at the broiler-fryer stage they taste the same.

Average weight per bird is the total dressed weight of all birds divided by the total number of birds. This index factors in weight lost both to excess fat and to uncontrollable inedible portions such as intestines, feathers, heads, feet, and blood. A good average for the edible portion is approximately 75 percent of a bird's live weight.

Readiness for Butchering

Your management style will influence the rate at which your meat birds grow. Furthermore, different breeds grow at different rates and reach readiness for butchering at different weights — a bantam, for instance, will not be ready to butcher at the same weight and age as a Jersey Giant. To assess a bird's readiness for butchering and the quality of its meat, examine it for these factors:

Feathering. If you plan to pick the bird rather than skin it, butcher when it has mature feathers only, no stubby broken feathers or pinfeathers.

Fleshing. A well-fleshed bird has a meaty breast, legs, and thighs.

Finish. The best flavor comes from a bird with a nice layer of fat beneath the skin. To assess finish, spread the breast feathers and examine the skin. A creamy or yellow color indicates good finish, while a reddish or bluish color indicates too little fat.

Conformation, or body shape. The ideal shape of a meat bird ready for butchering is blocky and rectangular, not narrow and triangular.

If you're raising meat birds for your own family use, you needn't be too concerned about freedom from defects. You might, in fact, cull defective birds from your flock by putting them into the freezer. But if you're raising meat birds for market, they must be free of such defects as crooked breastbones, crooked or hunched backs, deformed legs and wings, bruises, cut or torn skin, breast blisters, and calluses.

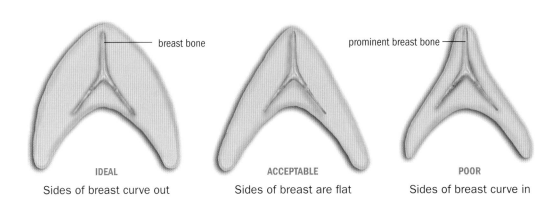

IDEAL
Sides of breast curve out

ACCEPTABLE
Sides of breast are flat

POOR
Sides of breast curve in

One way to assess the readiness of a meat bird for butchering is to determine if its shape more closely resembles a rectangle (right) or a triangle (below).

14

MEAT ON THE TABLE

When your chickens are plump enough to butcher, catch a few and examine them for pinfeathers. Chickens with emerging pinfeathers won't look clean no matter how carefully you pick them, so let them continue to grow until they pass the pinfeather stage. When the time comes, confine the birds overnight with plenty of water but without feed.

To move them to a holding area, catch them at night, when they're least active, to minimize struggling. Struggling depletes energy they need to relax muscles and keep the meat tender. Put them in a wire- or slat-floored coop where they cannot eat feathers or litter.

The goal of holding them without feed is to allow their intestines to empty, minimizing contamination during butchering. They need water, however, to prevent dehydration, which turns the skin dark, dry, and scaly, and causes meat to get tough and stringy.

Killing

The worst part about killing chickens is the reflex reaction that causes them to flap and twitch for a few moments after they're dead. During these death throes, blood gets spattered around. If butchering is going to upset you, flapping and bleeding are the most likely aspects to do it. That goes double for children, who may not understand that, even though a bird is still moving, it's dead and isn't suffering. Nearly every family has at least one member (or a neighbor) who can handle killing chickens.

Set up a killing place away from the rest of the flock, where remaining birds won't get upset, and where you can clean up easily afterward to avoid attracting flies, wasps, and four-legged predators. Have on hand a trash can or large bucket lined with a sturdy plastic bag, so you won't be tempted to toss discarded parts on the ground.

Killing Methods

Kill the chickens quickly and humanely with a minimum of stress, since stress reduces meat quality. Of several possible methods for killing chickens, the one most commonly used in backyards is the least suitable.

Using an ax. Most people's image of killing a chicken is to lay its neck on a stump and chop off its head. Using an ax is not the best idea, however. An ax severs the bird's spinal cord, tightening the feathers, and severs the jugular vein at the same time as the windpipe, letting blood into the lungs and contaminating the meat.

Serving a chicken you grew yourself is satisfying, in addition to furnishing a meal that tastes better and is better for you than the supermarket option.

By hand (or foot). A more suitable, time-honored method is to dislocate the bird's neck, which kills instantly. Hang the bird upside down by holding both legs in one hand. With the other hand, grasp the head with your thumb behind the comb and your little finger beneath the beak. Tilt the head back, and pull steadily until the head snaps free of the neck. Continue holding the head until struggling stops.

If you have trouble snapping a chicken's neck by hand, use your feet. Grasp the bird by the legs and lay its neck on the ground. Place a broom or rake handle across the neck. With one foot on each side of the neck, stand on the handle and firmly pull the bird upward.

After the neck has been snapped, hang the dead bird by its legs with a piece of twine. Cut the neck on both sides and let it bleed out.

Using a knife. If you have a lot of chickens to kill at one time, using a sharp knife lets you work fast. Suspend each bird from twine tied around its legs. With one hand, grasp the beak in front of the eyes and pull down hard to hold the bird steady. With a sharp knife in the other hand, make a 2-inch (5 cm) cut just behind the jaw and into the base of the skull on both sides of the neck, severing the jugular vein.

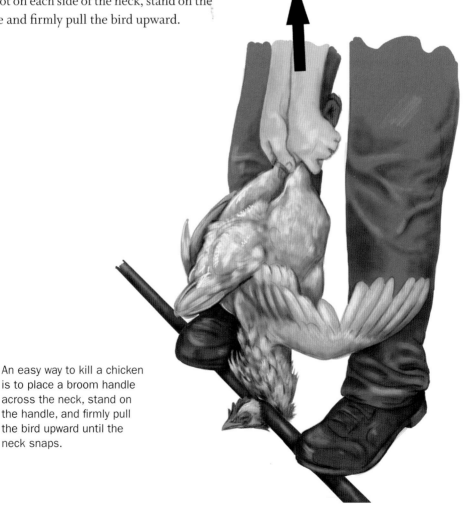

An easy way to kill a chicken is to place a broom handle across the neck, stand on the handle, and firmly pull the bird upward until the neck snaps.

Debraining helps loosen feathers for hand picking. Debraining is done with a sticking knife, which has a dagger-type blade that's usually sharp on both sides; any knife with a sharp, narrow blade will do. After cutting the vein for bleeding, insert the knife into the mouth, its sharp edge toward the groove at the roof. Push the knife toward the back of the skull and give it a one-quarter twist. The trick is to avoid sticking the front of the brain, which causes feathers to tighten instead of loosen. You can tell your knife has hit home when reflex causes the dead bird to shudder and utter a characteristic squawk.

Using a gun. A .22 handgun makes a fast, clean job of it but is a suitable option only if you are familiar with guns and live in a rural area where shooting is legal and may be done safely. Hang the bird by its legs and shoot a round into the back of the head just above the eyes. This method kills quickly, causes nerves to relax, and loosens feathers for easy removal. Immediately after shooting, bleed by cutting the jugular vein.

Stainless-steel killing cones come in various sizes to snugly fit the size of the chicken; a traffic cone, or an empty bleach or laundry detergent jug with the top and bottom cut off, makes a workable substitute.

BLEEDING

Bleeding is an important part of the killing process to ensure no trace of blood remains in the meat for the best flavor, appearance, and keeping quality. It must be done while blood still flows freely. Be ready to deal with the several tablespoons of blood that will flow from each chicken, either by working where it may be easily flushed away or by having a bucket or other collection container on hand.

If you plan to compost the blood, a little water in the container will keep it from coagulating. Have additional water handy to rinse away any spills.

Use a knife to slit the jugular vein. To keep blood from spattering, you have at least three choices:

· Keep a tight hold on the bird until flapping and bleeding stop.

· Attach a hook to the lower beak and hang a 1-pound (0.5 kg) weight from it.

· Confine each bird in a killing cone, a funnel-shaped device (also called a holding funnel) into which the chicken is inserted, head first, to keep it still.

Picking

Once the bird is dead, you have to decide whether to remove the head before or after picking. The correct thing to do is leave the head on until the feathers are removed, which keeps the neck clean for making soup, broth, or gravy. Because I don't like having a chicken stare accusingly at me while I relieve it of its feathers, I behead first.

The process of removing feathers is called *picking* in the industry and *plucking* in other circles. For sanitary reasons, it makes good sense to kill and pluck in one area, then rinse and pack in another. If you have a helper, the operation goes quickly if one does the picking and the other does the packing.

Not being one to follow the conventional wisdom that picking is less messy if you do it outdoors, I prefer to work indoors, where it's cool and I don't have to fight off flies and yellow jackets. I kill four birds at once and bring them inside for picking. Four is as many as I can do before the last one starts getting stiff and difficult to manage. When I'm done, I go back and get four more.

Picking Methods

How you pick will depend on how often you butcher and how many birds you have. Picking may be done in four ways:

Hand picking involves pulling feathers out by hand. It may be done by either the dry-pick method or the scald-and-pick method discussed further on. With either method, using rubber finger tips from an office supply or rubberized knit gloves from a hardware store will help improve your grip to make the job go more swiftly.

Dry picking is suitable when you have only a few birds. It's easy when chickens are freshly killed and still warm, especially if they've been debrained. But it's not so easy after they start to cool. The best plan is to strip away the feathers quickly while the birds are hung for bleeding. If you're allergic to feathers or dander, wear a dust mask during dry picking.

After the main feathers have been removed, some immature feathers will remain. These pinfeathers are less numerous in a commercial-broiler strain than in other breeds. They may be removed using your fingers, with the occasional stubborn one pulled out with tweezers. A pinfeather scraper, or pinning knife, with a rounded 3- to 4-inch (7.5 to 10 cm) blade is easier on the fingers; a suitable substitute is a paring knife with a dull blade. If you regularly pluck chickens of a breed with lots of pinfeathers, you might consider using hot wax.

Wax picking follows rough hand picking as a fast way to get birds clean. It involves dunking the whole bird in heated picking wax paraffin, then dunking it in cold water to harden the wax. Peel away the wax, and the feathers and pinfeathers alike come right off. When you're all done, let the remaining wax cool in the pot. Feather debris will settle to the bottom, and you can save the clean wax on top for reuse. If you melt the wax in a kettle of steaming hot water, the paraffin will float at the top and you will need less of it; allow one pound per four birds. Wax picking is used less often for chickens than for ducks and geese, which are more difficult to pick clean.

Machine picking saves time if you're butchering a lot of chickens. Picking machines come in two styles, both of which require chickens to be scalded first. A *tabletop picker* has a rotating drum with rubber fingers against which you hold one bird at a time while the feathers are flailed off. A *tub picker* has rubber fingers lining a rotating drum into which you drop one chicken or more; the feathers are flailed off as the drum spins.

Picking machines don't come cheap. You can reduce the cost by making your own, using plans included in the book *Anyone Can Build a Tub-Style Mechanical Chicken Plucker* by Herrick Kimball. If local laws allow, you might defray the cost of buying or building a mechanical picker by doing custom plucking on the side or by letting others use your machine on a rental basis.

A mechanical picker can cut down plucking time to as little as half a minute, which sounds nifty until you learn that a University of Arkansas study found mechanically picked chickens are 2½ times tougher than hand-plucked birds. If you butcher older chickens, which tend to be tough anyway, hand picking is the way to go.

Custom picking is no longer common in many areas, but if the idea of killing your own chickens seems too messy or distasteful, you might look for a local custom butcher or picker. You may even find someone with a licensed mobile unit. Otherwise, you'll have to transport your live chickens to the facility.

As an alternative, seek out someone who cleans game birds for hunters. Make sure the person knows how to deal with live birds, though: one fellow I know who regularly picked dead birds for hunters happily killed a bunch of chickens for a friend but didn't know he was supposed to bleed them.

Scald and Pick

Chickens are easier to pick if you first loosen the feathers by dipping each bird in scalding water. The dunking kettle must be large enough so when you dip and slosh, the rising water won't overflow. A standard-size enamel canning pot two-thirds full is just right for most chickens. A large stockpot is also a good size for all but the biggest breeds.

NO MORE PLUCKING

Poultry is the only meat harvested with the skin intact, and with today's concern about fat and cholesterol, few people eat the skin anyway. So why go through all the mess and bother of plucking chickens? For one thing, the meat may be more moist and tender cooked with the skin on and then skinned before eating. But lots of recipes produce chicken that's moist and tender without the skin. You can skin a young bird in about half the time needed for hand plucking, and you won't have to delay butchering day until your broilers pass the pinfeather stage.

Another option is to wait until featherless chickens come to the United States. Developed in Israel from a Naked Neck (Turken) crossed with a normal broiler, these chickens have only a few tufts of feathers on their pink skin. They don't waste a lot of energy making feathers that get thrown out anyway, and compared to regular broilers they grow faster in hot climates, where the growth rate is normally limited because nutritional energy must be restricted to keep broilers from overheating under all that plumage. But don't think about raising these chickens in a cold climate, where they require the expense of a heated facility, or in an unshaded pasture, as they have no protection from sunburn.

You might heat the kettle on a gas stove, over a propane burner designed for camping, or on a grill over a wood fire. Some folks heat the water in a turkey fryer. If you plan to pick a lot of birds, you might invest in a thermostatically controlled dipping vat, or make your own from plans in the book *Anyone Can Build a Whizbang Chicken Scalder,* also by Kimball.

Scalding a chicken without tearing the skin or cooking the meat requires a combination of proper water temperature and appropriate scalding time. Young birds with tender skin need a lower temperature and shorter scalding time than older birds with tough skin. Unfortunately, scalding at a high temperature stiffens muscle tissue, making tough meat even tougher. A high temperature also causes the outer layer of skin to come loose. When you're butchering for your own use, appearance may not be a problem, but it becomes an issue when chickens are butchered for sale.

A *semiscald*, or so-called *slack scald*, at 125 to 130°F (52 to 54°C) keeps the outer layer of skin intact. A *subscald*, at 138 to 140°F (59 to 60°C) loosens the outer skin layer but also loosens the feathers better, making them easier to pick. A *full scald*, or *hard scald*, at 140 to 150°F (60 to 66°C) speeds things up, but the skin tears more easily, causing the meat to dry out.

Some chicken pluckers are fussy about maintaining a specific water temperature, but after you pick a few you get a feel for the right combination of temperature and time. Hold the bird by the shanks, and completely immerse it, head first, until all the feathered parts are under water. Move the bird up and down and back and forth so the water evenly penetrates the feathers. A squirt of dishwashing detergent in the water helps with penetration, especially for densely feathered breeds. Lifting the chicken out of the water and dipping it back in a time or two also helps improve penetration.

After about 30 seconds of dipping, pull a large tail or wing feather. If it does not slip right out, quickly dip the bird again.

A scalded chicken is ready for mechanical or hand picking. Some hand pickers hang a bird by its legs or lay it on a table during picking. Others pluck under a running faucet, which has the advantage that feathers wash away instead of clinging, but the disadvantage is that it uses a lot of water.

I pick a scalded chicken in a basin of cold tap water, which cools it down so the hot feathers neither burn my fingers nor continue heating up the skin. I periodically rinse off clinging feathers, and refill the basin with fresh water for each bird.

Working as rapidly as possible, pull handfuls of feathers, rubbing with the base of your thumb as you work. Feathers should strip off easily and uniformly. If they don't, you didn't scald long enough, the water didn't penetrate evenly, or your water isn't hot enough. Torn skin and patches of feathers coming off with attached skin are signs that the bird was dipped too long or the water is too hot. Pluck feathers in the sequence that works best for you. I first get the large wing and tail feathers out of the way, then pick both wings clean, work down the breast and legs, and finally move down the back to the tail.

The world hand-picking record is 4.4 seconds per bird. The USDA claims an experienced picker shouldn't take more than 5 minutes for each chicken. Call me slow, but even after decades of experience, my hand-picking time has never been much better than 15 minutes per chicken.

Once the main feathers are off, squeeze out pinfeathers by holding the bird under a running faucet and scraping the skin with a pinning knife or a dull-bladed paring knife. Stubborn pins may need to be tweezed out or pulled between the knife blade and your thumb. When the skin is clean, rinse off loose feathers, pat the bird dry, and set it aside while you pluck the next.

Singeing

If you raise a utility breed and you serve chicken with the skin on, you'll need to singe away the hairlike feathers that pop up after birds have been plucked and dried. Hairs singe off readily when passed over an open flame, which might be from a gas burner or a propane torch. Singeing torches, similar to a handyman's propane torch, are used commercially.

If you're going to peel off the skin before cooking the meat, you needn't worry about the hairs — no one will ever know your chickens are hairy. If you raise a Cornish cross, you don't need to singe, anyway, since hybrid broilers have had their hairs genetically eliminated.

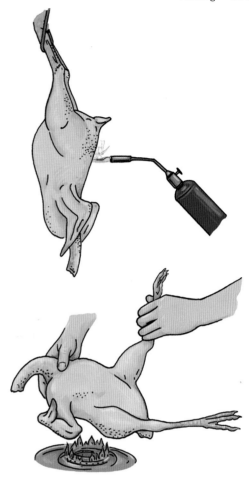

Use a gas flame or propane torch to singe off hairlike feathers.

TERMINOLOGY

Eviscerating a chicken is not as complicated as you might infer from the terminology used in government publications and other manuals. Knowing what all those strange words mean makes the whole thing sound as simple as it is.

Viscera — internal organs

Entrails — another word for viscera

Giblets — edible organs
(liver, heart, gizzard)

Offal — inedible organs
(crop, intestines, and so forth)

Eviscerating — removing the organs

Drawing — another word for eviscerating

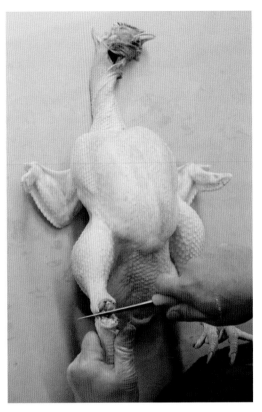

1. Remove the feet by cutting through the leg just below the hock joint.

Evisceration

Some people eviscerate their chickens right after plucking them. Others chill the chickens first, finding internal organs easier to remove from a cooled bird. Try both ways, and see which you prefer.

If you butcher regularly, a boning knife or a pair of boning shears comes in handy. Or you could use a pair of clean, sharp pruning shears.

First Steps

Use the knife or shears to remove the feet at the hock joint. If you're cleaning an older bird — roaster, stewing hen, or capon — avoid cutting the tough tendons so you can pull them from the drumsticks: cut the skin along the back of the shank to expose the tendon, insert a hook, twist the hook to get a good hold on the tendon, and pull. A tie-wire twister, designed for twisting concrete tie wires, makes a dandy tendon remover.

To remove the feet, cut through the leg just below the hock joint. Unless you've already cut the back part to remove the tendons, work

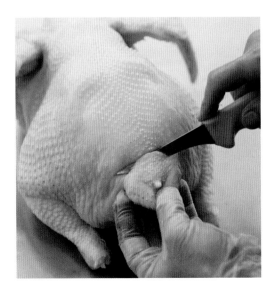

2. Remove the oil gland with a deep, wedge-shaped cut.

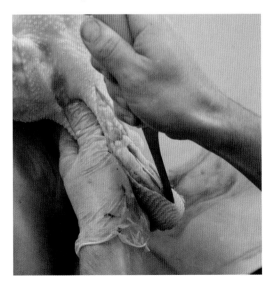

3. Slit the skin along the back of the neck.

from the front toward the back to leave a little flap of skin that keeps the meat from shriveling away from the bone during cooking. Some people save the feet to make soup stock or fried snacks, for which recipes may be found online. If that idea appeals to you, declaw the feet, then wash and skin them — you *know* what those feet have been walking in.

With the feet removed, turn the bird over to cut off the oil gland on the tail. One inch (2.5 cm) above the gland's nipple, make a cut deep enough to reach the tailbone. Since the gland goes quite deep, cut with a scooping motion as you move the knife along the bone to the tip of the tail. Be sure to get the whole gland out so the nasty-tasting oil doesn't spread into the surrounding meat.

If you haven't already done so, remove the bird's head next. Use your knife or shears, and with a twisting motion cut between the head and the first neck vertebra.

For a bird that will be stuffed and roasted, leave a flap of neck skin to hold the stuffing in and keep it from drying out. Insert the knife

through the skin at the back, near the shoulders. Slit the skin by guiding the knife up the back of the neck. Pull the skin away from the neck down to the crop, and cut below the crop to remove the crop and the windpipe. If your birds were taken off feed for at least 12 hours prior to butchering, the crop should be empty of sour contents that might otherwise spill out.

Use your kitchen shears to cut the neck all the way around, as close as possible to the shoulders. Grasp the neck and twist it off. If you can't get a good grip, wrap the neck in a dry paper towel. Although it doesn't have much meat, you might save the neck to simmer for soup stock or gravy.

Fold the flap of neck skin toward the back, turn the wings so the tips cross at the back, and use the wing tips to hold the neck flap. If you're going to cut up the bird for cooking, you won't need this flap, in which case cut off the skin at the base of the neck.

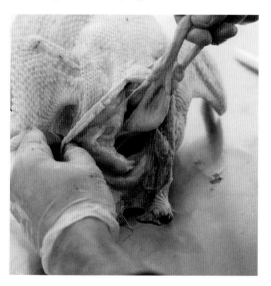

4. Pull out the crop and the windpipe.

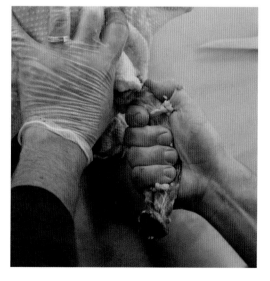

5. Cut through the neck at the base and twist it off.

Opening the Abdomen

The abdomen may be opened to remove the internal organs in one of two ways. Which way you use will depend on the bird's conformation and whether or not you plan to roast it whole.

A *bar cut* offers a natural trussing method for whole birds by leaving a horizontal strip of skin across the abdomen for the legs to be tucked into, keeping the breast meat from drying out. A bar cut doesn't work for birds with large deposits of fat, which loosen the abdominal skin too much to hold the legs. Neither does it work for birds with legs that are too short to tuck in.

Use a sharp, thin knife to separate the vent from the tail, halfway around. Do so by inserting the knife between the vent and tail and working upward, one-quarter around until you reach the pointed pinbone at the side of the vent. Then work upward in the other direction toward the other pinbone, taking care not to cut into the intestine.

Insert your finger into the opening as a guide, and use your shears to continue cutting the skin all the way around the vent. Grasp the vent, and pull it out a little so it can't drop into the cavity and release fecal matter inside.

OPENING THE ABDOMEN

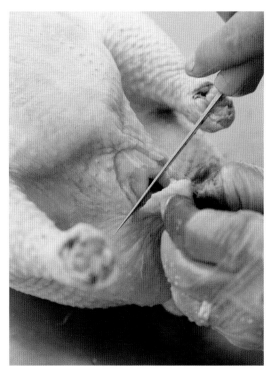 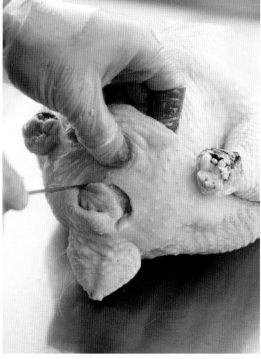

Take care when cutting around the vent, as you do not want to break into the colon. Even when a bird has gone overnight without eating, feces may remain in the intestine. Try to cut through the skin surrounding the vent, so the vent can be removed along with the internal organs.

About halfway between the vent opening and the breastbone, pinch the skin and insert your knife to make a horizontal cut about 3 inches (7.5 cm) across. The strip of skin between this cut and the vent opening is the skin bar that holds the legs in place during roasting.

A *midline cut* is a vertical opening running from the vent straight to the breastbone. This cut is used for small-size roasters, as well as for broiler/fryers. Roasters opened this way must be trussed with string or skewers to hold in stuffing or to keep the breast meat of an unstuffed bird from drying out.

Stretch the abdomen skin with one hand, and use a knife in the other hand to cut through the skin. Start at the keel, and slowly work the knife toward the vent, taking care not to cut deep enough to break into the intestine. When you reach the vent, insert your index finger into the opening and lift up on the intestine while continuing to cut around the vent beneath your finger.

Drawing

Once the abdomen is properly opened, the next step is to remove the internal organs or viscera. Insert a hand into the opening and work it around the inside wall, as far as you can reach on both sides, to break the attaching membranes. When you come to the gizzard (it feels hard in comparison to other organs), cup your hand around the bundle of organs and pull gently.

Although the idea is to bring out all the organs at once, if you miss any, you can always reach back in for them. When you gently pull them out, the bundle of organs may remain attached. For the moment, leave them that way.

If you're cleaning a cockerel, be sure to remove the testicles — two soft, white,

Remove the internal organs in one bundle.

Save the gizzard, heart, and liver (with bile sac removed) before discarding the offal.

oval-shaped organs along the backbone. If you're cleaning a pullet or hen, remove the mass of undeveloped eggs located in the same area.

In both cases, remove the lungs — pink, spongy organs pressing against the upper back. Some people enjoy the lungs in soup, provided the lungs were not contaminated by improper bleeding during killing. You can easily remove the lungs by inserting an index finger and lifting one and then the other with a scraping motion.

Giblets

In the bundle of organs are three things you may want to save: the liver, the heart, and the gizzard. Collectively, they are known as giblets

RESPONSIBLY DISPOSING OF WASTE

In most areas, regulations prohibit the disposal of butchering refuse in dumpsters or landfills. If you have a place to bury it away from water sources, bury it at least 3 feet (1 m) deep. Otherwise, compost it. Your local Extension office may have information on responsibly composting chicken offal, feathers, and blood. Information is also available from the National Sustainable Agriculture Information Service (also known as ATTRA).

Chicken	Yield	Waste
Broiler/fryer	75%	25%
Stewing hen, heavy breed	70%	30%
Stewing hen, light breed	68%	32%
Roaster or capon	76%	24%

and may be cooked and eaten or used as the basis for soup, stock, or gravy.

The liver from homegrown chickens is delicious, provided you don't contaminate it by breaking the gallbladder — the small green sac nestled into one of its folds. You'll know the gallbladder has broken if the liver becomes stained with green bile. When that happens — as it will from time to time — throw out the liver, as it will taste bitter.

Some people can remove the gallbladder by pinching it between their thumb and index finger. I've never been able to do that without squeezing out bile. Instead, I carefully insert a sharp knife tip under the connective tissue and slice upward, pressing the end of the bladder between my thumb and the blade to keep it from spilling.

Our favorite way to serve chicken livers is to fry them hot and fast and serve them with sweet onions over rice. When we have enough at one time, we make a tasty pâté.

Liver doesn't keep well, so it is best served fresh. If you have too many livers at once, pack them into plastic freezer containers large enough to hold one meal each. Cover the liver with a piece of plastic wrap or waxed butcher paper, snap on the lid, and freeze.

The gizzards and hearts may be chopped or ground and added to gravy or stuffing or used to make pizza, spaghetti sauce, tacos, or chili. In various parts of the country, grilled, stewed, fried, or pickled gizzards are enjoyed as a snack. Gizzards and hearts freeze well for later use.

Remove the membrane surrounding the heart. Trim off the top, slit the heart halfway to open it out flat, and rinse it.

Cut away the gizzard where it attaches to the stomach and intestines. Cut into the large end, aiming toward the center until you come to the tough lining. In an old bird, you can separate the gizzard from the lining without cutting into the latter. If the lining is tender enough to tear easily, cut into it, rinse away the grit, and peel off the lining with your fingers.

When butchering whole chickens for roasting, some people wrap all the giblets from each bird in a little plastic bag and place it in the body cavity for later use in stuffing or gravy. These days a lot of people don't bother to save the giblets, but if you clean many chickens, they can add up to a heap of good meat. Even when I don't have enough to serve at the table, I cook up the gizzards and hearts for my appreciative cat.

With the giblets removed, break any attachments connecting the remaining organs. Drop this refuse into a large bucket or cardboard box lined with a sturdy plastic bag for disposal (see box on facing page).

Cooling

As soon as the birds are cleaned, rinse them thoroughly in running water and cool them quickly to remove remaining body heat and check bacterial growth. USDA guidelines call for chickens to be chilled to 40°F (4°C) or less

CLEANING THE GIZZARD

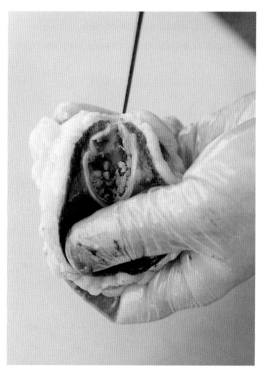

Cut into the large end of the gizzard until you reach the tough lining.

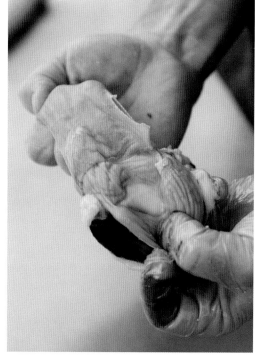

Peel away the lining and discard it, along with its gritty contents.

within four hours of slaughter. The chickens may be water cooled or air cooled.

For water cooling, put them in a clean container full of ice water, and change the water or add ice as often as necessary until the birds are chilled down. Then remove them from the water, and either prop them on paper towels or hang them by a wing to drain for about 20 minutes before wrapping them. Chickens cooled by this method may absorb as much as 12 percent of their weight in water. Most of the water is absorbed by skin, which makes crisping the skin difficult — a disadvantage for people who enjoy eating crisply cooked skin.

Even though our family doesn't eat the skin, we prefer air cooling, which is feasible because we have a spare refrigerator for the purpose, and by the time we get done cleaning

and rinsing the chickens under running water, most of their body heat has dissipated. We cover the chickens with clean cotton sheets, which allow air circulation while keeping the meat from drying out. Air chilling requires a cold-enough unit with sufficient space to spread out the chickens and hasten cooling to the requisite temperature.

Although you can't avoid bacteria altogether — the little critters are everywhere — proper cooling lets you avoid a major source of contamination found in store-bought chicken meat. Most commercial birds are cooled in vats of water laden with rinsed-away fecal matter spilled from intestines that have been torn open by eviscerating machines. The meat soaks up the foul water and bacteria, which is why the USDA now suggests not rinsing store-bought chickens before cooking them — to avoid spreading bacteria in your kitchen — and advises cooking the meat to 165°F (75°C) — to destroy the soaked-up bacteria.

Cutting Up

Broilers are cut into quarters or smaller pieces for frying, halved or quartered for grilling, or left whole for open-pit roasting. Roasters and capons are generally left whole. Old hens, being too tough to roast, are cut up to be stewed.

You may occasionally run across directions for eviscerating and cutting up a bird in one operation, but cutting up a chicken soon after it has been killed causes its muscles to bunch up, making them dense and tough. A chicken that is eviscerated and aged before being cut up will be more tender. Besides, chilled meat is easier to handle.

Some folks package all the parts of each chicken together. I prefer to sort the meatier pieces (breasts, thighs, and drumsticks) from

AGING

Unless a freshly killed chicken is rushed to the stove, muscle protein coagulates (*rigor mortis*), causing the meat to become tough. Like all meat, chicken meat must be aged to give the muscles time to relax. An aged chicken tastes better and is more tender than a chicken cooked or frozen a few hours after being killed. The older the chicken, the longer the meat needs to age.

After freshly killed chickens have cooled, wrap them loosely, and age them in the refrigerator for one to three days, leaving enough space around each for cool air to circulate. If you plan to cook a chicken fresh, you may refrigerate it for up to five days. If the chicken will be canned or frozen, process it no later than three days after dressing.

the bonier pieces (back, neck, and wings). Or if I'm cutting up enough chickens to make a batch of buffalo wings, I'll wrap and label them separately. I package and freeze the meaty pieces for meals, then boil the bony parts and strip the meat for canning or making soup.

Light and Dark Meat

I'm fascinated by the preferences different people have for different parts of a chicken. To me the tastiest meat is next to the bone; my husband prefers boneless chicken on his dinner plate. I like the lower half, or bottom quarters (legs and back); my husband likes the breasts. More people seem to side with his preferences.

Homegrown chicken has firmer breasts and darker legs than supermarket chicken. A bird uses breast muscles for flying and leg muscles for walking, but chickens do more walking than flying. Active muscles need oxygen, and oxygen is carried in blood cells. The more active the muscles, the more blood they require, and the more blood they require, the darker the meat.

Muscles get their energy from fat stored within the muscle cells. The more the muscles are exercised, the more energy they need, so the more fat they store. Since leg muscles get more exercise than breast muscles, the leg muscles contain more fat, which is why a chicken's hindquarters have more flavor, but the less fatty breasts are considered more healthful. The breast of a fast-growing broiler has less fat and flavor than that of other breeds.

Dark meat is denser than light meat and therefore takes longer to cook. Commercial-strain broilers don't do much more than sit by the feeder waiting for the next meal, so their meat cooks faster than that of a more active breed that forages between sit-down meals.

COMPONENTS OF A 4-LB (2 KG) BROILER/FRYER	
Component	**Approximate Percentage**
Head	2%
Feet	4%
Blood	4%
Feathers	8%
Offal	10%
Giblets	10%
Dressed bird	62%
Usable Parts of a Dressed Bird	
Fat	2.5%
Back and neck	13%
Wings	13.5%
Breast	34%
Legs and thighs	37%

Similarly, the older the chicken, the greater the difference between the density of the active leg muscles and the less active breast muscles.

Because of this difference between light and dark meat, when I fry a chicken, for instance, I put the dark pieces into the pan first and add the breasts after the first pieces have heated up. Another plan of action is to separate the light from the dark pieces into different freezer packages at the time of butchering.

All parts of a chicken consist of short-fiber meat that is easily digested by children, older people, and anyone with digestive issues. The nutritional value varies with the part and the method by which it is cooked. Dark and light meat are similar in vitamin and mineral content, but light meat has slightly more protein, and dark meat has more fat and calories. Frying in oil adds fat and calories to all parts.

HOW TO CUT UP A CHICKEN

To cut up a chicken, use a sharp, heavy knife and follow these steps:

1. Cut the skin between the thighs and the body.

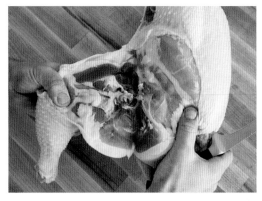

2. Grasp a leg in each hand, lift the bird, and bend the legs back until the hip joints pop free.

3. Cut a leg away by slicing from the back to the front at the hip, as close as possible to the backbone.

4. If you wish to separate the thigh from the drumstick, cut through the joint between them, which you can locate by flexing the leg and thigh to identify the bending point.

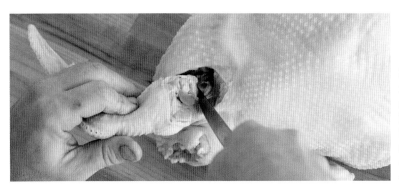

5. On the same side, remove the wing by cutting along the joint inside the wingpit, over the joint, and down around it. Turn the bird over, and remove the other leg and wing.

6. If you plan to make buffalo wings (named after Buffalo, New York, where the dish originated), separate the lower two bony sections of each wing, collectively known as the *wingette* or *flat*, from the upper, meatier portion, called the *drummette*.

7. To divide the body, stand the bird on its neck and cut from the tail toward the neck, along the end of the ribs on one side. Cut along the other side to free the back. Bend the back until it snaps in half, and cut along the line of least resistance to separate the ribs from the lower back.

8. With the breast on the cutting board, skin side down, cut through the white cartilage at the V of the neck.

9. Grasp the breast firmly in both hands, and bend each side back, pushing with your fingers to snap the breastbone. Cut the breast in half lengthwise alongside the bone. For boned breasts, place the breast on the cutting board skin side up. Insert the knife along one side of the keel and cut the meat away from the bone. Repeat for the other side.

The Fat of It

The fat in chickens varies more than any other nutritional element. Although some fat is in the muscle and a lot is in the skin, abdominal fat is the most variable fat deposit and therefore the main measure of a chicken's fatness.

Chickens evolved with the ability to deposit large quantities of abdominal fat — the so-called *fat pad* — to use as reserve energy during times when forage is scarce. Usually, the fat pad of a young bird may be easily peeled from the abdominal wall. Most young chickens, especially active pastured birds, have a relatively thin fat pad. An exception is commercial-strain broilers fed for rapid growth; any excess feed they don't convert into muscle (meat) metabolizes into fat. Excess abdominal fat represents wasted feed and wasted money.

Older birds generally accumulate more fat than younger ones, and females have more fat than males. Old hens, especially if they've been fed too much grain, can have enormous quantities of fat, to the point that the abdominal cavity is virtually filled with fat.

Chicken fat is an essential ingredient for matzo balls and a few other traditional recipes, and a little may be used to flavor chicken soup. But chicken fat contains about 20 percent saturated fat and lots of calories, so most people these days (except folks who can accumulate enough chicken fat to convert into biodiesel) consider it a waste product of butchering.

With the fat pad removed, chicken as a source of protein is among the lowest in calories, fat, saturated fat, and cholesterol (see the chart below). The fat and calorie difference between similar cooking methods, with or without the skin, is negligible if you don't eat the skin.

Storing Poultry

Cleaning a chicken only when you plan to cook it is not an economical use of time. Because of the mess involved, butchering several birds at once makes more sense. Freshly butchered chicken meat keeps for up to five days in a refrigerator with a temperature between 29 and 34°F (−2 and 1°C). To store it longer, either freeze it or can it within three days of butchering.

NUTRITIONAL COMPOSITION				
Serving 3.5 oz (100 g)	Calories	Protein	Saturated Fat	Cholesterol
Light meat				
Stewed with skin	184	27.4 g	2.0 g	75 g
Roasted without skin	173	30.9 g	1.3 g	85 g
Roasted with skin	197	29.8 g	2.2 g	89 g
Fried, flour coated	222	31.8 g	2.5 g	89 g
Fried, batter dipped	260	24.8 g	3.5 g	85 g
Dark meat				
Stewed with skin	233	23.6 g	3.6 g	82 g
Roasted without skin	205	27.4 g	2.7 g	93 g
Roasted with skin	253	25.7 g	4.0 g	91 g
Fried, flour coated	254	26.8 g	3.9 g	94 g
Fried, batter dipped	273	21.8 g	4.3 g	90 g

Freezing

Freezing preserves the quality of fresh meat and retards the growth of bacteria. But freezing does not stop bacterial growth, so strict cleanliness is essential in all aspects of butchering and packing. The quality of frozen chicken depends on:

- How fresh it is when you freeze it
- The temperature under which it remains frozen
- The length of time it remains frozen

Chicken may be frozen raw or cooked, whole or in pieces. If you package cut-up chicken, place just enough pieces in each packet for one meal. If you freeze whole birds, don't stuff them before freezing them. Dense stuffing slows freezing, giving bacteria more time to multiply.

Fat goes rancid fairly rapidly, even in the freezer, so whether you freeze birds whole or in pieces, trim away excess fat. Cut off sharp bones that may pierce the wrapping, exposing the meat and causing grayish leathery spots known as freezer burn. This condition occurs when food is not adequately wrapped, allowing air to circulate over the exposed surface and dry it out. Freezer-burned areas are tough and tasteless, and cutting them away gives the chicken parts an unsightly appearance.

CHICKEN MEAT STORAGE SAFETY

		Degrees Fahrenheit	Degrees Centigrade	
		212	100	Water boils
		180	82	Whole chickens and chicken parts cooked done
		170	77	Chicken breast cooked done
		158	70	Maximum incubation temperature for thermophilic bacteria
		140	60	Minimum safe holding temperature for hot chicken
		104	40	Minimum incubation temperature for thermophilic bacteria
		98	37	Incubation temperature for mesophilic bacteria
Spoilage Zone*	Bacteria double every — 30 minutes	90	28	
	1 hour	70	21	Incubation temperature for psychrophilic bacteria
	2 hours	50	10	
	6 hours	40	4	Safe refrigerator storage temperature
	20 hours	32	0	
		0	−18	Maximum safe storage temperature for frozen chicken
		−20	−29	Quick frozen
		−40	−40	
		−60	−51	All water in chicken freezes
		−100	−73	

*Spoilage zone — bacteria and food toxins develop fast.

Source: USA Poultry & Egg Export Council

Heavily freezer-burned chicken isn't worth cooking. Freezer burn may be avoided by:

- Using a vacuum pump and plastic bags designed for freezer use (provided the vacuum holds)
- Packing chicken in bags made of Cryovac, a kind of plastic that shrinks when dipped in boiling water
- Double-wrapping the chicken, first with plastic wrap, then with either heavy foil or waxed butcher paper sealed with freezer tape

On each packet, jot down the date (so you'll use the oldest first) and the contents (so you won't keep the freezer door open while you try to guess). Spread fresh packets around in the freezer, leaving space between them for the air to circulate until they freeze hard, which takes at least 12 hours. After the chicken is frozen, you can stack it any way you like.

During freezing, ice crystals form from moisture in the meat. Quick freezing produces small ice crystals. Slow freezing causes big crystals that damage meat tissue. To ensure quick freezing, don't add more at one time than 3 pounds (1.5 kg) of unfrozen meat per cubic foot (30 L) of freezer capacity.

Storing chicken in the freezer for any appreciable length of time is an option only if you have a dedicated freezer. The freezer compartment of most refrigerators isn't cold enough to hold chicken for more than a couple of weeks.

To monitor the quality and storage time of your frozen chickens, keep a reliable thermometer in your freezer. At freezer temperatures above 0°F, chicken deteriorates, even though it may remain cold and frozen hard. Chicken stored at 0°F (–18°C) or lower remains safe indefinitely but eventually deteriorates in quality. Every now and then, I find a packet of chicken that got lost in my freezer for several years, and as long as the wrapper is intact, it cooks up fine, although it isn't always quite as tasty as fresher chicken.

SAFE-STORAGE GUIDE FOR POULTRY

	Time in Refrigerator 35–40°F (1.5–4.5°C)	Time in Freezer 0°F (–18°C)
Raw		
Whole	5 days	12 months
Pieces	2 days	9 months
Giblet	1–2 days	3 months
Cooked		
Slices or pieces	1–2 days	1 month
Slices or pieces in gravy or broth	1–2 days	6 months
Casserole	1–2 days	6 months
Fried	1–2 days	4 months
Gravy or broth	1–2 days	3 months

Freezing Cooked Chicken

Freezing fully cooked chicken is a handy way to have quick meals later. If you freeze a roasted, stuffed bird, remove the stuffing and freeze it separately from the meat. Frozen cooked meat eventually takes on a rancid taste, but packing the cooked meat in broth or gravy keeps out air and lengthens the storage time.

Chicken cooked in a liquid or semiliquid base may be frozen in heat-sealed boilable plastic bags. To ensure a complete seal, cool the chicken before filling the bags, and fill them so they're no more than 1 inch (2.5 cm) thick laid flat. When you're ready to reheat the chicken-in-a-pouch, drop the whole pouch into boiling water.

To freeze a chicken casserole, cook the dish in an ovenproof or microwaveable container lined with foil. Cool the casserole, cover it, and freeze it — container and all. When the food is fully frozen, remove the casserole from its dish, wrap it, and freeze it. When you're ready to reheat it, unwrap the casserole and drop it back into its original container.

Any time you freeze cooked food, cool it first so it won't heat up the inside of your freezer. Pack the cooked food for freezing, then let it cool at room temperature for 30 minutes before popping it into the freezer. Better yet, chill it for a few hours in the refrigerator before putting it into the freezer.

Power Failure

If your power goes out, you can safely refreeze chicken if

- It still contains ice crystals or
- The temperature has been above freezing (but below 40°F [4°C]) for no more than two days

When the power goes out, the first thing to do is avoid opening the freezer, even if it means preparing something other than the meal you had planned. If you keep the door closed, a loaded freezer will stay cold enough to preserve chicken and other foods for one or two days, depending on its size. Even a partially loaded freezer should keep meat frozen for a day. In a heated room, you can delay thawing within the freezer by wrapping the unit in comforters, quilted furniture pads, bed pads, or anything else that's large and thick.

If you live in an area where power outages are frequent, a generator makes a good investment. As an alternative, look in advance for a place where you can buy dry ice in a hurry. Dry ice is carbon dioxide, solidified at a temperature of −220°F (−140°C). As it warms, it turns into a gas that evaporates, leaving no puddle the way wet ice does. A 50-pound (25 kg) block will keep a 20-cubic-foot (0.5 cu m) freezer going for up to four days if it's fully loaded, three days if it's at least half full. To avoid burning your fingers when handling dry ice, wear gloves and leave the ice in its brown-paper wrapper.

Although refrozen chicken is safe to eat, it won't taste as good as meat that hasn't been thawed and refrozen. To preserve the remaining flavor, cook the chicken before refreezing it. Even raw meat that's too far thawed to refreeze, but is still safe to eat, may be cooked and refrozen. Cooked chicken that has thawed should not be refrozen but may be eaten within two days.

Thawing

Frozen chicken pieces may be cooked without being thawed, but they'll cook faster if thawed first. A whole chicken should always be thawed before being roasted. Thawing chicken on a counter at room temperature is dangerous because the outside layer may thaw and start to spoil while the inside is still frozen. Chicken may be safely thawed in one of three ways:

In the refrigerator chicken pieces will thaw overnight, and a 4-pound (2 kg) chicken will thaw in about a day. This method is the safest, since you don't run the risk of forgetting about the chicken and letting it get too warm for too long. Put the package of frozen chicken on a plate or tray to catch drips, and place the plate in the refrigerator until the meat is pliable.

In cold water pieces will thaw in about an hour, and a 4-pound (2 kg) chicken will thaw in two to three hours. Seal the frozen chicken in a plastic bag and submerge it in cold tap water until the meat is pliable. To speed up the defrosting of pieces, separate them as soon as they have thawed enough to break them apart. In a warm room, change the water every half hour to make sure it stays cold; in a cool room, the thawing chicken will likely keep the water colder than cold tap water. Cook the chicken as soon as it has thawed.

In a microwave oven chicken may be thawed in a matter of minutes, although some spots may begin to cook while others have just barely thawed. Chicken defrosted in a microwave should therefore be cooked immediately after thawing. Every microwave oven is different, so follow your microwave manufacturer's directions for thawing frozen foods.

If you start to thaw chicken in the refrigerator and it isn't completely thawed by the time you're ready to cook it, speed things along by putting the packet in cold water or in the microwave. Another way to shorten thawing time is by freezing pieces with wax paper or freezer wrap between them, so you can separate the pieces easily for thawing.

Don't be alarmed if frozen chicken happens to look dark near the bone after it's cooked. Darkening is a reaction to slow freezing that normally occurs in home-frozen chicken.

Canning

Home-canned chicken has the distinct advantage that storage does not require electricity. To safely can chicken, you need a pressure canner; for pint (0.5 L) and half-pint (0.25 L) jars, you may use a pressure saucepan. No method that processes without pressure is safe for any kind of meat.

Chicken, like other meat, may contain bacteria that cause botulism, a form of food poisoning. The bacteria are destroyed by processing at 240°F (115°C) for a specific length of time that depends on the volume. If the temperature is lower or the time is shorter than recommended, the risk of botulism occurs.

If you are not familiar with the use of a pressure pot, do not attempt to can chicken based on the suggestions offered here. Read the manual that came with your canner, consult a reliable canning guidebook, or get information from your state or county Extension home economist.

A pressure canner operated at sea level at 10 pounds of pressure reaches a temperature of 240°F (115°C). If you live above sea level, you'll have to adjust the pressure for your altitude.

Chicken meat is easier to handle and will be more tender if it has been chilled for 6 to 12 hours before being canned, but there's nothing wrong with canning it as soon as the body heat is gone. Remove as much fat as possible. Do not can excessively fatty pieces. Add salt for flavor, if you wish. Work as quickly as possible, and process the jars as soon as they are filled.

Canned Chicken Safety

Stored in a cool, clean, dry pantry, canned chicken will keep for up to five years. In my experience, the limiting factor is the quality of the lids, which don't always have enough coating on the inside to prevent them from rusting through. Another potential problem is that a bit of fat or meat on the jar rim might prevent a good seal. To ensure the safety of canned chicken, observe these precautions:

- Keep all equipment clean.
- Carefully wipe jar rims before sealing.
- Meticulously follow processing times and temperatures.
- Let jars cool before washing them and storing them in a cool, dry place.
- Boil home-canned meat at a rolling boil for 15 to 20 minutes, stirring constantly, before serving it.

If you wish to use canned chicken for cold salads or sandwiches, chill the meat in the refrigerator after it's been boiled. Signs of spoilage include

- Bulging or leaking jar lids
- Gas bubbles inside a jar
- Liquid spurting out when a jar is opened
- Off-colors or off-odors
- Foaming while the food is boiling

If a jar shows any sign of spoilage, dispose of it as you would any toxic substance. Don't use it. Don't even taste it. At a home-canning seminar I attended, another attendee claimed she had canned something that later developed botulism. When asked how she knew, she said it looked funny, so she dipped the tip of her finger into the jar and tasted it, whereupon she was told, "If that had been botulism, you would be dead."

Hot Pack

Chicken may be hot-packed with or without the bones, but deboned canned chicken is easier to use, and you can get more meat into each jar. Deboning lets you use meat from bony

PROCESSING TIMES FOR CANNING CHICKEN

Chicken	Jar Size	Minutes
Without bones	Half pints (0.25 L)	60
	Pints (0.5 L)	75
	Quarts (1 L)	90
With bones	Pints	65
	Quarts	75
Broth	Pints	20
	Quarts	25

Pressure: For a dial-gauge canner, use a pressure of 11 psi at an altitude of up to 2,000 feet above sea level; 12 psi for 2,001 to 4,000 feet; 13 psi for 4,001 to 6,000 feet; and 14 psi for 6,001 to 8,000 feet. For a weighted-gauge canner, use a pressure of 10 pounds for up to 1,000 feet and 15 pounds above 1,000 feet.

parts to make salads and sandwiches. A regular meal at our house is taco shells filled with canned chicken heated together with a little chopped onion and spiced with chili powder and cumin.

To hot-pack boned chicken, simmer the bony pieces, covered in water, for 30 to 45 minutes or just until the meat starts to fall off the bone. Remove all the bones and skin, and pack the meat loosely into clean glass jars. Keep the broth simmering.

If you wish, add salt: ¼ teaspoon per half-pint (0.25 L), ½ teaspoon per pint (0.5 L), 1 teaspoon per quart (1 L). Cover the meat with simmering broth, leaving 1¼ inch (3.5 cm) of headspace. Wipe the jar rims to rid them of fat and meat particles. Seal jars with clean lids, and process.

When I have broth left over, I can it for use in making soups later on. Cool the broth, and remove the fat that accumulates at the top. Reheat the broth to boiling, fill jars, leaving 1 inch (2.5 cm) of headspace, seal and process.

Raw Pack

If you wish to can chicken pieces bones and all, raw packing is easier than hot packing. For a raw pack, canning the meaty pieces (thighs, breasts, and drumsticks) makes more sense. Since the breastbone and drumstick take up lots of room, at least bone the breasts and saw drumsticks short. Trim off any fat, and pack pieces loosely into quart (1 L) jars.

Place thighs and drumsticks with their skin next to the glass. Fit breasts into the center. Use smaller pieces to fill up the remaining space. Leave 1¼ inch (3.5 cm) of headspace at the top of the jar. If you wish, add 1 teaspoon of salt. Do not add liquid—raw meat generates its own juice while it cooks.

Wipe the jar rims to rid them of fat and meat particles. Seal the jars with clean lids, and process.

Sanitation and Safety

One of the big advantages to growing your own chicken is the safety factor. You are bound to take care in raising, butchering, and storing your meat, knowing you and your family will eat it. A *Consumer Reports* survey found that 83 percent of fresh, whole broilers bought in stores nationwide harbor *Campylobacter* or *Salmonella*, the leading bacterial causes of foodborne disease. Other bacteria found in chicken include *Staphylococcus* and *Listeria*.

Even though these bacteria may be present in your chickens, you have the advantage of using better sanitation. Much of the widely publicized precautions for handling and cooking chicken are designed to ensure that the excrement absorbed into the meat and skin of commercially processed chicken is thoroughly cooked before you eat it.

It stands to reason that any time an intestine or other organ is broken and spills its contents during butchering, you should immediately rinse away the spillage. After butchering is complete, thoroughly rinse the chickens in running water. Minimize bacterial growth by cooling and refrigerating the meat as quickly as possible.

Cleanup

Cleaning chickens is a messy affair and is best done somewhere other than in the kitchen. No matter where you do it, after handling raw meat, thoroughly clean countertops, sinks, knives, and any other utensils used.

Begin by scraping bits of meat and blood from counters and cutting boards. Then wash your hands, knife, counter, cutting board, and other utensils with hot soapy water, and rinse well in running water. Finish up by sanitizing all the cleaned surfaces by wiping them with a clean cloth soaked in white vinegar.

Kitchen Safety

Any bacteria present on raw or undercooked chicken will multiply at temperatures between 40 and 140°F (5 and 60°C), which is most likely to occur if chicken is taken out of the refrigerator too long before it is cooked. To ensure safe eating, follow these tips:

- Check the temperature of your refrigerator and freezer with an appliance thermometer; the refrigerator should be 40°F (5°C) or lower and the freezer 0°F (–18°C) or lower.

- Store chicken at 40°F or below. If you won't use it for a couple of days, can it or freeze it; freezing does not destroy bacteria, but it does significantly slow its growth.

- Before cooking frozen meat, thaw it in the refrigerator or in a plastic bag under cold running water, never on the countertop at room temperature.

- If you're going to stuff a whole chicken, do so just before you pop the bird into the oven.

- Cook chicken to a safe internal temperature, as indicated in the Doneness Guide for Cooked Chicken table that follows.

- Never ever return cooked chicken to the container in which it was carried while raw.

- Separate the chicken meat from stuffing or gravy before storing leftovers.

DONENESS GUIDE FOR COOKED CHICKEN

To check temperature, insert a thermometer into the thickest part of the thigh without touching bone or, when cooking breasts, into the thickest part of the breast.

Chicken	Temperature
Whole	180°F / 82°C
Leg, drumstick, thigh	180°F / 82°C
Breast, bone in	170°F / 77°C
Breast, boneless	160°F / 70°C
Ground	165°F / 75°C
Stuffing in whole chicken	165°F / 75°C
Casseroles and combination dishes	165°F / 75°C
Reheated leftovers	165°F / 75°C

abdominal capacity. Total abdominal depth and width.

abdominal depth. The distance between the pelvic bones and the keel bone.

abdominal width. The distance between the two pelvic bones.

acute. Description of a disease with rapid development of severe symptoms, often measured in hours and ending in death or recovery; opposite of chronic.

American Standard of Perfection. A book published by the American Poultry Association describing each breed recognized by that organization.

amino acid. One of the building blocks of protein, of which eight (the essential amino acids) cannot be manufactured in the body.

anthelmintic. An antiworm drug.

aragonite. Calcium carbonate derived from seashells.

avian. Pertaining to birds.

bantam. A miniature chicken, about one-fourth the size of a regular-size chicken.

Bantam Standard. A book published by the American Bantam Association describing each of the bantam breeds recognized by that organization.

banty. Affectionate word for a bantam breed.

barnyard chicken. A chicken of mixed breed.

beak. The hard, protruding portion of a bird's mouth, consisting of an upper beak and a lower beak.

beard. The feathers (always found in association with a muff) bunched under the beak of such breeds as Antwerp Belgian, Faverolle, and Houdan.

bedding. Straw, wood shavings, shredded paper, or anything similar scattered on the floor of a chicken coop to absorb moisture and manure.

biddy. Affectionate word for a hen.

bill out. To use the beak to scoop feed out of a trough onto the floor.

biosecurity. Disease-prevention management.

bleaching. The fading of color from the beak, shanks, and vent of a yellow-skin laying hen.

blood feathers. Pinfeathers, so called because the quills are filled with blood to nourish the growing feather.

bloom, egg. The moist, protective coating on a freshly laid egg that dries so fast you rarely see it.

blowout. Vent damage caused by laying an oversize egg.

brail. To temporarily restrict flight by wrapping one wing so it can't open.

bran. A by-product of milling consisting of the outer coating, or husks, of the grain kernels.

break up. To discourage a hen from setting.

breed. A group of chickens that are alike and capable of reproducing more of the same; also, pairing a rooster and hen for the purpose of obtaining fertile eggs.

breeders. Mature chickens from which fertile eggs are collected; also, a person who manages such chickens.

breed true. The ability to produce offspring exactly like the parents.

broiler. A young, tender meat chicken; also called a fryer.

brood. To care for a batch of chicks; also, the chicks themselves.

brooder. A mechanical device used to imitate the warmth and protection a mother hen gives her chicks.

broody. A mother hen.

candle. To examine the contents of an intact egg with a light.

candler. A light used to examine the contents of an intact egg.

cannibalism. The bad habit chickens have of eating each other's flesh, feathers, or eggs.

cape. The short, narrow feathers between a chicken's shoulders, where the neck joins the back, and growing underneath the hackle. As a group, these feathers grow in the shape of a cape.

capon. A cockerel, raised for meat, that has been castrated so it will grow larger and more tender.

carrier. An apparently healthy individual that transmits disease to other individuals; also, a container used to transport chickens.

cecum. A blind pouch at the juncture of the small and large intestine that resembles the human appendix; plural: ceca.

cestode. Tapeworm.

chalazae. Two white cords on opposite sides of a yolk that keep the yolk properly positioned within the egg white; singular: chalaza.

check. An egg with a cracked shell but with the shell membrane still intact.

chronic. Description of a disease having long duration, measured in days, months, or years, and being somewhat resistant to treatment; opposite of acute.

class. A group of chickens competing against each other at a show.

classification. The grouping of purebred chickens according to their place of origin (such as Asiatic) or according to shared characteristics (such as rose comb or clean legged).

clean legged. Having no feathers growing down the shanks.

clinical. Having signs of disease that may be readily observed.

cloaca. The chamber just inside the vent where the digestive, reproductive, and excretory tracts come together.

cluck. The sound a hen makes to comfort her chicks; also, the hen herself.

clucker. Affectionate word for a mother hen.

clutch. A batch of eggs that are hatched together, either in a nest or in an incubator (from the Old Norse word *klekja*, meaning "to hatch"), also called a setting; also, all the eggs laid by a hen on consecutive days, before she skips a day and starts a new laying cycle.

coccidiasis. Infection with coccidial protozoa that produces no symptoms.

coccidiosis. A parasitic protozoal infestation that causes clinical disease, usually occurring in damp, unclean housing.

coccidiostat. A drug used to keep chickens from getting coccidiosis.

cock. A male chicken; also called a rooster.

cocker. A person who maintains fighting cocks.

cockerel. A male chicken younger than 1 year old.

cocking. Cockfighting.

complete protein. Description of a feedstuff containing all eight essential amino acids.

condition. A chicken's state of health and cleanliness.

conformation. A chicken's body structure.

contagious. Description of a disease that's readily transmitted from one individual or flock to another.

coop. The house or cage in which chickens live.

crest. A puff of feathers on the heads of breeds such as Houdan, Silkie, or Polish; also called a topknot.

crop. A pouch at the base of a chicken's neck that bulges with feed after the bird has eaten; also, to trim a bird's wattles.

crossbreed. The offspring of a hen and a rooster of two different breeds.

crumbles. Crushed pellets.

cull. To eliminate a nonproductive or inferior chicken from a flock; also, the nonproductive or inferior chicken itself.

dam. Mother.

dam family. Sibling chickens that have the same dam and sire.

debeak. To remove a portion of a bird's top beak to prevent cannibalism.

defect. Any characteristic that makes a chicken less than perfect.

disinfectant. Anything used to destroy disease-causing organisms.

disqualification. A defect or deformity serious enough to bar a chicken from a show.

down. The soft, furlike fluff covering a newly hatched chick; also, the fluffy part near the bottom of any feather.

drench. To give liquid medication orally; also, the liquid medication itself.

droppings. Chicken manure.

dual purpose. Suitable for the production of both eggs and meat.

dub. To trim the comb.

dust. The habit chickens have of thrashing around in soft soil to clean their feathers and discourage body parasites.

electrolytes. Natural chemicals in the blood that help regulate fluid balance and other important physiological functions.

embryo. A fertilized egg at any stage of development prior to hatching.

enteritis. An inflammation of the intestine.

exhibition breeds. Chickens kept and shown for their beauty rather than their ability to lay eggs or produce meat.

faking. The dishonest practice of concealing a defect or disqualification from a potential buyer or show judge.

feather legged. Having feathers growing down the shanks.

fecal. Pertaining to feces.

feces. Manure.

fertile. Capable of producing a chick.

fertilized. Description of an egg that contains sperm and therefore is capable of developing into a chick.

finish. The amount of fat beneath the skin of a meat bird.

flighty. Easily excited.

flock. A group of chickens living together.

forced-air incubator. A mechanical device for hatching fertile eggs that has a fan to circulate warm air.

fowl. Domesticated birds raised for food; also, a stewing hen.

free choice. Available to chickens at all times.

free range. Describes chickens allowed to roam at will.

frizzle. Feathers that curl rather than lie flat.

fryer. A tender young meat chicken; also called a broiler.

gizzard. An organ that contains grit for grinding up the grain and plant fiber a chicken eats.

grade. To sort eggs according to their interior and exterior qualities.

grit. Sand and small pebbles eaten by a chicken and used by its gizzard to grind up grain and plant fiber.

ground fed. Free to move about outdoors, as opposed to being housed and fed within a building or a cage.

hackle. The collective group of feathers growing along the back and sides of the neck. Also any single one of those feathers.

hatch. The process by which a chick comes out of an egg; also, a group of chicks that come out of their shells at roughly the same time.

hatchability. Percentage of fertilized eggs that hatch under incubation.

hatching egg. A fertilized egg stored in a way that does not destroy its ability to hatch.

hen. A female chicken.

hen feathered. The characteristic of a rooster having rounded rather than pointed sex feathers.

host. A bird (or other animal) on or in which a parasite or an infectious agent lives.

hybrid. The offspring of a hen and rooster of different breeds, each of which might themselves be crossbred; often erroneously applied to the offspring of a hen and rooster of different strains within a breed.

immunity. Ability to resist infection.

impaction. Blockage of a body passage or cavity, such as the crop or cloaca.

incubate. To maintain favorable conditions for hatching fertile eggs.

incubation period. The time needed for a bird's egg to hatch; also, the time elapsed between exposure to a disease-causing agent and the appearance of the first signs.

incubator. A mechanical device for hatching fertile eggs.

infectious. Capable of invading living tissue and multiplying therein, causing disease.

infertility. Temporary or permanent inability to reproduce.

keel. The breastbone, which resembles the keel of a boat.

layer ration. A type of commercially prepared feed that provides adequate protein and calcium to meet the specific dietary needs of laying hens.

line. A strain of chickens bred to be as similar as possible to one another.

litter. Bedding.

mash. Feed that has been ground to various degrees of coarseness but is still recognizable so chickens can pick out what they like.

mate. The pairing of a rooster with one or more hens; a hen or rooster so paired.

mesophilic bacteria. Microorganisms that grow best in temperatures between 68 and 113°F (20 and 45°C).

mite. A tiny jointed-leg body parasite.

molt. The annual shedding and renewing of a bird's feathers.

mortality. Percentage killed by a disease.

nematode. Roundworm.

nest. A secluded place where a hen feels she may safely leave her eggs; also, the act of brooding.

nest egg. A wooden or plastic egg placed in a nest to encourage hens to lay there.

omphalitis. An often fatal inflamation of the navel in newly hatched chicks, caused by various species of staph bacteria.

oviduct. The tube inside a hen through which an egg travels when it is ready to be laid.

parasite. An organism that lives on or inside a host animal and derives food or protection from the host without giving anything in return.

pasting. Loose droppings sticking to the vent area.

pastured poultry. Chickens housed on grassland in movable shelters.

pathogenic. Capable of causing disease.

peck order. The social rank of chickens.

pellet. Mash that has been compressed into small tubular pieces, each having identical nutritional value.

pelvic bones. Two sharp, slender bones that end in front of the vent; also called pinbones.

pen. A group of chickens entered into a show and judged together; also, a group of chickens housed together for breeding purposes.

perch. The place where chickens sleep at night; the act of resting on a perch; also called roost.

pickout. Vent damage due to cannibalism.

pigmentation. The color of a chicken's beak, shanks, and vent.

pinbones. Pelvic bones.

pinfeathers. The tips of newly emerging feathers.

pip. The hole a newly formed chick makes in its shell when it is ready to hatch; also, the act of making the hole.

plumage. The complete set of feathers covering a chicken.

postmortem. Pertaining to or occurring after death.

poultry. Chickens and other domesticated birds raised for food.

predator. One animal that hunts another for food, or, as with dogs, for sport.

processor. A person or firm that kills, cleans, and packages meat birds.

producer. A person or firm that raises meat birds or laying hens.

prolapse. Slipping of a body part from its normal position, often erroneously used to describe an everted organ.

protein. A nutrient furnished by any feedstuff (such as eggs, meat, milk, nuts, seed germs, and soybeans) that is high in amino acids.

protozoa. Single-cell microscopic creatures that may be either parasitic or beneficial.

proventriculus. A chicken's stomach.

psychrophilic bacteria. Microorganisms that grow best in temperatures below 59°F (15°C).

pullet. A female chicken younger than 1 year old.

purebred. The offspring of a hen and a rooster of the same breed.

range. An open area on which chickens forage.

range fed. Description of chickens that are allowed to graze pasture.

ration. The combination of all feed consumed.

resistance. Immunity to infection.

roaster. A cockerel or pullet, usually weighing 4 to 6 pounds (2 to 2.7 kg), suitable for cooking whole in the oven.

roost. The place where chickens spend the night; the act of resting on a roost; also called perch.

rooster. A male chicken; also called a cock.

saddle. The part of a chicken's back just before the tail.

scales. The small, hard, overlapping plates covering a chicken's shanks and toes.

scratch. The habit chickens have of scraping their claws against the ground to dig up tasty things to eat; also, any grain fed to chickens.

set. To keep eggs warm so they will hatch; also called brood.

setting. A group of hatching eggs in an incubator or under a hen; the incubation of eggs by a hen (incorrectly called "sitting" by people who try too hard to be grammatically correct).

sex-link. A genetic trait carried on the sex chromosomes.

sexed. Newly hatched chicks that have been sorted into pullets and cockerels.

shank. The part of a chicken's leg between the claw and the first joint.

sickles. The long, curved tail feathers of some roosters.

sire. Father.

spent. No longer laying well.

sport. Cockfighting.

spur. The sharp pointed protrusion on a rooster's shank and sometimes on a hen's.

stag. A cockerel on the brink of sexual maturity, when his comb and spurs begin to develop.

standard. The description of an ideal specimen for its breed; also, a chicken that conforms to the description of its breed in the *American Standard of Perfection*; sometimes erroneously used when referring to large as opposed to bantam breeds.

started pullets. Young female chickens that are nearly old enough to lay.

starter. A ration for newly hatched chicks.

starve-out. Failure of chicks to eat.

station. The degree of upright stance or carriage that is desirable in a game breed.

steal. A hen's instinctive habit of hiding her eggs.

sterile. Permanent inability to reproduce.

sternum. Breastbone or keel.

still-air incubator. A mechanical device for hatching fertile eggs that does not have a fan to circulate air.

straightbred. Purebred.

straight run. Newly hatched chicks that have not been sexed; also called "unsexed" or "as-hatched."

strain. A flock of related chickens selectively bred by one person or organization for so long the offspring have become uniform in appearance or production; also called a line.

stress. Any physical or mental tension that reduces resistance.

stub. Down on the shank or toe of a clean-leg chicken.

thermophilic bacteria. Microorganisms that grow best between 106 and 252°F (41 and 122°C).

tidbitting. Repeatedly picking up and dropping a bit of food.

trace mineral salt. A blend of salt along with many other minerals the body needs in infinitesimally small amounts.

trachea. Windpipe.

treading. Rapid movements of a cock's feet while mating.

trematode. A parasitic fluke.

trio. A cock and two hens or a cockerel and two pullets of the same breed and variety.

type. The size and shape of a chicken that tells you what breed it is.

unthrifty. Unhealthy appearing and/or failing to grow at a normal rate.

urates. Uric acid (salts found in urine).

vaccine. A product made from disease-causing organisms and used to produce immunity.

variety. A subdivision of a breed based on color, comb style, beard, or leg feathering.

vent. The outside opening of the cloaca, through which a chicken emits eggs and droppings from separate channels.

virulence. The strength of an organism's ability to cause disease.

wattles. The two red or purplish flaps of flesh that dangle under a chicken's chin.

zoning. Laws regulating or restricting the use of land for a particular purpose, such as raising chickens.

Some of the publications listed below are regularly updated, so be sure to look for the latest editions of those with that designation.

American Bantam Association. *Bantam Standard.* Augusta, NJ: American Bantam Association (latest edition).
Pictorial guide for breeders and exhibitors of bantams.

American Poultry Association. *American Standard of Perfection* (latest edition).
Pictorial guide for breeders and exhibitors of fancy fowl.

Damerow, Gail. *The Chicken Health Handbook.* 2nd ed. North Adams, MA: Storey Publishing, 2015.
Comprehensive book on preventing, identifying, and treating diseases in backyard flocks.

———. *Hatching and Brooding Your Own Chicks.* North Adams, MA: Storey Publishing, 2013.
How to incubate, hatch, and brood baby chickens, ducklings, goslings, turkey poults, and guinea keets.

———. *The Chicken Encyclopedia.* North Adams, MA: Storey Publishing, 2012.
An illustrated A-to-Z reference guide on the terminology of everything chicken.

Ekarius, Carol. *Storey's Illustrated Guide to Poultry Breeds.* North Adams, MA: Storey Publishing, 2007.
Photographs and detailed description by breed, organized according to primarily egg producers, meat producers, or ornamental.

Foreman, Pat. *City Chicks.* Buena Vista, VA: Good Earth Publications, 2009.
How to maintain a small flock of urban chickens for egg laying, insect control, food waste disposal, garden fertilizer, and endless entertainment.

Heuser, G. F. *Feeding Poultry: The Classic Guide to Poultry Nutrition.* Blodgett, OR: Norton Creek Press, 2003.
A comprehensive guide first published in 1955, when most chickens were kept in small flocks and raised on homegrown feeds.

Hutt, F. B. *Genetics of the Fowl.* Blodgett, OR: Norton Creek Press, 2003.
A classic guide to poultry breeding and chicken genetics, first published in 1949.

Kimball, Herrick. *Anyone Can Build a Tub-Style Mechanical Chicken Plucker.* Moravia, NY: Whizbang Books, 2003.
Step-by-step guide to building a homemade mechanical chicken-plucking machine.

———. *Anyone Can Build a Whizbang Chicken Scalder.* Moravia, NY: Whizbang Books, 2005.
Step-by-step guide to building a thermostatically controlled scalding pot.

Lee, Andy, and Patricia Foreman. *Day Range Poultry: Every Chicken Owner's Guide to Grazing Gardens and Improving Pastures.* Buena Vista, VA: Good Earth Publications, 2002.
A guide to grazing chickens in gardens and pastures.

National Organic Program. *National Organic Program (NOP) Standards & Guidelines.*
Only available online: www.ams.usda.gov/nop

Schrider, Don, and Jeannette Beranger. *Chicken Assessment for Improving Productivity.* Pittsboro, NC: American Livestock Breeds Conservancy, 2007.
An outline for selecting desirable traits for meat qualities, egg production, and breeding stock; available in print or as a free download at www.albc-usa.org/downloads.html

Solomon, Steve. *Gardening When It Counts*. Gabriola Island, BC: New Society Publishers, 2006.
Methods for growing a highly productive, but inexpensive vegetable garden.

U.S. Department of Agriculture. *Regulations Governing the Voluntary Grading of Poultry Products, and U.S. Classes, Standards, and Grading*. 7 CFR Part 70. Washington, DC: U.S. Department of Agriculture. 2008.
Periodically updated regulations pertaining to standards for marketing dressed meat birds (may be available through your state poultry specialist or agricultural library).

U.S. Department of Agriculture Agricultural Marketing Service. *Agriculture Handbook #75: Egg-Grading Manual*. Washington, DC: U.S. Department of Agriculture Agricultural Marketing Service, 2000.
Illustrated guide and color wall chart; accompanying film available through your county Extension agent or the film library at your state land-grant university.

INDEX

Page numbers in *italics* indicate drawings and photographs.
Page numbers in **bold** indicate tables and charts.

METRIC CONVERSION CHARTS

VOLUME EQUIVALENTS

US	Metric
1 teaspoon	5 milliliters
1 tablespoon	15 milliliters
¼ cup	60 milliliters
½ cup	120 milliliters
1 cup	230 milliliters
1¼ cups	300 milliliters
1½ cups	360 milliliters
2 cups	460 milliliters
2½ cups	600 milliliters
3 cups	700 milliliters

TEMPERATURE CONVERSION

To convert Fahrenheit to Celsius, subtract 32 from Fahrenheit temperature, multiply by 5, then divide by 9.

Easy-to-Remember Equivalents

0°C = 32°F	30°C = 86°F
10°C = 50°F	40°C = 104°F
20°C = 68°F	Every 10°C = 18°F

VOLUME CONVERSION

To convert	to	multiply
teaspoons	milliliters	teaspoons by 4.93
tablespoons	milliliters	tablespoons by 14.79
fluid ounces	milliliters	fluid ounces by 29.57
cups	milliliters	cups by 236.59
cups	liters	cups by 0.24
pints	milliliters	pints by 473.18
pints	liters	pints by 0.473
quarts	milliliters	quarts by 946.36
quarts	liters	quarts by 0.946
gallons	liters	gallons by 3.785

WEIGHT EQUIVALENTS

US	Metric
0.035 ounce	1 gram
¼ ounce	7 grams
½ ounce	14 grams
1 ounce	28 grams
1¼ ounces	35 grams
1½ ounces	40 grams
1¾ ounces	50 grams
2½ ounces	70 grams
3½ ounces	100 grams
4 ounces	112 grams
5 ounces	140 grams
8 ounces	228 grams
8¾ ounces	250 grams
10 ounces	280 grams
15 ounces	425 grams
16 ounces (1 pound)	454 grams

LENGTH / AREA CONVERSION

To convert	to	multiply by
inches	centimeters	2.54
feet	meters	0.305
yards	meters	0.91
miles	kilometers	1.61
square feet	square meters	0.09
acres	hectares	0.40

WEIGHT CONVERSION

To convert	to	multiply
ounces	grams	ounces by 28.35
pounds	grams	pounds by 453.5
pounds	kilograms	pounds by 0.45

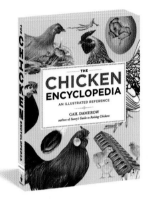
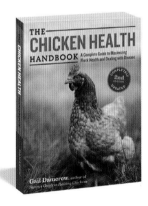
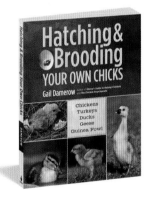

STOREY'S GUIDE TO RAISING

The Definitive Series for Essential Animal Husbandry Information

MORE THAN

2.2 MILLION

COPIES SOLD!

This best-selling series offers fledgling farmers and seasoned veterans alike what they most need to know to ensure both healthy livestock and profits. Each book includes information on selection, housing, space requirements, behavior, breeding and birthing, feeding, health concerns, and remedies for illnesses. They also cover business considerations and marketing products that come from the animals.

THE COMPLETE STOREY'S GUIDE TO RAISING LIBRARY INCLUDES:

Beef Cattle
by Heather Smith Thomas

Chickens
by Gail Damerow

Dairy Goats
by Jerry Belanger and Sara Thomson Bredesen

Ducks
by Dave Holderread

Horses
by Heather Smith Thomas

Keeping Honey Bees
by Malcolm T. Sanford and Richard E. Bonney

Llamas
by Gale Birutta

Meat Goats
by Maggie Sayer

Miniature Livestock
by Sue Weaver

Pigs
by Kelly Klober

Poultry
by Glenn Drowns

Rabbits
by Bob Bennett

Sheep
by Paula Simmons and Carol Ekarius

Training Horses
by Heather Smith Thomas

Turkeys **by Don Schrider**